Nonnitrogenous Organocatalysis

Organocatalysis Series

Series Editor

Sanjay Malhotra
Frederick National Laboratory for Cancer Research,
Maryland, USA

Nonnitrogenous Organocatalysis, *Andrew M Harned*

Nonnitrogenous Organocatalysis

Edited by
Andrew M. Harned

CRC Press is an imprint of the
Taylor & Francis Group, an **informa** business

CRC Press
Taylor & Francis Group
6000 Broken Sound Parkway NW, Suite 300
Boca Raton, FL 33487-2742

First issued in paperback 2021

ISBN 13: 978-1-03-209596-7 (pbk)
ISBN 13: 978-1-4987-1503-4 (hbk)

Publisher's Note
The publisher has gone to great lengths to ensure the quality of this reprint but points out that some imperfections in the original copies may be apparent.

Library of Congress Cataloging-in-Publication Data

Names: Harned, Andrew, author.
Title: Nonnitrogenous organocatalysis / Andrew Harned.
Description: Boca Raton, Florida : CRC Press, [2018] | Series:
Organocatalysis series | Includes bibliographical references and index.
Identifiers: LCCN 2017032468| ISBN 9781498715034 (hardback : alk. paper) |
ISBN 9781315371238 (ebook)
Subjects: LCSH: Organic compounds--Synthesis. | Catalysis. | Chemistry,
Organic.
Classification: LCC QD262 .H2697 2018 | DDC 547/.215--dc23
LC record available at https://lccn.loc.gov/2017032468

Visit the Taylor & Francis Web site at
http://www.taylorandfrancis.com

and the CRC Press Web site at
http://www.crcpress.com

Dedication

Dedicated to the memory of Professor Ronald Breslow, a giant with broad shoulders.

Contents

Preface

Why *nonnitrogenous organocatalysis*? The answer to this is as follows. At the time the idea of this book was conceived there seemed to be a perception, by some in the community, that an organocatalytic reaction was the one that involved the transient formation of a carbon–nitrogen bond and relied on the reactivity imparted by enamine or iminium ion intermediates (sometimes both). To be sure this concept has been a sea change in the area of catalysis and has demonstrated a remarkably broad utility. However, I think it is fair to say that when Prof. David MacMillan coined the term *organocatalysis* in the late 1990s; it was meant to encompass more than just enamine and iminium ions. Furthermore, when the general reactivity of nitrogen is considered you quickly find that a broader net needs to be cast in order to identify more reactions that could, in principle, be carried out by nonmetallic catalysts. For example, nitrogen will likely be a poor reactive center for an oxidation reaction. Similarly, amines and phosphines often demonstrate quite disparate reactivity toward the same functional groups.

In addition to a desire to show that organocatalytic transformations can involve activation modes other than enamines and iminium ions, it was also puzzling to see that most of the reactions being referred to as *organocatalytic* were asymmetric in nature. Yet, digging into the literature, one finds that there are numerous examples of nonasymmetric reactions that are catalyzed by small organic molecules. Should these not be classified as being organocatalytic, just because the product is racemic or lacks stereocenters?

With these concerns in mind, I have asked the authors to focus more on the reactivity of the various catalysts and to include discussions on nonasymmetric reactions. In doing so, I hope to alert the reader to how different reactive intermediates can be accessed in a catalytic fashion and give them some sense of how these intermediates behave. In doing so, they can then use this as an inspiration for devising new reactivity patterns.

In keeping with the title, many of the catalysts covered in this volume are devoid of nitrogen altogether. However, there are some that do contain nitrogen. This is most noticeable in our discussion of *N*-heterocyclic carbene (NHC) catalysts. At the same time, I have chosen not to include discussion of other nitrogen-containing catalysts, such as thioureas. It was difficult to decide where to draw the line. Ultimately, it was decided that NHC catalysts should be included because the catalytic cycle did not involve the formation of new sigma bonds to the nitrogen atoms that are present.

It is, of course, impossible to include everything between the pages of a single volume, especially the one that is somewhat broader in scope. I have asked the contributors to be somewhat selective in their coverage and limit their discussion to the most important reaction types that illustrate the overall reactivity of the catalysts in question. As an editor I have also had to make some choices about the overall coverage. Consequently, some topics that may be the obvious choices for inclusion are missing. Phase-transfer catalysis has not been included. This is mostly due to the nature of the ionic interactions that are involved with these catalysts. Boronic and borinic acids have not been included. Yes, these are nonmetallic organic catalysts, but it is, at times, difficult to separate the traditional Lewis acidic nature of the boron atom from their ability to activate substrates, mostly alcohols, by transient covalent bonds. Similarly, I made an editorial decision not to include Lewis base catalysis; though there is some discussion of this in the coverage of phosphine oxides. In addition, this topic is far too broad to adequately cover as a single chapter (as evidenced by the recent three-volume set edited by Prof. Scott Denmark and Prof. Edwin Vedejs). I had planned to include a chapter on the use of aryl iodides in organocatalytic oxidation reactions. Unfortunately, outside circumstances prevented its inclusion and there was not sufficient time to find an alternative contributor.

Finally, a project such as this is not completed easily and involves the assistance and input from many individuals whose names do not appear on the cover. First, I would thank Sanjay Malhotra for the invitation to contribute to his larger book series on the topic of organocatalysis. Second, I would thank all the contributing authors for their efforts and providing me with a high quality product. Third, I thank Hilary LaFoe, Cheryl Wolf, Marsha Pronin, and Natasha Hallard of the CRC Press for their tremendous efforts behind the scenes and for their encouragement from the start to the finish. I also thank the authors and the editorial team for their tremendous patience during some of the unfortunate delays that came up.

Andrew M. Harned
Texas Tech University

Editor

Andrew M. Harned was born in Fort Benning, Georgia in 1977 and received his BS in biochemistry (1999) from Virginia Tech, Blacksburg, Virginia. He earned his PhD (2005) from the University of Kansas, Lawrence, Kansas where he worked in the laboratories of Prof. Paul Hanson. While there he received an ACS Division of Organic Chemistry Nelson J. Leonard fellowship. After his graduate work, he spent time at Caltech as an NIH postdoctoral fellow in the laboratory of Prof. Brian Stoltz. He started his independent career in 2007 as an assistant professor in the Department of Chemistry at the University of Minnesota, Minneapolis, Minnesota. In 2015, he joined the Department of Chemistry and Biochemistry at Texas Tech University, Lubbock, Texas as an associate professor. His group's research interests focus on developing new methodologies, catalysts, and strategies for natural product synthesis and medicinal chemistry applications.

Contributors

Yi An Cheng
Department of Chemistry
National University of Singapore
Singapore, Singapore

Yonggui Robin Chi
Division of Chemistry and Biological Chemistry
Nanyang Technological University
Singapore, Singapore

Xinqiang Fang
Fujian Institute of Research on the Structure of Matter
Chinese Academic of Science
Fuzhou, China

Andrew M. Harned
Department of Chemistry & Biochemistry
Texas Tech University
Lubbock, Texas

Choon Wee Kee
Division of Chemistry and Biological Chemistry
Nanyang Technological University
Singapore, Singapore

Zhiqi Lao
Department of Chemistry
The University of Hong Kong
Hong Kong, People's Republic of China

Pavel Nagorny
Department of Chemistry
University of Michigan
Ann Arbor, Michigan

Choon-Hong Tan
Division of Chemistry and Biological Chemistry
Nanyang Technological University
Singapore, Singapore

Jia-Hui Tay
Department of Chemistry
University of Michigan
Ann Arbor, Michigan

Patrick H. Toy
Department of Chemistry
The University of Hong Kong
Hong Kong, People's Republic of China

Yunus Emre Türkmen
Department of Chemistry
UNAM-Institute of Materials Science and Nanotechnology
Bilkent University
Ankara, Turkey

Ying-Ying Yeung
Department of Chemistry
The Chinese University of Hong Kong
Hong Kong, People's Republic of China

1 Introduction

Andrew M. Harned

CONTENTS

1.1 A NEW ERA FOR CATALYSIS

The turn of the millennium saw the publication of several seminal reports that would eventually usher in a new era of chemical research. In the mid-1990s, several groups reported the use of small organic catalysts for the kinetic resolution of various secondary alcohols (Figure 1.1). Fu[1] and Fuji[2] reported that chiral 4-dimethylaminopyridine (DMAP) analogs **1** and **2** could be used to resolve acyclic and cyclic alcohols, respectively. It is true that the Fu catalyst does contain a metal, but the ferrocene portion of **1** likely has little to do with the reactivity and just serves as a source of chirality. Soon after, Miller[3] reported that the small, histidine-containing peptide **3** could function as a catalyst for the resolution of cyclic alcohols. The Miller group continued to develop these peptide-based catalysts into a very selective method for alcohol resolutions[4] and other asymmetric transformations.[5]

In 1998 and 1999, Jacobsen[6] and Corey[7] reported asymmetric Strecker reactions promoted by what would now be called hydrogen-bonding catalysts (Figure 1.2).[8] Jacobsen's catalyst was identified by parallel screening of a library of amino acid-derived Schiff base catalysts. Eventually, thiourea catalyst **4** was found to be optimal and could be applied to a range of different *N*-allyl imine substrates. In Corey's report, chiral C_2-symmetric guanidine **5** was found to be an effective catalyst for asymmetric Strecker reactions of *N*-benzhydryl imines. It was proposed that hydrogen bonds in a pretransition state assembly similar to **6** were responsible for organizing the reactants.

In 2000, List, Lerner, and Barbas reported that asymmetric aldol reactions could be carried out using catalytic amounts of the simple α-amino acid proline.[9] In the next two years, List and Barbas would extend the concept of proline catalysis by demonstrating its use in ketone aldol,[10] Mannich,[11] and Michael reactions (Figure 1.3).[12] The key intermediate in all of these reactions is the enamine (**7**) formed between proline and the starting ketone or aldehyde. Important contributions by MacMillan, Jørgensen, Ley, Hayashi, and many others have shown that the so-called *enamine catalysis* is an incredibly robust reaction manifold that can be extended to numerous reaction types.[13]

2 mol% 1
Ac_2O, Et_3N
Et_2O, rt

R_u = Unsaturated group (aryl, alkene)
R_a = Alkyl group

99.7%–92.2% ee
51%–67% conv
s = 14–52

(a)

Resolution conditions
5 mol% **2**, 1 equiv. 2,4,6-collidine,
0.7 equiv. isobutyric anhydride
toluene, rt

Resolved products obtained, R = p-$Me_2NC_6H_4$
(19%–25% recovery, 92%–99% ee, s = 4.7–10.1)

(b)

Resolution conditions
5 mol% **3**, 1 equiv. Ac_2O,
10 equiv. substrate
toluene, 0°C

Resolved products obtained
(14%–84% ee, s = 1.4–12.6)

(c)

FIGURE 1.1 Kinetic resolutions of chiral secondary alcohols reported by (a) Fu, (b) Fuji, and (c) Miller.

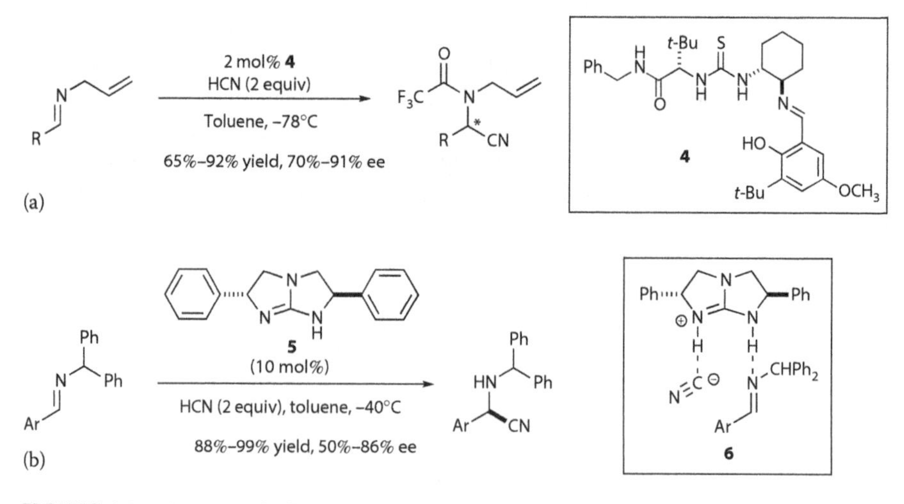

FIGURE 1.2 Asymmetric Strecker reactions reported by (a) Jacobsen and (b) Corey.

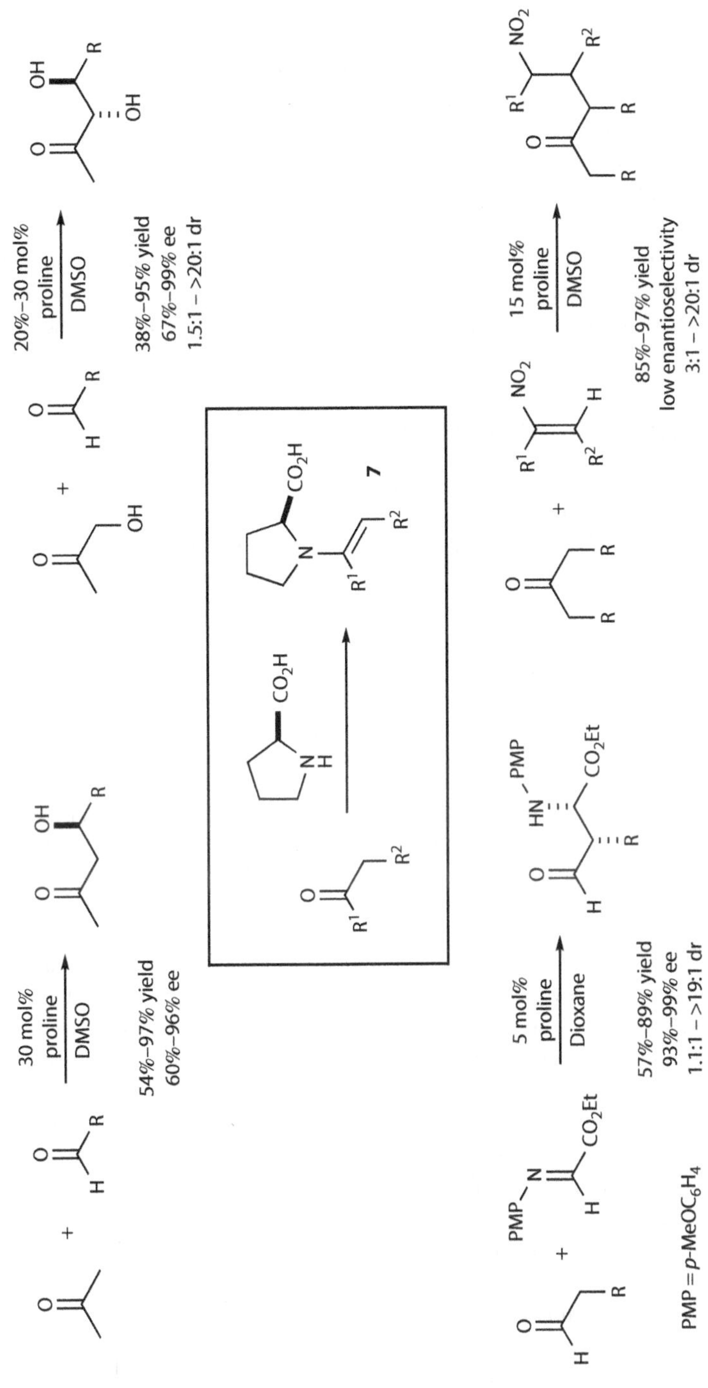

30 mol% proline, DMSO

54%–97% yield
60%–96% ee

20%–30 mol% proline, DMSO

38%–95% yield
67%–99% ee
1.5:1 – >20:1 dr

5 mol% proline, Dioxane

57%–89% yield
93%–99% ee
1.1:1 – >19:1 dr

PMP = p-$MeOC_6H_4$

15 mol% proline, DMSO

85%–97% yield
low enantioselectivity
3:1 – >20:1 dr

FIGURE 1.3 Early proline-catalyzed reactions.

A report from the MacMillan laboratory also appeared in 2000 concerning asymmetric Diels–Alder reactions that were catalyzed by a phenyl alanine-derived imidazolidinone catalyst.[14] This initial report was quickly followed by reports of other asymmetric cycloaddtions,[15] Friedel–Crafts alkylations,[16] and Mukaiyama–Michael reactions (Figure 1.4).[17] All of these reactions are predicated on the formation of an iminium ion (**8**) from the α,β-unsaturated aldehyde starting material and the imidazolidinone catalyst. Once formed, the iminium ion would have a lowest unoccupied molecular orbital (LUMO)-lowering effect similar to that of a Lewis acid–aldehyde complex. This activation manifold has come to be called *iminium ion catalysis* and, much like the proline-catalyzed reactions earlier, has proven to be applicable to numerous reaction types.[13] The orthogonal reactivity of the nucleophilic enamines and electrophilic iminium ions can even work in concert, thus allowing one to devise powerful cascade reactions.[18]

1.2 DEFINITIONS AND INTERPRETATIONS

Another notable detail of the initial MacMillan report[14] is its introduction of a new term to the chemical lexicon—organocatalysis. Interestingly, enough *organocatalysis* is not defined in that original paper. In the foreword of Berkessel and Gröger's 2004 book *Asymmetric Organocatalysis*, MacMillan offered these thoughts on how the term was coined:

> Inspired directly by the work of Shi, Denmark, Yang, Fu, Jacobsen, and Corey, I became convinced of the general need for catalysis strategies or concepts that revolved around small organic catalysts...During the preparation of our Diels–Alder manuscript I became interested in coining a new name for what was commonly referred to as "metal-free catalysis." My motivations for doing so were very simple I did not like the idea of describing as area of catalysis in terms of what it was not, and I wanted to invent a specific term that would set this field apart from other types of catalysis. The term "organocatalysis" was born and a field that had existed for at least 40 years acquired a new name.[19]

As alluded to by MacMillan, it is important to recognize that the groundbreaking work described previously was not done in a vacuum. Scattered reports of enamine and/or iminium ion catalysis can be found dating back to at least 1970s. The most famous of which are the reports of Hajos and Parrish[20] and Eder, Sauer, and Wiechert.[21] Discussion of a proline-catalyzed aldol reaction can also be found in Woodward's synthesis of erythromycin,[22] which is, perhaps, the first use of an organocatalytic reaction in natural product synthesis. The year 1970 also saw a report describing the enhanced reactivity of iminium ion dienophiles in Diels–Alder reactions.[23] The ability of hydrogen-bonding interactions to activate a substrate and organize the ensuing transition state in nonbiological transformations was recognized in the early 1980s[24,25] but was not appreciated as a general approach for catalysis until later. It would be careless to not mention that other examples of reactions are catalyzed solely by small organic molecules appeared between the years 1970 and 2000. In a 2008 *Nature* article, MacMillan offered an enlightening discussion of how these pioneering works inspired the development

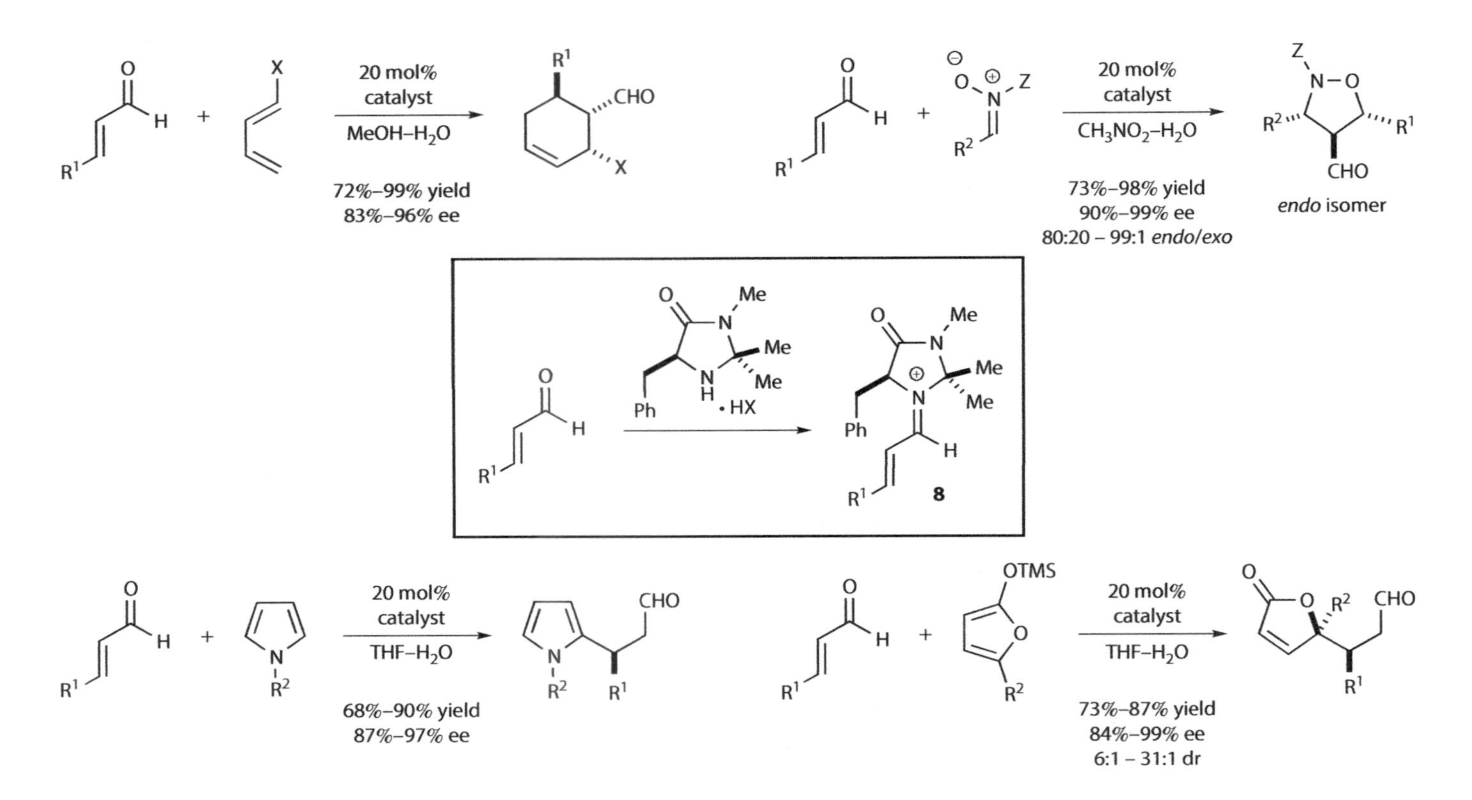

FIGURE 1.4 Early reactions promoted by MacMillan's imidazolidinone catalyst.

of the burgeoning field of organocatalysis.[26] Some of these important results are mentioned earlier, whereas others will be discussed next and in the following chapters.

Unfortunately, the full intent of MacMillan's definition of organocatalysis was not 100% clear at the time of its original usage. As a result, one finds that *organocatalysis* has mostly been used to refer to reactions that involve enamine intermediates, iminium ion intermediates, activation by hydrogen bond donors, or some combination of the three. But, should its use really be this limited? If one follows the spirit of MacMillan's definition,[14,19,26] it is clear that he meant the term *organocatalysis* to have a broad reach that could include many different catalysts and activation methods. For example, should the chiral phase transfer catalysts[25] developed by O'Donnell,[27] Corey,[28] and Maruoka[29] be included under the organocatalysis umbrella? When one sees the variety of different substrates and reactions these catalysts can be applied to,[30] it is easy to conclude that they should be.

Another curious aspect of organocatalysis is that the field seemingly did not have much of a nonasymmetric gestation period. This is in stark contrast with more traditional metal-mediated reactions. For example, transition metal-mediated hydrogenation reactions were known for many decades before the first truly useful asymmetric versions were developed. Similarly, asymmetric Heck reactions and asymmetric olefin metathesis reactions were realized many years after the nonasymmetric versions had been developed. There are, of course, many reasons why asymmetric metal-mediated reactions were slow to develop. Foremost among them being that widely available, commercial analytical methods for rapid determination of enantiomeric composition (i.e., chiral gas chromatography, GC, or high-performance liquid chromatography, HPLC) were not available until at least the 1980s.[31] Nowadays, many researchers have some access to a chiral GC or chiral HPLC column, if they are not housed in the investigator's own laboratory. Having this access significantly lowers the barrier toward developing new asymmetric transformations. Nevertheless, these specialized analytical techniques are only needed if asymmetric reactions are being developed. So the question must be asked: where are the nonasymmetric organocatalytic reactions? Once again, the answer may be in a perceived bias as to what one describes as an organocatalytic reaction or as an organocatalyst.

Many of the prototypical organocatalysts contain at least one nitrogen atom and for many of them, this nitrogen atom is intimately involved in a bond-forming step. In reactions catalyzed by proline (and its derivatives) or MacMillan-type catalysts, a C–N bond must be formed to generate the key enamine or iminium ion intermediate. With thiourea catalysts, the nitrogen is not directly involved in forming a new bond, but the multiple N–H bonds do serve to activate the substrate and organize the transition state. There are also many reactions that are catalyzed by various *Cinchona* alkaloid derivatives that rely on the basicity of the quinuclidine nitrogen to proceed. Together these catalysts exemplify the classic reactivity of nitrogen: nucleophile, base, and H-bond donor. It is quite remarkable how many different catalysts have been devised from just this behavior and how many publications have arisen from it. But by limiting ourselves to nitrogen, we miss out on a host of other organocatalytic reactions.

1.3 EARLY EXAMPLES OF NONNITROGENOUS ORGANOCATALYSIS

In 1968, Morita described a reaction between an aldehyde and an electron-deficient alkene (acrylates and acrylonitrile).[32] The reaction was catalyzed by tricyclohexylphosphine and afforded α-hydroxyalkylated products **9** (Figure 1.5). Several years later, Baylis and Hillman[33] reported that this same reaction could be carried out using a highly nucleophilic tertiary amine catalyst (DABCO). This reaction has since become known as the Morita–Baylis–Hillman (MBH) reaction and has been the subject of intense research in recent decades owing to the highly functionalized nature of the products. It should be mentioned that Rauhut and Currier described a related dimerization of acrylates in a 1963 patent.[34] This reaction was also promoted by a trialkylphosphine.

In 1973, Stetter reported that sodium cyanide promoted the conjugate addition of aldehydes to α,β-unsaturated ketones and esters, affording the corresponding 1,4-dicarbonyl compound (Figure 1.6a).[35] The mechanism of this transformation has some similarity to the benzoin condensation, another cyanide-catalyzed reaction, and one that has been known for more than 180 years (Figure 1.6b).[36] Should sodium cyanide be considered as an organocatalyst? Possibly. The use of cyanide in these reactions has largely been supplanted by thiazolium (e.g., **10**) and imidazolium salts, however. In the presence of mild bases, these salts can be deprotonated to form relatively stable and nucleophilic *N*-heterocyclic carbenes (NHC) (e.g., **11**).[37] Notably, the first report of a benzoin reaction promoted by a thiazolium catalyst appeared

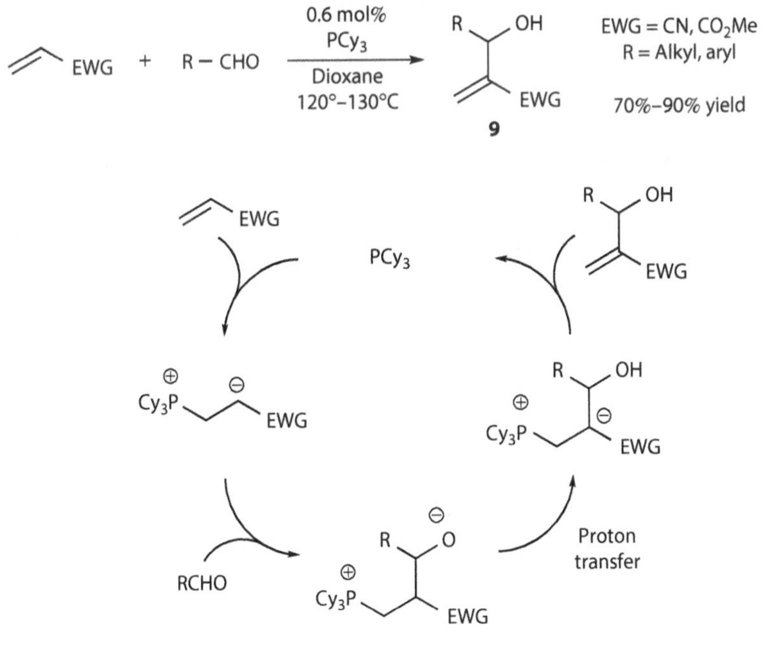

FIGURE 1.5 Morita's phosphine-catalyzed MBH reaction.

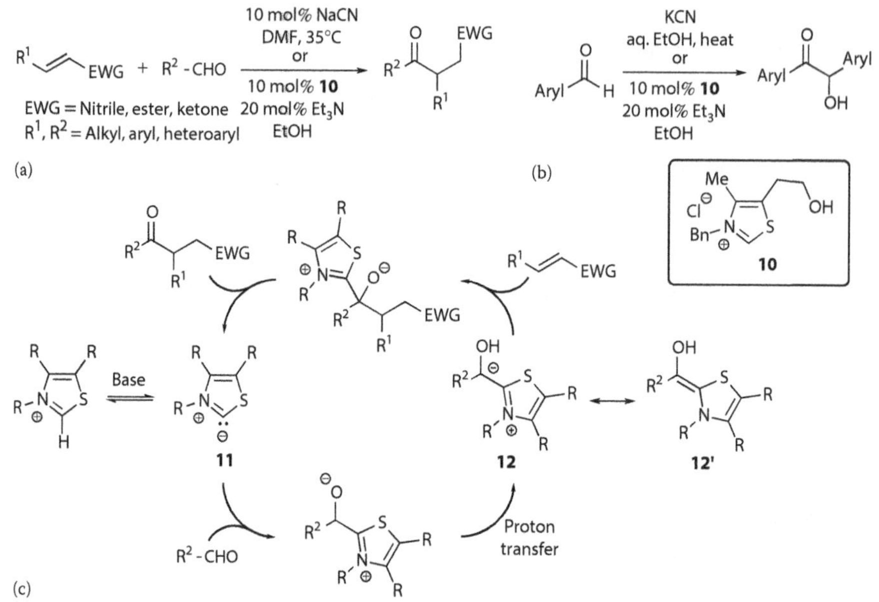

FIGURE 1.6 (a) The Stetter reaction. (b) The benzoin reaction. (c) The generally accepted mechanism of the carbene-catalyzed Stetter reaction.

in 1943,[38] and the first report of a Stetter reaction promoted by a thiazolium catalyst appeared in 1974.[39] This reactivity is quite analogous to the biological mechanism of thiamine (vitamin B1) proposed by Breslow.[40] Given that thiazolium and imidazolium catalysts share reactivity with a coenzyme required by all living organisms, it would be difficult to describe these catalysts as anything but organocatalysts.

The benzoin, Stetter, and MBH reactions were highlighted here not only because they are early (pre-1980) examples of organocatalytic reactions that can be run with an achiral catalyst but also because they proceed by mechanisms in which nitrogen is not intimately involved in a bond-forming step. In the case of the phosphine-catalyzed MBH reaction (Figure 1.5), in which the catalyst obviously does not contain nitrogen, the reaction is initiated by conjugate addition of the phosphine into the electron-deficient alkene, thereby forming a C–P bond. In the benzoin and Stetter reactions (Figure 1.6c), the addition of the carbene catalyst into the aldehyde generates what has now become known as the Breslow intermediate (**12**) and requires the formation of a C–C bond. To be fair, the presence of the nitrogen atom in thiazolium catalyst **10** is necessary to help stabilize both the NHC and the Breslow intermediate, but there is no new bond to this atom.

The Stetter and MBH reactions are also interesting because they employ essentially the same starting materials: an aldehyde and an electron-deficient alkene. It is the identity of the catalyst that determines which product is ultimately formed. One could imagine that if an enolizable aldehyde was used and a secondary amine was used as the catalyst, a third product could be formed. In all of these cases, the catalyst functions as a nucleophile, but all of the catalysts react at different locations. The phosphine preferentially reacts with the alkene, the carbene reacts with the aldehyde, and an amine catalyst would form an enamine. It is really the inherent reactivity of the main group atom (carbon, nitrogen, or phosphorus) involved in the initial bond formation that governs which product is being formed.

1.4 CONCLUSION

Hopefully, this discussion has shown that if we are willing to expand our sights into other main group elements, we will see that many other reaction types, and catalyst types, should be included under the umbrella of organocatalysis. The following chapters will describe the reactivity of several organocatalysts that are either devoid of nitrogen, do not involve formation of new nitrogen bonds in the catalytic cycle, or do not use N–H bonds as hydrogen-bond donors. Some of these catalysts can trace their roots to reactions discussed earlier. Others are still quite new and, therefore, very much in the developmental stage. Nevertheless, these new catalyst designs open the door to reactions nitrogen is not capable of achieving.

REFERENCES

1. (a) Ruble, J. C., Fu, G. C. *J. Org. Chem.* **1996**, *61*, 7230–7231. (b) Ruble, J. C., Latham, H. A., Fu, G. C. *J. Am. Chem. Soc.* **1997**, *119*, 1492–1493.
2. (a) Kawabata, T., Nagato, M., Takasu, K., Fuji, K. *J. Am. Chem. Soc.* **1997**, *119*, 3169–3170. (b) Kawabata, T., Yamamoto, K., Momose, Y., Yoshida, H., Nagaoka, Y., Fuji, K. *Chem. Commun.* **2001**, 2700–2701.
3. Miller, S. J., Copeland, G. T., Papaioannou, N., Horstmann, T. E., Ruel, E. M. *J. Am. Chem. Soc.* **1998**, *120*, 1629–1630.
4. (a) Jarvo, E. R., Copeland, G. T., Papaioannou, N., Bonitatebus, P. J., Jr., Miller, S. J. *J. Am. Chem. Soc.* **1999**, *121*, 11638–11643. (b) Copeland, G. T., Miller, S. J. *J. Am. Chem. Soc.* **2001**, *123*, 6496–6502.
5. Miller, S. J. *Acc. Chem. Res.* **2004**, *37*, 601–610.
6. Sigman, M. S., Jacobsen, E. N. *J. Am. Chem. Soc.* **1998**, *120*, 4901–4902.
7. Corey, E. J., Grogan, M. J. *Org. Lett.* **1999**, *1*, 157–160.
8. Doyle, A. G., Jacobsen, E. N. *Chem. Rev.* **2007**, *107*, 5713–5743.
9. List, B., Lerner, R. A., Barbas, C. F., III *J. Am. Chem. Soc.* **2000**, *122*, 2395–2396.
10. (a) Notz, W., List, B. *J. Am. Chem. Soc.* **2000**, *122*, 7386–7387. (b) List, B., Pojarliev, P., Castello, C. *Org. Lett.* **2001**, *3*, 573–575.
11. (a) Córdova, A., Watanabe, S.-I., Tanaka, F., Notz, W., Barbas, C. F., III *J. Am. Chem. Soc.* **2002**, *124*, 1866–1867. (b) List, B., Pojarliev, P., Biller, W. T., Martin, H. J. *J. Am. Chem. Soc.* **2002**, *124*, 827–833.
12. (a) List, B., Pojarliev, P., Martin, H. J. *Org. Lett.* **2001**, *3*, 2423–2425. (b) List, B., Castello, C. *Synlett* **2001**, 1687–1689.
13. (a) Mukherjee, S., Yang, J. W., Hoffman, S., List, B. *Chem. Rev.* **2007**, *107*, 5471–5569. (b) Jensen, K. L., Dickmeiss, G., Jiang, H., Albrecht, Ł., Jørgensen, K. A. *Acc. Chem. Res.* **2012**, *45*, 248–264. (c) Watson A. J. B., MacMillan, D. W. C. Enantioselective organocatalysis involving iminium, enamine, SOMO, and photoredox activation. In *Catalytic Asymmetric Synthesis*, 3rd ed., Ojima, I., (Ed.), Wiley: Hoboken, NJ, pp. 39–57, 2010.
14. Ahrendt, K. A., Borths, C. J., MacMillan, D. W. C. *J. Am. Chem. Soc.* **2000**, *122*, 4243–4244.

15. (a) Jen, W. S., Wiener, J. J. M., MacMillan, D. W. C. *J. Am. Chem. Soc.* **2000**, *122*, 9874–9875. (b) Northrup, A. B., MacMillan, D. W. C. *J. Am. Chem. Soc.* **2002**, *124*, 2458–2460.
16. (a) Paras, N. A., MacMillan, D. W. C. *J. Am. Chem. Soc.* **2001**, *123*, 4370–4371. (b) Austin, J. F., MacMillan, D. W. C. *J. Am. Chem. Soc.* **2002**, *124*, 1172–1173. (c) Paras, N. A., MacMillan, D. W. C. *J. Am. Chem. Soc.* **2002**, *124*, 7894–7895.
17. Brown, S. P., Goodwin, N. C., MacMillan, D. W. C. *J. Am. Chem. Soc.* **2003**, *125*, 1192–1194.
18. (a) Wang, Y., Mo, P.-F. Application of organocatalytic cascade reactions in natural product synthesis and drug discovery. In *Catalytic Cascade Reactions*, Xu, P.-F., Wang, W., (Eds.), Wiley: Hoboken, NJ, pp. 123–144, 2014. (b) Grondal, C., Jeanty, M., Enders, D. *Nat. Chem.* **2010**, *2*, 167–178. (c) Volla, C. M. R., Atodiresei, I., Rueping, M. *Chem. Rev.* **2014**, *114*, 2390–2431.
19. Berkessel, A., Gröger, H. Forward. *Asymmetric Organocatalysis.* Wiley–VCH: Weinheim, Germany, 2004; pp. 13–14 (Forward written by David MacMillan).
20. (a) Hajos, Z. G., Parrish, D. R. Asymmetric synthesis of optically active polycyclic organic compounds. German Patent DE 2,102,623, July 29, 1971. (b) Hajos, Z. G., Parrish, D. R. *J. Org. Chem.* **1974**, *39*, 1615–1621.
21. (a) Eder, U., Sauer, G., Wiechert, R. Optically active 1,5-indanone and 1,6-napthalenedione. German Patent DE 2,014,757, October 7, 1971. (b) Eder, U., Sauer, G., Wiechert, R. *Angew. Chem. Int. Ed. Engl.* **1971**, *10*, 496–497.
22. Woodward, R. B., Logusch, E., Nambiar, K. P., Sakan, K., Ward, D. E., Au-Yeung, B.-W., Balaram, P. et al. *J. Am. Chem. Soc.* **1981**, *103*, 3210–3213.
23. Baum, J. S., Viehe, H. G. *J. Org. Chem.* **1976**, *41*, 183–187.
24. (a) Hiemstra, H., Wynberg, H. *J. Am. Chem. Soc.* **1981**, *103*, 417–430. (b) Oku, J.-I., Inoue, S. *J. Chem. Soc., Chem. Commun.* **1981**, 229–230.
25. Dolling, U.-H., Davis, P., Grabowski, E. J. J. *J. Am. Chem. Soc.* **1984**, *106*, 446–447.
26. MacMillan, D. W. C. *Nature* **2008**, *455*, 304–308.
27. O'Donnell, M. J., Bennett, W. D., Wu, S. *J. Am. Chem. Soc.* **1989**, *111*, 2353–2355.
28. (a) Corey, E. J., Zhang, F.-Y. *Org. Lett.* **1999**, *1*, 1287–1290. (b) Corey, E. J., Bo, Y., Busch-Petersen, J. *J. Am. Chem. Soc.* **1999**, *120*, 13000–13001. (c) Corey, E. J., Xu, F., Noe, M. C. *J. Am. Chem. Soc.* **1997**, *119*, 12414–12415.
29. (a) Ooi, T., Kameda, M., Maruoka, K. *J. Am. Chem. Soc.* **1999**, *121*, 6519–6520. (b) Ooi, T., Takeuchi, M., Kameda, M., Maruoka, K. *J. Am. Chem. Soc.* **2000**, *122*, 5228–5229. (c) Ooi, T., Kameda, M., Maruoka, K. *J. Am. Chem. Soc.* **2003**, *125*, 5139–5151. (d) Kitamura, M., Shirakawa, S., Maruoka, K. *Angew. Chem. Int. Ed.* **2005**, *44*, 1549–1551.
30. (a) Tan, J., Yasuda, N. *Org. Proc. Res. Dev.* **2015**, *19*, 1731–1746 (b) Shirakawa, S., Moteki, S. A., Maruoka, K. Asymmetric phase-transfer catalysis. In *Modern Tool for the Synthesis of Complex Bioactive Molecules*, Cossy, J., Arseniyadis, S., (Eds.), Wiley: Hoboken, NJ, pp. 213–242, 2012. (c) Ooi, T., Maruoka, K. *Angew. Chem. Int. Ed.* **2007**, *46*, 4222–4266. (d) Maruoka, K., Ooi, T. *Chem. Rev.* **2003**, *103*, 3013–3028.
31. Beesley, T. E., Scott, R. P. W. *Chiral Chromatography.* Wiley: West Sussex, UK, 1998, pp. 2–3.
32. Morita, K., Suzuki, Z., Hirose, H. *Bull. Chem. Soc. Jpn.* **1968**, *41*, 2815.
33. Baylis, A. B., Hillman, M. E. D. Acrylic compounds. German Patent DE 2,155,113, May 10, 1972.
34. Rauhut, M. M., Currier, H. Preparation of dialkyl-2-methylene glutamates. U.S. Patent 3,074,999, January 22, 1963.
35. (a) Stetter, H., Schreckenberg, M. *Angew. Chem. Int. Ed. Engl.* **1973**, *12*, 81. (b) Stetter, H. *Angew. Chem. Int. Ed. Engl.* **1976**, *15*, 639–647.

36. (a) Wöhler, F., Liebig, J. *Ann. Pharm.* **1832**, *3*, 249–282. (b) Lapworth, A. J. *J. Chem. Soc., Trans.* **1903**, *83*, 995–1005. (c) Lapworth, A. J. *J. Chem. Soc., Trans.* **1904**, *85*, 1206–1215.
37. Dove, A. P., Pratt, R. C., Lohmeijer, B. G. G., Li, H., Hagberg, E. C., Waymouth, R. M., Hedrick, J. L. N-Heterocyclic carbenes as organic catalysts. In *N-Heterocyclic Carbenes in Synthesis*, Nolan, S. P., (Ed.), Wiley-VCH: Weinheim, Germany, pp. 275–296, 2006.
38. Ukai, T., Tanaka, R., Dokawa, T. *J. Pharm. Soc. Jpn.* **1943**, *63*, 296–300.
39. Stetter, H., Kuhlmann, H. *Angew. Chem. Int. Ed. Engl.* **1974**, *13*, 539.
40. Breslow, R. *J. Am. Chem. Soc.* **1958**, *80*, 3719–3726.

2 Alcohols and Phenols as Hydrogen Bonding Catalysts

Yunus Emre Türkmen

CONTENTS

2.1 INTRODUCTION

Hydrogen bonding catalysis has emerged as an exciting subfield of homogeneous catalysis within the last two decades and has found widespread applications in synthetic organic chemistry [1–7]. In this type of catalysis, a small-molecule hydrogen bond donor activates a substrate through one or more hydrogen bonding interactions, whereas in the closely related area of Brønsted acid catalysis, a substrate is activated via full proton transfer [8–12]. These two approaches, which form a continuum and thus complement each other, can be utilized effectively in the catalysis of a broad range of transformations. In this chapter, selected examples of hydrogen bonding catalysts based on alcohols and phenols will be covered with an emphasis on detailed mechanistic investigations.

Hine and coworkers reported a series of studies between 1984 and 1987 that constituted a major breakthrough in the area of hydrogen bonding catalysis [13–19]. In this pioneering work, 1,8-biphenylenediol was demonstrated to be a highly effective dual hydrogen bond donor to bind to oxygen-containing functional groups such as carbonyls, epoxides, and phosphoramides. The initial studies by Hine and coworkers, published in 1984, reported the crystal structures of the parent 1,8-biphenylenediol (**1**) complexed with hexamethylphosphoramide (HMPA), 1,2,6-trimethyl-4-pyridone, and 2,6-dimethyl-γ-pyrone (Figure 2.1) [13]. In all three crystal structures, biphenylenediol was found to make dual hydrogen bonds to the Lewis basic oxygen atoms present in each hydrogen bond acceptor. The distances between two oxygens (O-H•••O) in all three complexes were between 2.54 Å and 2.61 Å, which indicate the presence of two strong

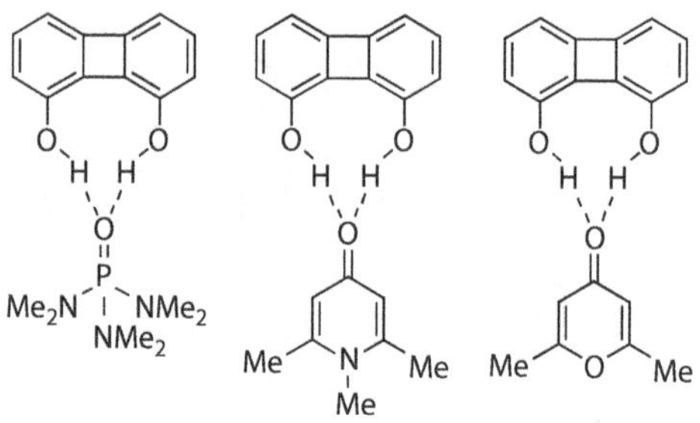

FIGURE 2.1 Complexes of 1,8-biphenylenediol (**1**) with various hydrogen bond acceptors.

hydrogen bonds in each complex [20]. In addition, all the O-H•••O angles were found to be between 168° and 178°, which are close to the ideal angle of 180° for strong hydrogen bonds.

Following the initial discovery, 1,8-biphenylenediol was used by Hine and coworkers in 1985 as a hydrogen bonding catalyst in the ring-opening reaction of phenyl glycidyl ether with diethylamine [14,16]. Besides biphenylenediol **1**, a series of other phenols were tested as catalysts, and the kinetics of the ring opening reactions were investigated thoroughly for the determination of the catalysis constants (k_c) for each catalyst (Scheme 2.1). Although all the mono-phenolic compounds tested exhibited a good correlation between their catalytic activities and their pK_a values in water, biphenylenediol **1** showed a remarkably higher activity, presumably due to its double hydrogen bonding nature with the substrate, phenyl glycidyl ether. Subsequently, the Hine group described the synthesis of 4,5-dinitro-1,8-biphenylenediol (**11**) in 1987 [18,19]. In accordance with the expectations due to the electron-withdrawing effect of the nitro groups, diol **11** was found not only to have higher acidity compared to biphenylenediol **1** but also to have higher association constants with various Lewis bases as hydrogen bond acceptors [18].

In 1990, Kelly and coworkers used a biphenylenediol derivative as a dual hydrogen bonding catalyst in the Diels–Alder (DA) reaction of various dienes with

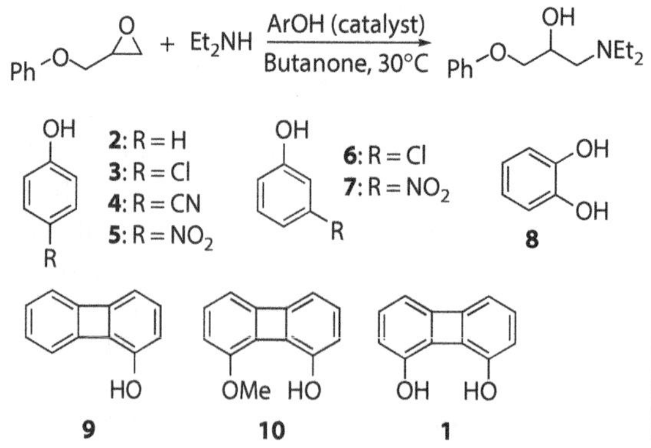

Catalyst	10^5k_c ($M^{-2}s^{-1}$)	pK_a in water
2	6.0	9.98
3	7.7	9.38
4	15.3	7.95
5	17.0	7.15
6	8.2	9.02
7	14.3	8.40
8	11.9	9.49
9	11.5	8.64
10	7.3	9.15
1	75	8.00

SCHEME 2.1 Ring-opening reaction of phenyl glycidyl ether catalyzed by hydrogen bond donors.

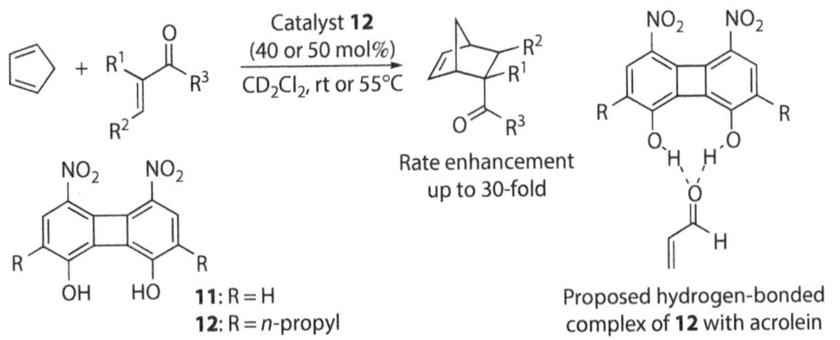

Catalyst **12** (40 or 50 mol%) CD_2Cl_2, rt or 55°C

Rate enhancement up to 30-fold

11: R = H
12: R = *n*-propyl

Proposed hydrogen-bonded complex of **12** with acrolein

SCHEME 2.2 Diels–Alder reactions catalyzed by biphenylenediol **12**.

α,β-unsaturated carbonyl compounds [21]. Dinitro-substituted biphenylenediol scaffold was selected as a potential catalyst candidate due to its higher hydrogen bond donating ability compared to the parent biphenylenediol **1** [19]. However, since the dinitro-substituted biphenylenediol **11** showed insufficient solubility in noncoordinating solvents such as CH_2Cl_2, the more soluble, dipropyl-substituted diol **12** was prepared and used in catalytic studies. In the presence of 40 or 50 mol% of diol **12**, the rates of the DA reactions of cyclopentadiene with several α,β-unsaturated aldehydes and ketones were shown to be enhanced significantly up to 30-fold with reference to the background reaction (Scheme 2.2). Although other dienes such as 2,3-dimethyl-1,3-butadiene and 1-methoxy-1,3-butadiene were found to be potentially useful dienes in this catalytic process, the use of α,β-unsaturated esters did not exhibit a significant rate enhancement. In accordance with the findings of Hine [13], a doubly hydrogen-bonded complex formation between diol **12** and α,β-unsaturated carbonyl compound was proposed by the authors. The inferior results obtained by the use of *p*-nitrophenol and 4-nitro-3-(trifluoromethyl)phenol as monophenol-containing catalysts in control experiments supported this hypothesis.

2.2 ALCOHOLS AS HYDROGEN BONDING CATALYSTS

The field of hydrogen bonding catalysis using alcohols and phenols has witnessed a renaissance at the beginning of this century, fueled particularly by the development of chiral alcohols for the catalysis of enantioselective reactions. The spark for the discovery of chiral, alcohol-based hydrogen bond donors originated from a solvent-screening study for a cycloaddition reaction in 2002 [22]. In this work, Rawal and coworkers observed that the rates of the hetero-Diels–Alder (HDA) reaction between 1-amino-3-siloxybutadiene (**13**) and *p*-anisaldehyde (**14**) did not exhibit a correlation with the dielectric constants of the solvents studied (Scheme 2.3). For instance, the rate of formation of the HDA adduct **15** was 10 and 30 times faster in chloroform than the more polar aprotic solvents acetonitrile and tetrahydrofuran (THF), respectively. In protic solvents such as *tert*-butanol and isopropanol the rate enhancement was even more pronounced where a relative rate value as high as 630 was obtained in isopropanol compared to THF. The possibility of catalysis by trace amount of

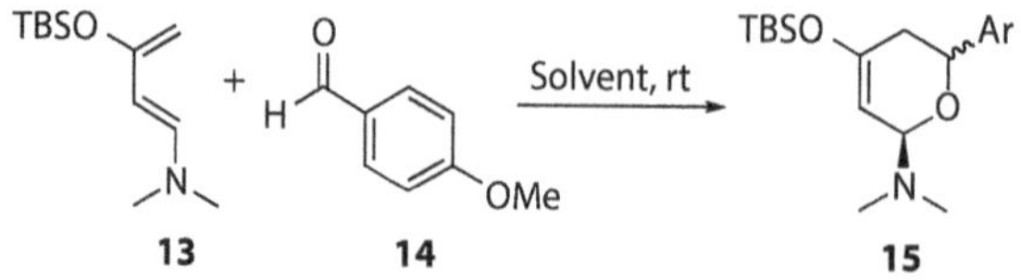

Solvent	Relative rate
THF-d_8	1
Benzene-d_6	1.3
Acetonitrile-d_3	3.0
Chloroform-d	30
tert-butanol-d_{10}	280
Isopropanol-d_8	630

SCHEME 2.3 Effect of solvents on the rate of hetero-Diels–Alder reaction.

acid that might be present in chloroform was excluded by the control experiments in which the addition of a catalytic amount of HCl or triethylamine as an acid scavenger did not influence the reaction rate. The higher rates observed in alcohols and chloroform were therefore attributed to hydrogen bond-making abilities of these solvents and the activation of the C=O group of *p*-anisaldehyde (**14**) by such hydrogen bonds. The authors demonstrated a successful application of this observation in the more challenging HDA reactions of diene **13** with unactivated ketones where the products were obtained in high yields at room temperature when 2-butanol was used as a solvent [22].

The fundamental observation that high-rate enhancement values could be obtained when solvents with hydrogen bond-donating abilities were used in HDA reactions paved the way for the discovery that chiral alcohols could be used to impart enantioselectivity during the catalysis of such reactions. In the groundbreaking work of Rawal and coworkers reported in 2003, α,α,α′,α′-tetraaryl-2,2-disubstituted 1,3-dioxolane-4,5-dimethanol (TADDOL) derivative **16** was identified as a highly effective hydrogen bonding catalyst for the HDA reaction between diene **13** and various aldehydes (Scheme 2.4) [23]. The resulting dihydropyrone products were obtained in good to high yields (52%–97%) and with excellent levels of enantioselectivity (93:7 to >99:1 enantiomeric ratio). Importantly, the reaction was found to tolerate electron-rich and electron-deficient aromatic aldehydes as well as *trans*-cinnamaldehyde and cyclohexanecarboxaldehyde. Through this work, it was shown for the first time that high levels of enantio-induction could be achieved in a cycloaddition reaction using a chiral hydrogen bonding catalyst.

The scope of cycloaddition reactions that could be catalyzed by TADDOL-based hydrogen bond donors was extended to *all*-carbon DA reaction by Rawal and coworkers [24]. An initial catalyst screening in the DA reaction of aminosiloxydiene **13** with methacrolein (**17**) revealed a strong dependence of the reaction outcome on the nature of the aryl groups in TADDOL catalysts (Scheme 2.5). Interestingly,

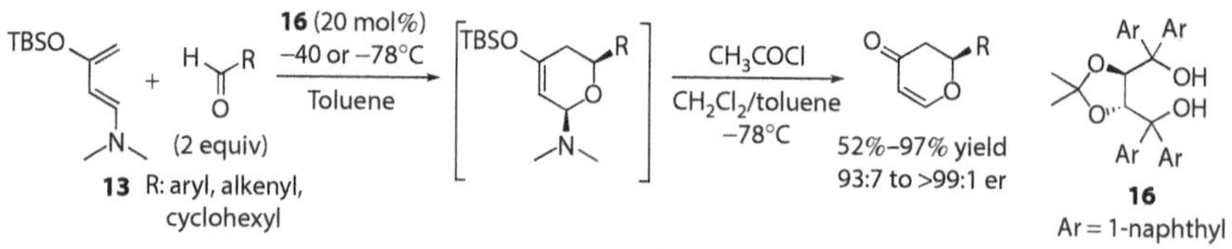

SCHEME 2.4 TADDOL-catalyzed enantioselective HDA reactions.

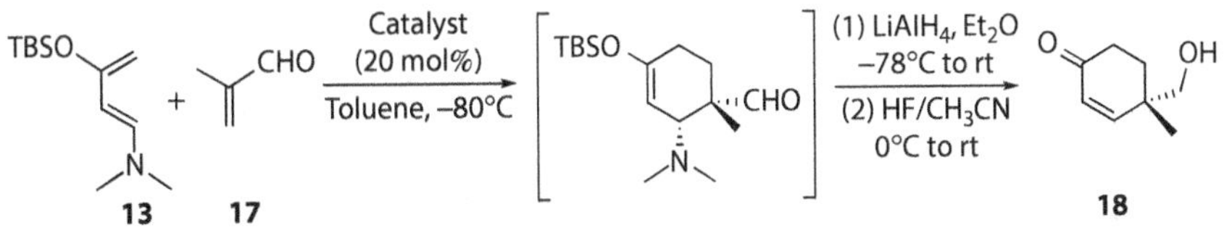

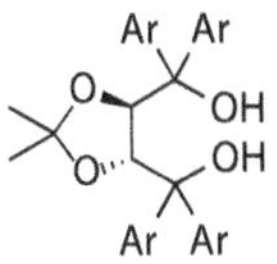

Catalyst	Ar	Yield (%)	ee (%)
19	Ph	30	31
20	2-naphthyl	45	33
16	1-naphthyl	83	91

SCHEME 2.5 Catalyst optimization for enantioselective Diels–Alder reactions.

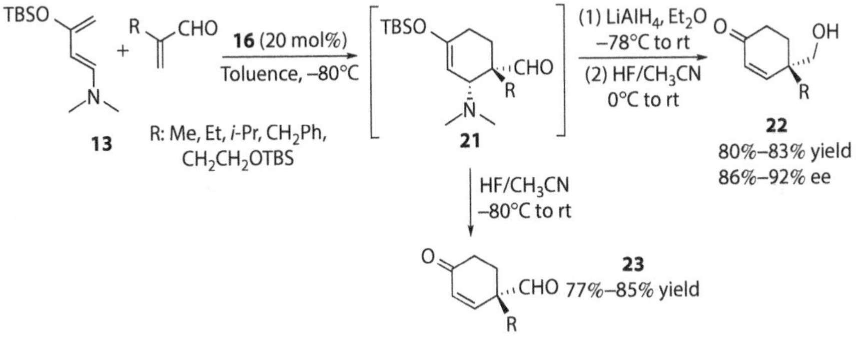

SCHEME 2.6 TADDOL-catalyzed enantioselective Diels–Alder reactions.

not only enantioselectivity but also reactivity was found to be affected significantly with the nature of the aryl groups such that 1-naphthyl-substituted TADDOL (**16**) afforded the cyclohexenone product (**18**) with the highest yield and enantiomeric excess (ee) values (83% yield, 91% ee). The scope of the reaction was investigated with respect to the dienophile component using various substituted α,β-unsaturated aldehydes (Scheme 2.6). When the initial *endo*-DA adducts (**21**) were subjected to reduction by $LiAlH_4$ followed by treatment with HF in acetonitrile, the resulting alcohol products (**22**) were obtained in high yields (80%–83%) and with high enantioselectivities (86%–92% ee). The diminished enantioselectivity with acrolein (73% ee) indicates the importance of the 2-substituent on enal substrates for attaining high levels of selectivity. The preparation of aldehyde products (**23**) was also shown to be feasible by the direct treatment of initial DA adducts (**21**) with HF/CH_3CN at low temperature. Overall, this methodology introduced a highly practical and convenient way for the asymmetric synthesis of cyclohexenone derivatives bearing *all*-carbon stereogenic centers.

In order to explain the mechanism of action of TADDOL derivatives as hydrogen bonding catalysts, a model was proposed by Rawal and coworkers [24]. This model was based on earlier crystal structures of TADDOLs obtained by the Seebach and Toda groups [25,26]. TADDOL derivatives are known to have a rigid,

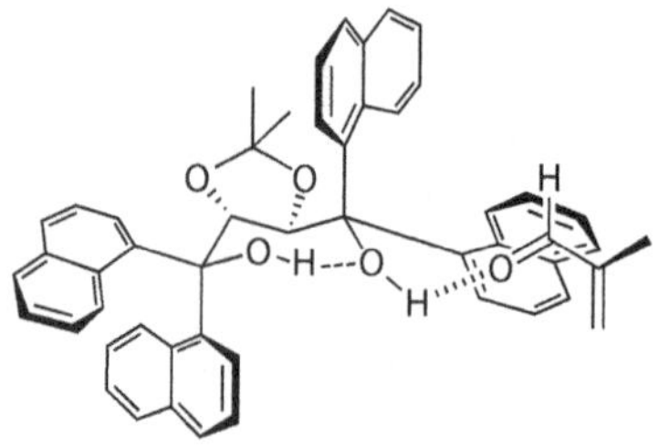

FIGURE 2.2 Proposed H-bonded complex between TADDOL **16** and methacrolein.

seven-membered ring conformation with an intramolecular hydrogen bond between the two hydroxyl groups. This type of intramolecular hydrogen bond is expected to increase the hydrogen bond-donating ability of the second hydroxyl, which remains free to interact with a suitable hydrogen bond acceptor. The proposed structure of the hydrogen-bonded complex of TADDOL **16** with methacrolein proposed by Rawal and coworkers is shown in Figure 2.2 [24]. In this working model, a π–π interaction between methacrolein and one of the naphthyl groups of **16** was proposed to be present in addition to the main hydrogen bonding interaction. This way, one of the faces of methacrolein is expected to be blocked by the naphthyl group, which allows the approach of the diene from the other face leading to high enantioselectivities in the DA reactions. In 2006, a crystal structure of a hydrogen-bonded complex between *rac*-**16** and *p*-anisaldehyde was reported by the Rawal group [27].

The ability of TADDOL-based alcohols to function as hydrogen bonding catalysts was utilized by the Yamamoto and Rawal groups in 2005 for the development of binaphthalene-derived alcohol catalysts possessing axial chirality [28]. Chiral 1,1′-biaryl-2,2′-dimethanol (BAMOL) derivatives were developed as hydrogen bond donors for the catalysis of HDA reactions of electron-rich diene **13** with aromatic and aliphatic aldehydes (Scheme 2.7). An initial catalyst screening revealed BAMOLs **25** and **26** as the best-performing catalysts in terms of both reactivity and selectivity.

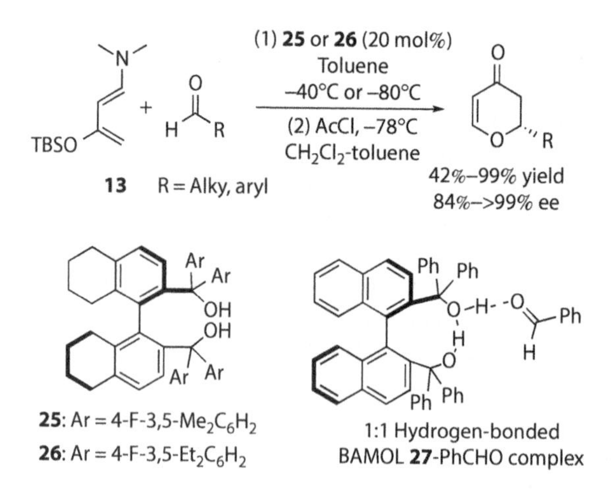

SCHEME 2.7 BAMOL-catalyzed enantioselective HDA reactions.

The resulting dihydropyrone products were obtained in good to high yields and with high levels of enantioselectivity. It is worth noting that a broad range of aliphatic aldehydes could be used as substrates in this methodology. The mechanism of action of the BAMOL catalysts was proposed by the authors to be similar to TADDOL-based hydrogen bonding catalysts. This proposal was supported by the crystal structure of 1:1 complex of BAMOL **27** with benzaldehyde (Scheme 2.7). The X-ray crystallographic analysis of this 1:1 complex revealed an intramolecular hydrogen bond between the two hydroxyls of BAMOL **27** resulting in the formation of a nine-membered ring. A second, intermolecular hydrogen bond is present between the free hydroxyl of BAMOL **27** and the carbonyl oxygen of benzaldehyde.

The reaction space of enantioselective TADDOL-catalyzed reactions was expanded further by the Rawal group to Mukaiyama aldol reactions [27,29–31]. In a study reported in 2006, the Mukaiyama aldol reaction between *O*-silyl-*N, O*-ketene acetal **28**, and a wide range of aldehydes was shown to be promoted by hydrogen bonding catalysis in a highly diastereo- and enantioselective manner (Scheme 2.8) [27]. An initial screening of various TADDOLs allowed the identification of cyclohexylidene-TADDOL derivative **29** as the optimized catalyst for achieving both high diastereo- and enantioselectivity values. The reaction was found to have a broad substrate scope with respect to the aldehyde component tolerating a high number of electron-deficient and electron-rich aromatic aldehydes as well as butanal as an aliphatic aldehyde. When TADDOL **29** was used in 10 mol% catalyst loading, β-hydroxyamide products were obtained in good to high yields (47%–94%), with high distereoselectivities (*syn*/*anti*: 2:1 to >25:1) and excellent enantioselectivities for the *syn* products (87%–98%). In an attempt to demonstrate the synthetic utility of this methodology, the authors reported the successful transformation of three amide aldol products to the corresponding aldehydes using Schwartz's reagent (Scheme 2.8). Under the optimized conditions, synthetically useful β-siloxyaldehyde products were obtained in high yields (84%–88%) and with little or no epimerization of the α-stereocenters, starting from TBS-protected β-hydroxyamides.

The electrophile scope of TADDOL-catalyzed Mukaiyama aldol reactions of *O*-silyl-*N, O*-ketene acetals was extended to acyl phosphonates by Rawal and coworkers [30]. Among various catalysts tested including TADDOL derivatives, BINOL, cinchona alkaloids, Takemoto's thiourea [32], and sugar-derived alcohols,

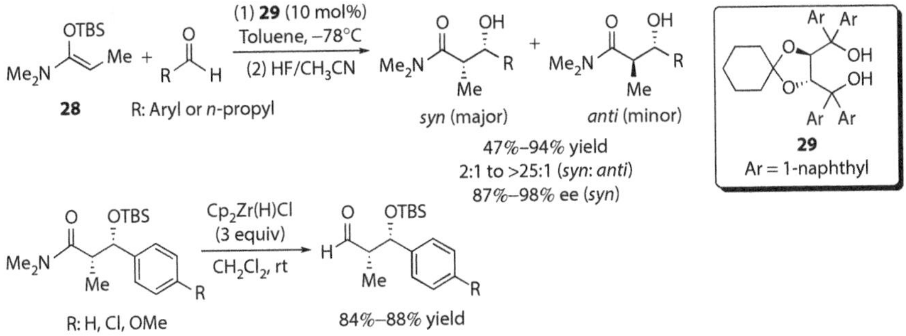

OTBS Me

(1) **29** (10 mol%) Toluene, –78°C; (2) HF/CH_3CN

28 — R: Aryl or *n*-propyl

syn (major) + *anti* (minor)

47%–94% yield
2:1 to >25:1 (*syn*: *anti*)
87%–98% ee (*syn*)

29
Ar = 1-naphthyl

$Cp_2Zr(H)Cl$ (3 equiv), CH_2Cl_2, rt

R: H, Cl, OMe

84%–88% yield

SCHEME 2.8 TADDOL-catalyzed enantioselective Mukaiyama-aldol reactions.

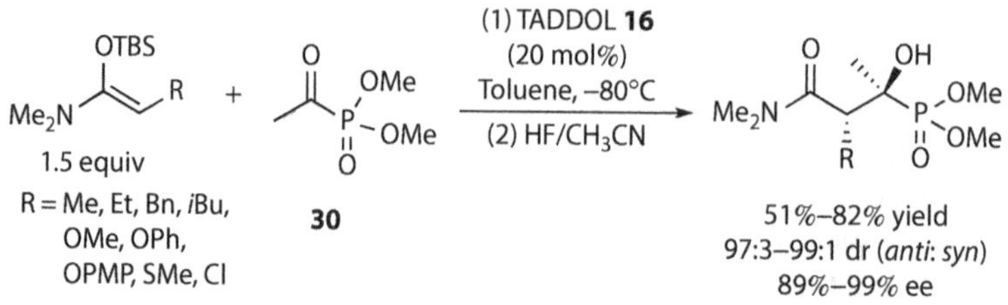

OTBS
Me2N
R
1.5 equiv
R = Me, Et, Bn, iBu, OMe, OPh, OPMP, SMe, Cl
+
O
OMe
P
OMe
O
30
(1) TADDOL 16
(20 mol%)
Toluene, –80°C
(2) HF/CH3CN
O
OH
Me2N
OMe
P
OMe
R
O
51%–82% yield
97:3–99:1 dr (anti: syn)
89%–99% ee

SCHEME 2.9 TADDOL-catalyzed enantioselective Mukaiyama-aldol reactions of *O*-silyl-*N*, *O*-ketene acetals with acetyl dimethyl phosphonate.

TADDOLs **16** and **29** were proved to be the best catalysts in terms of yield as well as diastereo- and enantioselectivity. Between these two catalysts, the commercially available TADDOL **16** was selected due to its easier accessibility. The Mukaiyama aldol reactions of a broad range of *O*-silyl-*N*, *O*-ketene acetals with acetyl dimethyl phosphonate (**30**) afforded the resulting chiral α-hydroxy-phosphonate products in good to high yields (51%–82%) and with excellent diastereo- (97:3-99:1 dr, *anti*:*syn*) and enantioselectivities (89%–99% ee, Scheme 2.9). Generation of two contiguous stereogenic centers one of which is quaternary and the successful use of heteroatom-substituted ketene acetals as substrates are noteworthy features of this transformation.

In 2005, Yamamoto and coworkers reported an elegant utilization of two different hydrogen bonding catalysts to achieve *O*- and *N*-regioselectivity in the enantioselective aldol-type reactions of achiral enamines with nitrosobenzene (**31**) [33]. When TADDOL **16** was used as a catalyst with 30 mol% loading, *N*-nitroso aldol products were obtained in good to high yields (63%–91%) and enantioselectivities (65%–91% ee) (Scheme 2.10). In contrast, complete *O*-selectivity was attained in the nitroso aldol reactions when (*S*)-1-naphthyl glycolic acid (**32**) was used as a catalyst (Scheme 2.11). Importantly, the regioselectivity of the process was found to be dependent not only on the type of the catalyst used but also on the structures of the enamine reactants. In order to provide a rationale for the different regioselectivities obtained, the

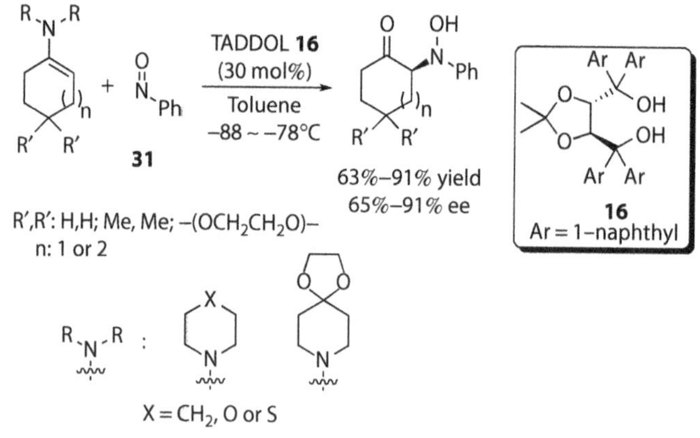

R N R
R' R'
n
+
O
N
Ph
31
TADDOL 16
(30 mol%)
Toluene
–88 ~ –78°C
O
OH
N
Ph
n
R' R'
63%–91% yield
65%–91% ee
Ar Ar
OH
OH
Ar Ar
16
Ar = 1-naphthyl
R',R': H,H; Me, Me; –(OCH2CH2O)–
n: 1 or 2
R N R :
X
N
O O
N
X = CH2, O or S

SCHEME 2.10 *N*-Selective nitroso aldol reactions catalyzed by TADDOL **16**.

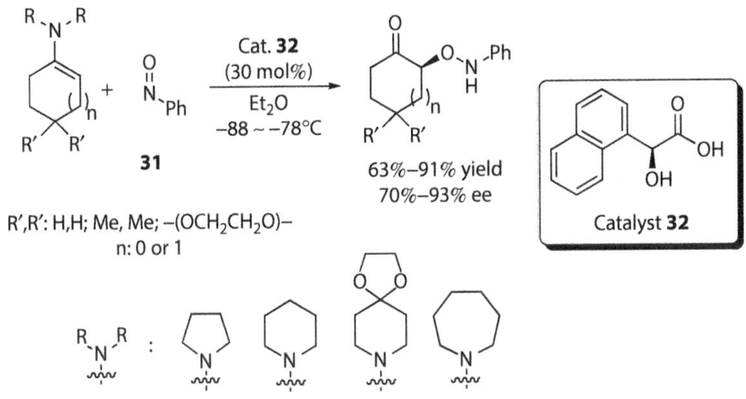

SCHEME 2.11 *O*-Selective nitroso aldol reactions catalyzed by (*S*)-1-naphthyl glycolic acid (**32**).

mechanism of the nitroso aldol reaction explained earlier was investigated computationally by Akakura et al. [34].

The cooperative behavior of the two alcohol units in TADDOL and BAMOL-type hydrogen bonding catalysts was utilized elegantly by Imahori and coworkers for the development of a photoswitchable catalyst system in 2012 [35]. In this design, BAMOL catalyst scaffold and a photoresponsive azobenzene unit have been merged leading to azobenzene-tethered bis(trityl alcohol) derivatives as photoswitchable hydrogen bonding catalysts (Scheme 2.12). It was anticipated that the two alcohol groups of **33** in its *trans* form would be positioned away from each other and therefore would not exhibit cooperativity as they would act independently as hydrogen bond donors. On the other hand, *cis*-**33** was expected to exhibit a cooperative behavior due to the spatial proximity of the two alcohols. In order to test this hypothesis, the authors first investigated the *cis–trans* isomerization of the bis(trityl alcohol) derivative **33**. In accordance with the expectations, the *trans*-to-*cis* isomerization was accomplished by the irradiation of a sample of *trans*-**33** with UV light at 365 nm

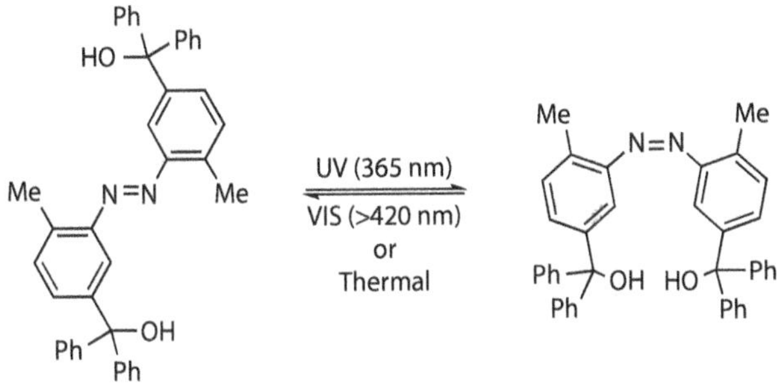

***trans*-33**
Catalytically OFF state

***cis*-33**
Catalytically ON state

SCHEME 2.12 *Cis–trans* isomerization of diol **33**.

leading to a *cis*/*trans* ratio of 95:5. Similarly, it was shown that the *cis*-isomer could be converted back to the *trans* isomer by irradiation using visible light (>420 nm) or thermally by heating at 150°C.

Following the successful demonstration of the photoswitchable behavior of the bis(trityl alcohol) **33**, Imahori and coworkers tested the catalytic activity of this diol in the Morita–Baylis–Hillman (MBH) reaction of 3-phenyl-1-propanal (**34**) and 2-cyclopenten-1-one (**35**) (Scheme 2.13) [35]. In the absence of a hydrogen bond donor, the background reaction afforded the MBH product **36** in 26.5% yield when $n\mathrm{Bu_3P}$ was used as a Lewis base catalyst. Not surprisingly, *trans*-**33** showed only a modest increase in yield (37%). On the other hand, the *cis*-isomer of **33** (*cis:trans* = 95:5) induced a high-rate enhancement in the MBH reaction increasing the yield to 78%, which corresponds to 81% yield when extrapolated for pure *cis*-isomer. The notable difference in chemical yields obtained for the *trans*- and *cis*-isomers of **33** is in agreement with the original hypothesis of the authors that the two –OH groups would exhibit a cooperative behavior in the *cis* form. This was further supported by the observation that trityl alcohol (**37**) afforded the MBH adduct **36** in 36% yield when used in 40 mol% catalyst loading. The similar chemical yields obtained by *trans*-**33** (20 mol%) and trityl alcohol (**37**, 40 mol%) indicated that the two alcohol units in *trans*-**33** acted independently of each other as hydrogen bond donors. Additional control experiments revealed diols **38**, **39**, and (*S*)-BINOL to be inferior catalysts compared to *cis*-**33** resulting in lower yields for the formation of the MBH adduct **36** (47%, 37%, and 61% yield, respectively).

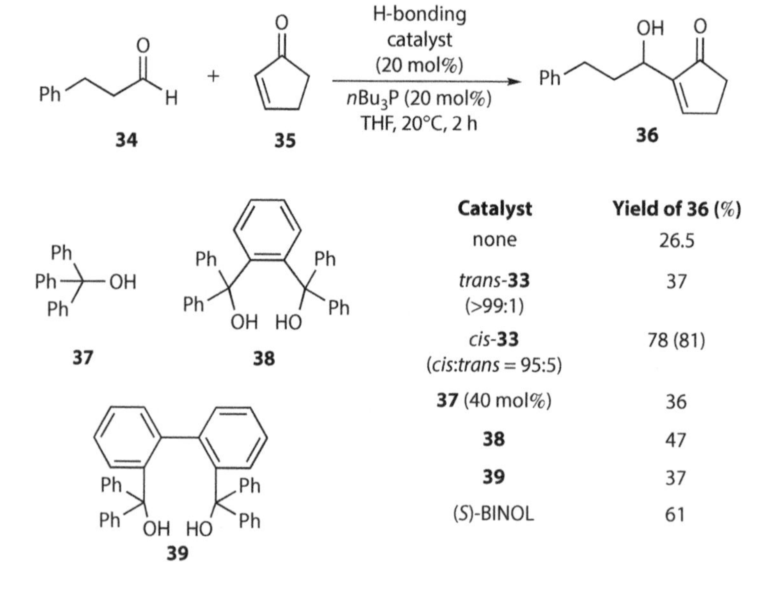

SCHEME 2.13 Morita–Baylis–Hillman reaction catalyzed by photoswitchable diol **33**.

2.3 FLUORINATED ALCOHOLS AS HYDROGEN BONDING CATALYSTS

Development of more effective catalysts to promote reactions with inherent low reactivity is a major theme in the area of hydrogen bonding catalysis. One strategy to achieve this goal is to use dual hydrogen bond donors to activate a substrate via two-point hydrogen bonding [13]. Another effective strategy that has been frequently employed is to increase the acidity of the hydrogen bond donor by the incorporation of electron-withdrawing groups [36,37]. In this respect, fluorination has been proven to have a positive effect in many cases on the activities of alcohols both as catalysts and reaction solvents due to the increased hydrogen bonding ability of the –OH groups and the decreased nucleophilicity of alcohol oxygens. Indeed, a variety of reactions have been shown to proceed in higher yields and with better selectivities in fluorinated alcohol solvents such as 2,2,2-trifluoroethanol (TFE) and 1,1,1,3,3,3-hexafluoro-2-propanol (HFIP) [38–40]. In recent years, fluorinated alcohols started to find widespread use as hydrogen bonding catalysts in a broad range of reactions.

A seminal work in this area reported by Hedrick and coworkers in 2009 described the use of dual hydrogen bond donor catalysts containing fluorinated alcohols for the ring-opening polymerization (ROP) reaction of lactides [41]. Among the various alcohols and diols screened as catalysts for the ROP of L-lactide (**40**), fluorinated hydrogen bond donors **41**, **42**, and **43** were found to be the most effective catalysts (Table 2.1). The polylactide (PLA) products were obtained with high conversion and molecular weight (M_n) values and with narrow polydispersities (low PDI). (–)-Sparteine was used as a cocatalyst base in this reaction, whereas benzyl alcohol (BnOH) acted as the alcohol initiator. Interestingly, fluorinated alcohol derivative **43** exhibited a comparable catalytic activity to the dual hydrogen bond donors **41** and **42**. On the other hand, nonafluoro-*tert*-butyl alcohol

TABLE 2.1
Ring-Opening Polymerization of L-Lactide Catalyzed by Fluorinated Alcohols

H-bonding catalyst, BnOH, (–)-sparteine

40 **40**: Catalyst:BnOH:sparteine = 200:5:1:1

41: R = Vinyl
42: R = H
43

Catalyst	Time (h)	Conversion (%)	M_n (g/mol)	PDI
41	52	73	25,900	1.07
41	91	95	27,500	1.08
42	47	72	19,000	1.06
43	50	75	22,400	1.09

did not promote the ROP of L-lactide even with an extended reaction time (143 h). The lack of catalytic activity was attributed to the high steric bulk of this alcohol. Finally, the use of HFIP as a less bulky fluorinated alcohol led to a high conversion value (91%) albeit with a lower molecular weight (12,900 g/mol) and higher PDI (1.28). These results suggest the possibility that HFIP can participate in the reaction in addition to acting as a catalyst due to its lower steric bulk.

The hydrogen bond formation between the fluorinated alcohol derivatives and carbonyl compounds were investigated by the Hedrick group both experimentally and computationally (Figure 2.3) [41]. Titration studies using ^{13}C-NMR spectroscopy were first performed to investigate the hydrogen bonding between fluorinated diol **44** and valerolactone (VL). The binding stoichiometry for the hydrogen-bonded complex between these two compounds was found to be 1:1, whereas alcohol **43** interacts with VL in a 2:1 ratio. Similarly, when diol **44** was titrated with VL in C_6D_6, ^{1}H-NMR spectra indicated strong hydrogen bonds between the diol and the lactone derivatives. The hydrogen-bonded complex formation was further supported by computational studies in which a dual hydrogen bond formation was observed between fluorinated diol **42** and L-lactide (**40**).

In 2015, Gilbert, Detrembleur, Jerome, and coworkers developed a solvent-free carbon dioxide fixation methodology for the synthesis of cyclic carbonates from epoxides using fluorinated alcohols as hydrogen bonding catalysts [42]. The conversion of 1,2-epoxydodecane (**45**) to the corresponding cyclic carbonate **46** was investigated in the presence of CO_2 (80 bar), nBu$_4$NI catalyst, and various alcohol and diol hydrogen bond donors at 80°C (Scheme 2.14). In the absence of a hydrogen bond donor, the reaction was found to be extremely slow (k = 0.0005 min^{-1}) and required more than five days to reach >98% conversion. The relative rate of the reaction exhibited a direct correlation with the extent of fluorination on the catalysts investigated. For instance, although *tert*-butanol (**47**) showed only a modest increase in the reaction rate, a relative rate value as high as 165 was obtained using perfluoro-*tert*-butanol (**49**) as the hydrogen bonding catalyst. In addition, the authors tested diol **50** and its fluorinated analog **42** as bidentate hydrogen bond donors. Not surprisingly,

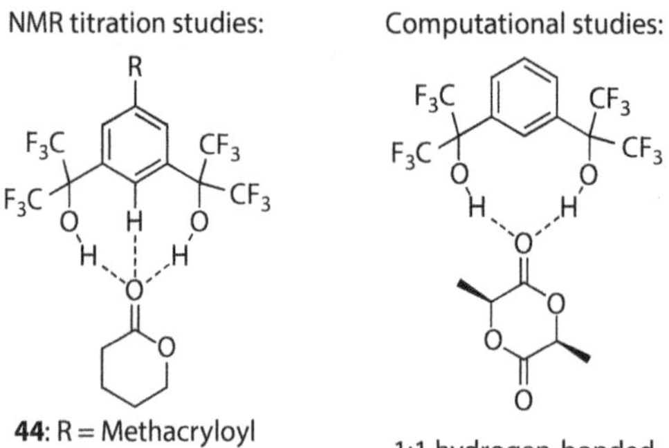

FIGURE 2.3 NMR titration and computational studies on the formation of H-bonded complexes.

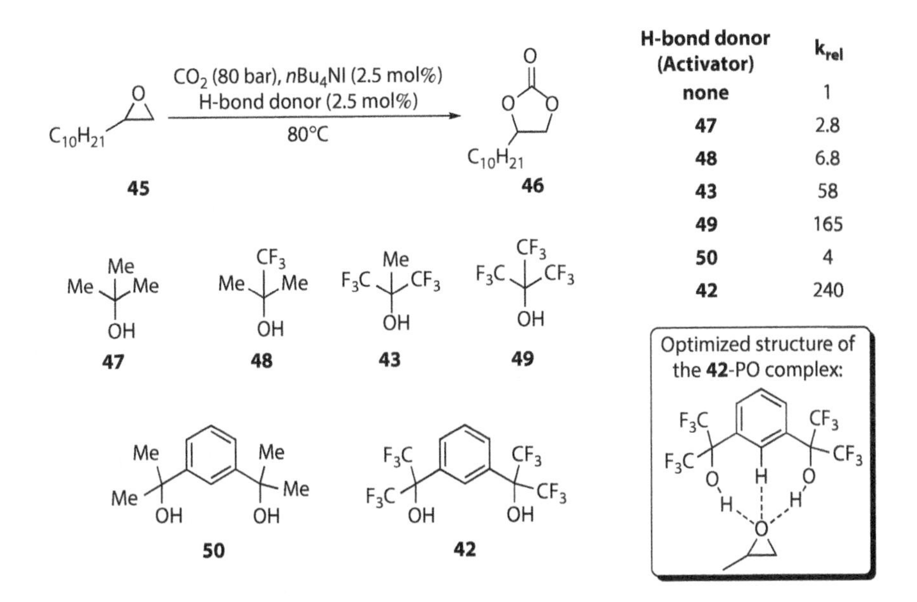

SCHEME 2.14 Fixation of carbon dioxide into epoxides catalyzed by fluorinated alcohols.

fluorinated diol **42** proved to be a much more effective catalyst compared to the nonfluorinated diol **50** (k_{rel} = 240 and 4, respectively). Moreover, the higher rate enhancement obtained by diol **42** compared to fluorinated alcohols **43**, **48**, and **49** might be due to its double hydrogen bond activation ability for the epoxide **45**. In order to shed light on the activation of epoxide substrates by the hydrogen bond donors, the complex formation was first investigated by ^{1}H- and ^{13}C-NMR titration studies. Indeed, a correlation was observed between the activating abilities of alcohols **42**, **43**, and **47–50** and the magnitudes of the chemical shift changes induced by complexation in titration studies. The nature of the interaction between fluorinated diol **42** and propylene oxide (PO) was also investigated by density functional theory (DFT) calculations (Scheme 2.14). A 1:1 hydrogen-bonded complex formation was observed in which the two OH groups and the C-H proton at the *ortho* position to the alcohols participate in hydrogen bonding to the epoxide oxygen. The ΔG of this complex formation was calculated to be –3.88 kcal mol^{-1}, which is indicative of a strong interaction between the two components.

Berkessel and coworkers reported in 2012 the syntheses of several new polyfluorinated TADDOL derivatives, α,α,α′,α′-tetrakis-(perfluoroalkyl/aryl)-2,2-dimethyl-1,3-dioxolane-4,5-dimethanols (TEFDDOLs), with the aim of developing more effective hydrogen bond donors [43]. The synthesis of the tetrakis–CF_3 TEFDDOL **52** starting from the acid chloride **51** was accomplished by using Ruppert's reagent [44] (CF_3-TMS) as the nucleophilic CF_3 anion source due to the instability of CF_3Li (Scheme 2.15). All the other TEFDDOL derivatives (**53–57**) were synthesized by the treatment of the acid chloride **51** with the perfluorinated organolithium reagents generated via lithium–halogen exchange. The conformations of all TEFDDOL derivatives were investigated thoroughly both in solid state using X-ray crystallography

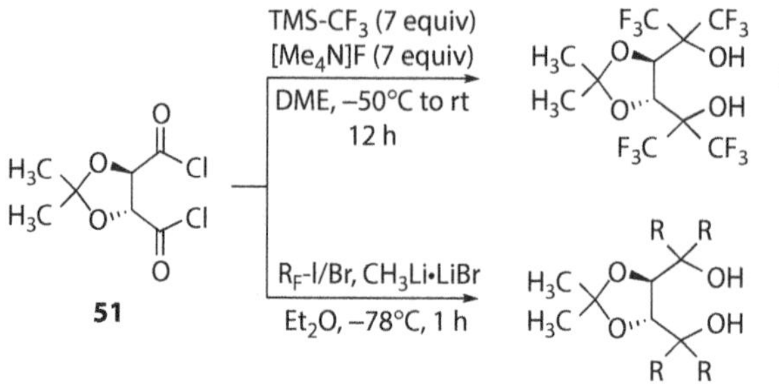

TMS-CF_3 (7 equiv)
$[Me_4N]F$ (7 equiv)
DME, −50°C to rt
12 h

52, 20%
pK_a = 5.7 in DMSO

51

R_F-I/Br, $CH_3Li{\cdot}LiBr$
Et_2O, −78°C, 1 h

	R	Yield (%)	pK_a in DMSO
53:	C_2F_5	55	2.4
54:	n-C_3F_7	25	
55:	n-C_4F_9	26	
56:	n-C_6F_{13}	15	
57:	C_6F_5	70	ca. 11

SCHEME 2.15 Syntheses of TEFDDOL derivatives **52–57**.

and in solution state using diffusion-ordered spectroscopy (DOSY) and nuclear overhauser effect spectroscopy (NOESY) techniques. Moreover, determination of the pK_a values of some TEFDDOL derivatives revealed that these fluorinated diols were highly acidic with pK_a values as low as 2.4 in dimethyl sulfoxide (DMSO) (Scheme 2.15) [45]. Due to their enhanced acidities, TEFDDOLs offer many opportunities in the context of hydrogen bonding catalysis, which have yet to be developed.

The computational and experimental investigation of the epoxidation reaction of alkenes with H_2O_2 in hexafluoroisopropanol (HFIP) by Berkessel and coworkers in 2006 revealed the involvement of more than one HFIP units in the transition state (Figure 2.4) [46]. This discovery paved the way for the development of polar dendritic polymers decorated with polyfluoroalcohol head groups by the Berkessel group in 2013 [47]. In this design, it was envisaged by the authors that the cooperative behavior of multiple HFIP units in an epoxidation reaction could be achieved by the increased local concentration of polyfluoroalcohols on a soluble dendritic polymer. Hyperbranched polyglycerol **58** (hPG-OH) with a molecular weight of 10 kDa was chosen for this purpose as a dendritic scaffold (Scheme 2.16). The conversion of –OH groups of hPG-OH (**58**) to azides ($-N_3$) followed by azide–alkyne click reaction using appropriate fluoroalcohol-containing alkynes afforded catalysts **59** and **60** in high yields. The results for the polyfluoroalcohol-catalyzed epoxidation of cyclooctene are shown in Scheme 2.16. The loading values of the catalysts were carefully arranged so as to provide 20 mol% of *HFIP equivalents*. Dendritic fluoroalcohols **59** and **60** were found to be highly active catalysts giving rise to quantitative yields of the epoxidation product **61** in 24 h. On the other hand, the activities of the monomeric fluoroalcohols **62** and **63** as well as HFIP proved to be much lower compared to the polymeric catalysts affording the product in only up to 14% yield. These findings supported the initial catalyst design and the hypothesis that the cooperative behavior of multiple HFIP units could be achieved by the use of a dendritic polymer. The scope of the epoxidation reaction catalyzed by **59** and **60** was investigated using various alkenes. Although cycloalkenes such as cyclohexene, cyclooctene, and 1-phenylcyclohexene were epoxidized in excellent yields (90%-quant.), open-chain alkenes such as 1-octene and styrene gave lower yields (28%–35%). Finally, dendritic fluoroalcohols **59** and **60** were shown to be reusable catalysts as no noticeable loss in yields was observed when they were reused twice in the epoxidation of cyclooctene.

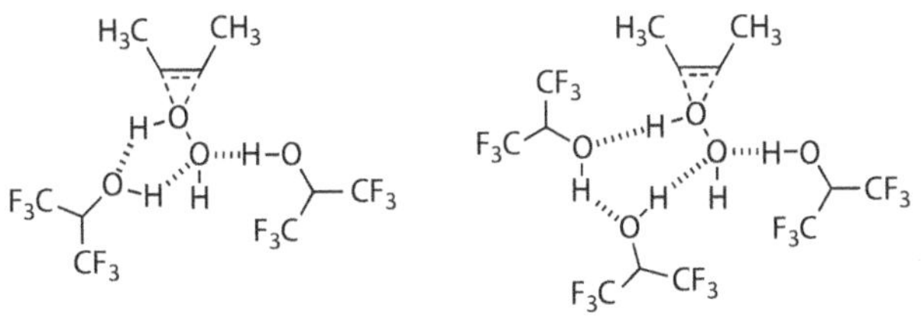

FIGURE 2.4 Proposed transition-states for the epoxidation of alkenes with H_2O_2 in HFIP.

Catalyst	Yield (%)
59	quant.
60	quant.
62	11
63	<10
HFIP	14

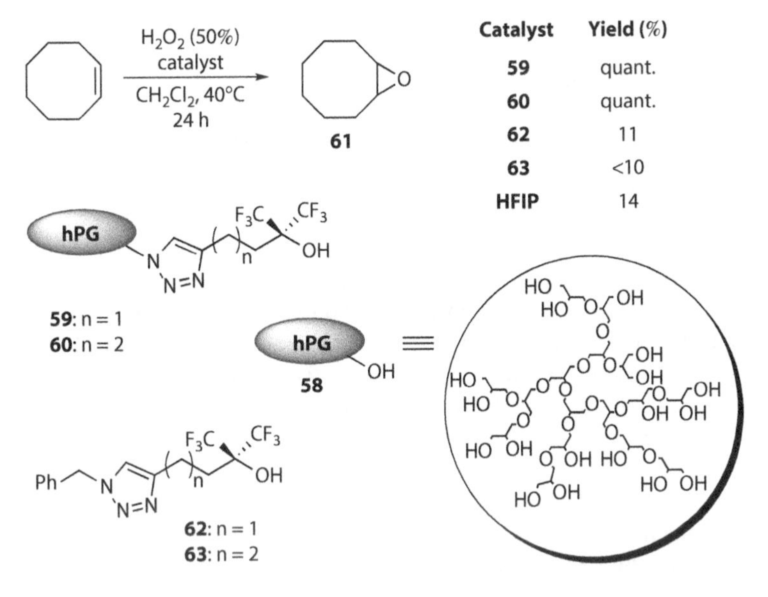

SCHEME 2.16 Epoxidation of cyclooctene catalyzed by dendritic fluoroalcohols **59** and **60**.

2.4 PHENOLS AS HYDROGEN BONDING CATALYSTS

The pioneering work of Hine and coworkers on biphenylenediols [14,15] stimulated further research for the development of bidentate phenol-containing hydrogen bonding catalysts in which two phenol groups are linked by a suitable spacer. In a study reported by Braddock and coworkers in 2003, 4,12-dihydroxy[2.2]paracyclophane (PHANOL) derivatives were utilized as dual hydrogen bond donors to activate α,β-unsaturated aldehydes and ketones in DA reactions [48,49]. The results obtained from the DA reaction between cyclopentadiene (**64**) and crotonaldehyde (**65**) using various catalysts are shown in Scheme 2.17. Although there is almost no background reaction (<1%) after 15 h in the absence of any additives, PHANOL derivatives **66** and **67** showed significant levels of conversion (17% and 37%, respectively) at 10 mol% catalyst loading. The higher conversion obtained by using PHANOL **67** can be attributed to the increased hydrogen bond strength of the phenols due to the presence of

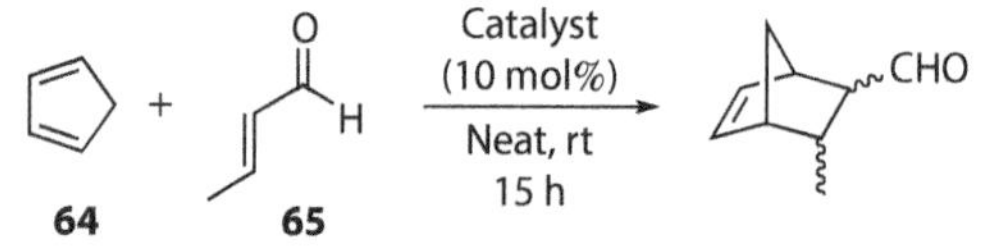

Catalyst	Conversion (%)	*endo/exo*
none	<1	1.9:1
66	17	3.8:1
67	37	1.9:1
68	<1	-
69	3	1.9:1
70	9	3.7:1

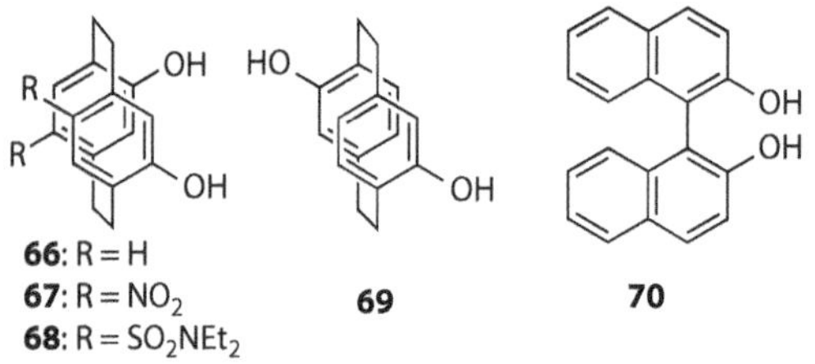

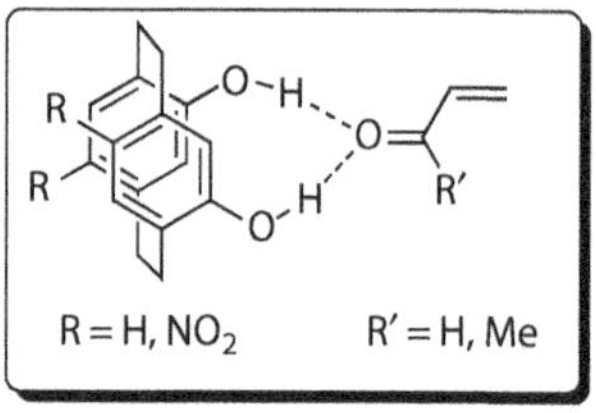

SCHEME 2.17 Catalysis of Diels–Alder reactions by PHANOL derivatives.

two electron-withdrawing $-NO_2$ groups. On the other hand, sulfonamide-substituted PHANOL derivative **68** exhibited no rate acceleration compared to the background reaction due to its low solubility in the reaction mixture. In a control experiment where diol **69** was tested as a catalyst, the conversion was found to be only 3%. Based on these results, it was proposed that PHANOL derivatives activated the α,β-unsaturated aldehydes and ketones by lowering the lowest unoccupied molecular orbital (LUMO) energies of these dienophiles through dual hydrogen bonds between the phenols and the two lone pairs of the carbonyl oxygens (Scheme 2.17). The observation that diol **69** in which the two hydroxyls are located at opposite ends of the paracyclophane moiety was ineffective as a catalyst further supported this proposal. Furthermore, 1,1′-bi-2-naphthol (BINOL) (**70**) was investigated for comparison with the PHANOL catalysts and found to be only moderately effective (9% conversion). Various other dienophiles such as acrolein, methacrolein, *trans*-cinnamaldehyde, and methyl vinyl ketone were also tested in DA reactions catalyzed by PHANOLs **66** and **67** where significant rate enhancements were generally observed. Finally, in addition to DA reactions, PHANOL derivatives were shown to be effective catalysts for the ring-opening reactions of epoxides with secondary amines.

In a study reported in 2013 by Rawal and coworkers, diarylacetylene diols were investigated as dual hydrogen bond donor catalysts for DA reactions [50]. The inspiration for the use of this scaffold in hydrogen bonding catalysis stemmed from the pioneering work of Wuest and coworkers on multidentate Lewis acids [51–53]. Phenol-containing diols **71** and **72** were initially designed by these researchers in order to serve templates for the development of aluminum- and titanium-based multidentate Lewis acids (Figure 2.5). In the absence of any additives, diphenol **71** was found to exhibit a sharp signal at 3502 cm^{-1} in its infrared (IR) spectrum. This observation led to the proposal that diphenol **71** adopted the conformation shown in Figure 2.5, in which the two O-H groups were involved in intramolecular hydrogen bonds with the internal alkyne moiety. On the other hand, diphenol **71** was shown to be capable of acting as a two-point (dual) hydrogen bond donor

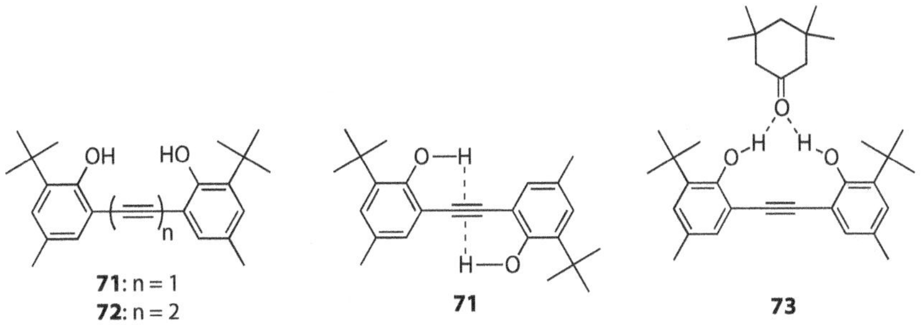

FIGURE 2.5 Structures of diols **71**, **72** and the hydrogen-bonded complex **73**.

in the presence of a ketone by Saied et al. [52]. A co-crystallization experiment using a 1:1 mixture of 3,3,5,5-tetramethylcyclohexanone and diphenol **71** resulted in the formation of a hydrogen-bonded adduct (**73**), analyzed by X-ray crystallography. According to this crystal structure, diphenol **71** was found to bind to the carbonyl oxygen via two moderately strong hydrogen bonds. The IR spectrum of this 1:1 adduct showed a broader signal in the O-H stretching region at 3329 cm^{-1}. More importantly, the dual hydrogen bonding of the diol to the cyclohexanone derivative induced a significant red-shift in the C=O stretching frequency of the ketone from 1717 cm^{-1} to 1670 cm^{-1}, which is indicative of a strong interaction in solid state. However, no change in the O-H and C=O frequencies was observed when the IR spectra of diphenol **71**, the free ketone, and their 1:1 mixture were recorded in CH_2Cl_2 (0.2 M).

The potential of diarylacetylenediols to act as dual hydrogen bonding catalysts was explored by Rawal and coworkers in the DA reaction of cyclopentadiene (**64**) with methyl vinyl ketone (**74**) [50]. First, the binding of several diols to formaldehyde was investigated computationally. In accordance with the aforementioned crystal structure, calculations have shown that the unsubstituted diphenylacetylenediol **75** was capable of binding to the carbonyl oxygen by two hydrogen bonds. Next, diphenols **71** and **72** as well as other diarylacetylenediols bearing electron-withdrawing $-CF_3$ groups were tested as catalysts in the DA reaction of diene **64** and dienophile **74** in d^8-toluene at −30°C (Scheme 2.18). Toluene was selected as a nonpolar reaction solvent in order to increase the associate constants of the hydrogen-bonded complexes between diols and methyl vinyl ketone. Relative rate constant values ($k_{rel} = k_{observed}/k_{background}$) under these reaction conditions were determined for each catalyst with 10 mol% loading. Not surprisingly, $-CF_3$ substituted diols **76** and **81** proved to be the most effective catalysts resulting in k_{rel} values as high as 21.8 with diol **81**. Acetylene-linked catalysts **71** and **76** were found to give higher k_{rel} values compared to the diacetylene-linked diols **72** and **77** (3.5 and 16.7 compared to 1.3 and 10.0, respectively), which was consistent with the original hypothesis. In addition, the catalytic activities monophenolic compounds **78**, **79**, **80**, and **82** were investigated in control experiments with 20 mol% catalyst loading. In all cases, these control catalysts were found to afford inferior k_{rel} values compared to their diphenol counterparts. These results supported the

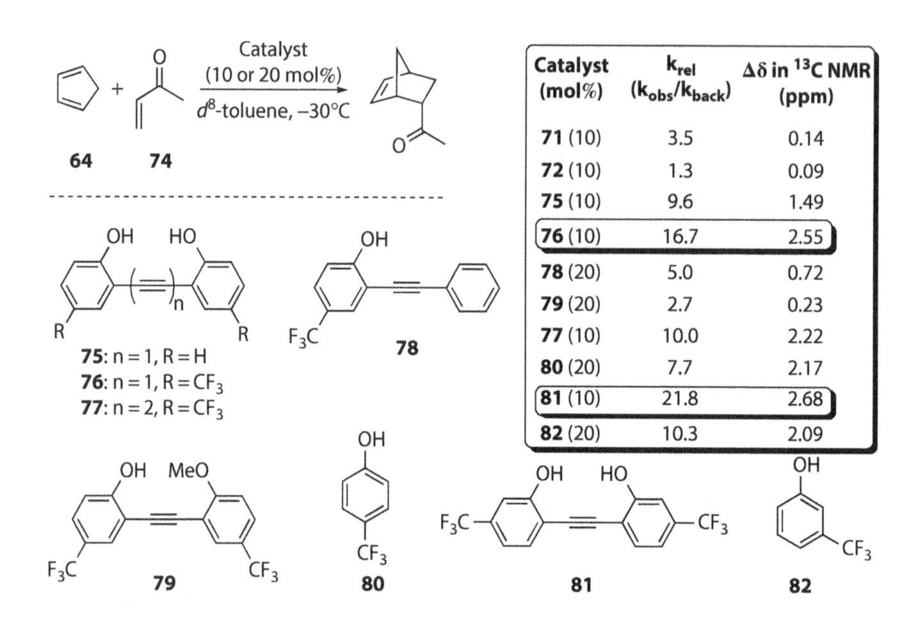

Catalyst (mol%)	k_{rel} (k_{obs}/k_{back})	Δδ in ^{13}C NMR (ppm)
71 (10)	3.5	0.14
72 (10)	1.3	0.09
75 (10)	9.6	1.49
76 (10)	16.7	2.55
78 (20)	5.0	0.72
79 (20)	2.7	0.23
77 (10)	10.0	2.22
80 (20)	7.7	2.17
81 (10)	21.8	2.68
82 (20)	10.3	2.09

SCHEME 2.18 Catalysis of Diels–Alder reactions by diarylacetylenediol derivatives.

potential of diarylacetylenediol derivatives to act as dual hydrogen bond donors for the activation of carbonyl compounds.

Finally, the binding strengths of the catalysts tested to methyl vinyl ketone (**74**) were investigated by ^{13}C-NMR spectroscopy (Scheme 2.18) [50]. In each case, a 0.06 M, 1:1 mixture of diol and **74** or a 2:1 mixture of mono-phenol and **74** was prepared in d^8-toluene, and ^{13}C-NMR spectrum was recorded. The chemical shift values of the C=O carbon of **74** found in these experiments were compared to the C=O signal obtained from blank solution of methyl vinyl ketone. Indeed, hydrogen bonding was found to induce chemical shift changes (Δδ) up to 2.68 ppm. More importantly, these Δδ values exhibited a very good correlation with the observed k_{rel} values.

Phenol-containing hydrogen bond donors have successfully been used as cocatalysts in MBH reactions within the last two decades. A pioneering study in this field, reported by Ikegami and coworkers in 2000, demonstrated that 1,1′-bi-2-naphthol (BINOL, **70**) was a highly effective cocatalyst in the MBH reactions of α,β-unsaturated carbonyl compounds with aliphatic aldehydes [54]. When Bu_3P was used as a Lewis basic phosphine catalyst (20 mol%), MBH adducts were obtained in good to excellent yields in the presence of *rac*-BINOL **70** (10 mol%, Scheme 2.19). Compounds 2-cyclopenten-1-one, 2-cyclohexen-1-one, methyl acrylate, and ethyl acrylate were found to be compatible substrates in this methodology. It is worth noting that other phenol derivatives such as 2-naphthol, phenol, and monomethylated BINOL were also capable of acting as effective catalysts. Finally, the MBH product of the reaction between 2-cyclopenten-1-one and 3-phenyl-1-propanal was obtained in low enantioselectivity (<10% ee) when (*R*)-**70** was used as a chiral hydrogen bonding catalyst. However, the calcium complex of (*R*)-**70** was found

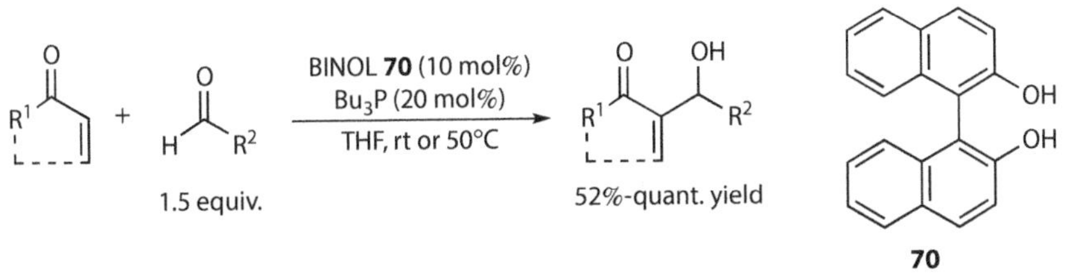

SCHEME 2.19 Morita–Baylis–Hillman reactions catalyzed by Bu_3P and BINOL.

to be a promising Lewis acid catalyst for this transformation affording the MBH adduct in 62% yield and 56% ee.

Schaus and coworkers reported in 2003 an enantioselective MBH reaction of cyclohexenone with aldehydes catalyzed by chiral hydrogen bond donors [55]. This work also represents the first highly enantioselective reaction catalyzed by a chiral phenol derivative as a hydrogen bonding catalyst. During the reaction optimization, the screening of various BINOL-derived hydrogen bond donor catalysts showed that saturation of the BINOL derivatives and substitutions at the 3,3′-positions had a positive effect on the catalytic activity. Indeed, partially saturated BINOL derivatives **83** and **84** with bulky aryl groups at the 3,3′-positions were found to be the most effective catalysts for obtaining high enantioselectivities and yields (Scheme 2.20). It should also be noted that the use of Me_3P and $n\text{Bu}_3P$ as the phosphine derivative gave inferior results in terms of enantioselectivity compared to Et_3P. The substrate scope was investigated using a wide range of aliphatic aldehydes, and the MBH products were obtained in high yields with high enantioselectivities. On the other hand, the use of benzaldehyde and cinnamaldehyde gave lower yields and enantioselectivities.

The proposed mechanism of the hydrogen bonding-catalyzed MBH reaction is shown in Scheme 2.21. The initial 1,4-conjugate addition of Et_3P to cyclohexenone is expected to form the enolate intermediate **85**, which is stabilized through hydrogen bonding to the BINOL-derived catalyst (B-H). In the chirality-inducing step, the attack of the enolate **85** to the aldehyde will form the alkoxide **86**, which will form the final MBH product upon the elimination of Et_3P and protonation. Further mechanistic studies revealed valuable information on the role of the phenolic –OH groups

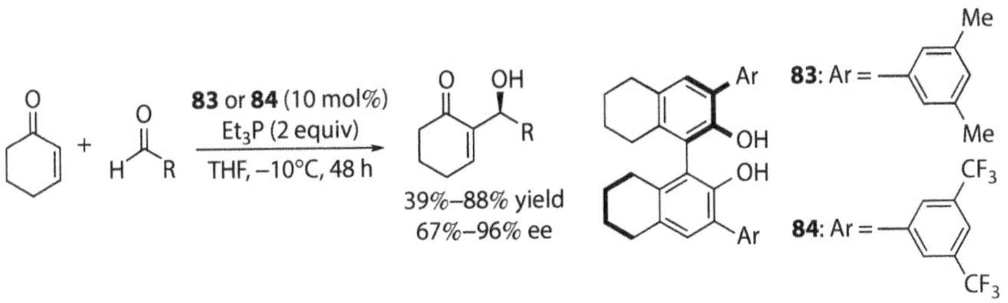

SCHEME 2.20 Enantioselective MBH reactions catalyzed by BINOL derivatives **83** and **84**.

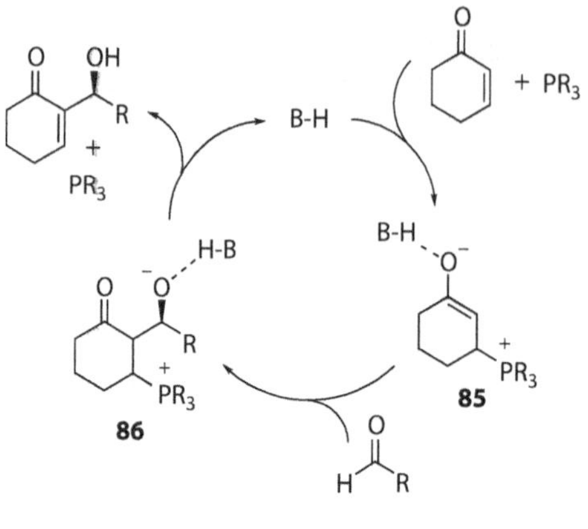

SCHEME 2.21 Proposed mechanism of the Morita–Baylis–Hillman reaction.

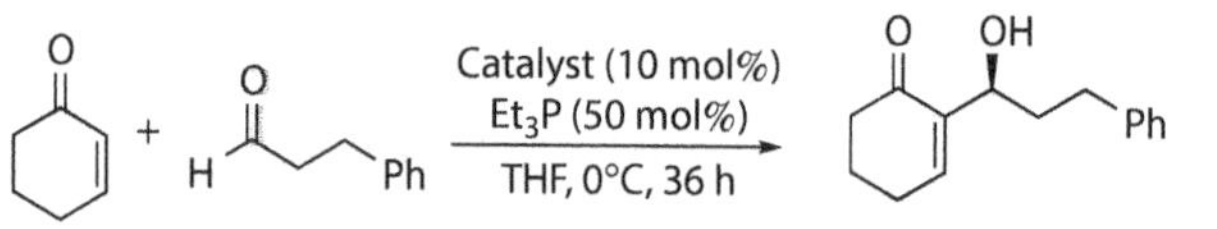

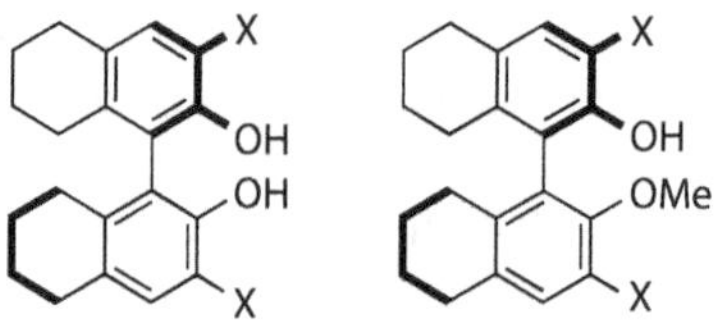

Catalyst	Yield (%)	ee (%)
87	73	48
88	73	79
89	43	3
90	15	3

SCHEME 2.22 Control experiments for the BINOL-catalyzed MBH reaction.

of the BINOL-derived catalysts. The results of the MBH reaction between cyclohexenone and 3-phenylpropanal using catalysts **87–90** are shown in Scheme 2.22. The lower yields and the lack of enantioselectivity with the use of –OMe-substituted catalysts **89** and **90** clearly demonstrate the importance of the presence of the two phenolic –OH groups for the high catalytic efficiency observed in this study.

A highly effective bifunctional organocatalyst containing phenols has been developed by Sasai and coworkers in 2005 for the enantioselective aza-MBH reaction between *N*-tosylimines and α,β-unsaturated carbonyl compounds [56]. In this catalyst design, a Lewis basic aminopyridine moiety is linked to the chiral (*S*)-BINOL unit whose naphthol groups act as hydrogen bond donors (**91**, Scheme 2.23). The initial catalyst screening revealed **92** as the optimal catalyst in terms of enantioselectivity and yield (87% ee, 93% yield) for the aza-MBH product (**93**) between methyl vinyl ketone (**74**) and phenyl *N*-tosylimine (**94**). The reaction was found to have a broad substrate scope with respect to the *N*-tosylimine component tolerating both electron-rich and electron-deficient aromatic groups. Methyl vinyl ketone, ethyl vinyl ketone,

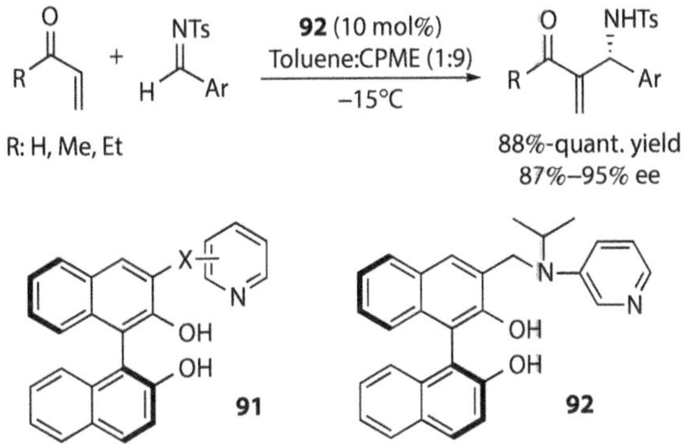

SCHEME 2.23 Enantioselective MBH reactions using bifunctional catalyst **92**.

and acrolein were tested as the α,β-unsaturated carbonyl component affording the aza-MBH products in high yields (88% to quantitative) with high enantioselectivities (87%–95% ee). In a control experiment, a mixture of (*S*)-BINOL (10 mol%) and 3-DMAP (10 mol%) was used as the catalytic system. Even though this combination of two catalysts was found to promote the aza-MBH reaction, the product was obtained almost as a racemic mixture (48% yield, 3% ee). This result underscores the importance of using a bifunctional catalyst for obtaining a more ordered transition state leading to higher enantioselectivity values.

In order to gain more mechanistic insight into the functions of each structural subunit of the catalyst, various additional control experiments were performed (Figure 2.6) [56]. First, compounds **95** and **96** bearing –OMe groups at the 2- and 2′-positions, respectively, were prepared and tested as catalysts to understand the effect of each phenolic –OH groups. The –OH group at the 2′-position appeared to have a significant effect on the catalytic activity as the aza-MBH product **93** was obtained only in 5% yield and with 24% ee when **95** was used as catalyst. In contrast, catalyst **96** exhibited only a slight decrease in activity affording product **93** in 85%

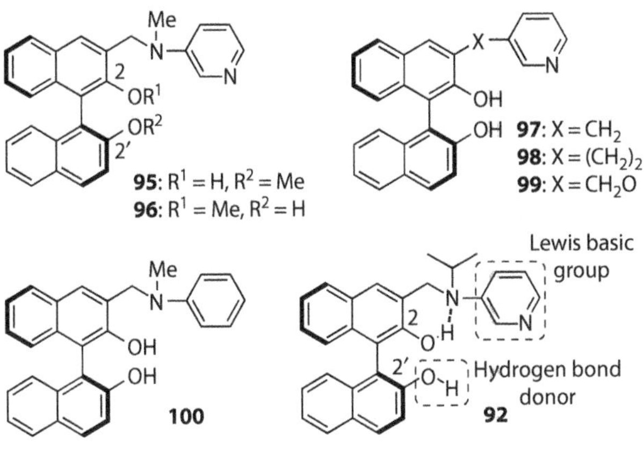

FIGURE 2.6 Mechanistic studies on the enantioselective MBH reaction.

yield and 79% ee. Furthermore, the effects of the pyridine moiety and the linker were investigated via the use of compounds **97–100** as catalysts. Surprisingly, none of these catalysts promoted the aza-MBH reaction, which point out to the crucial roles of the specific linker in catalyst **92** along with the 3-aminopyridine unit. Based on these results, the authors proposed that the phenolic –OH group at the 2-position and the amino group make an intramolecular hydrogen bond that keeps the catalyst in a fixed conformation. On the other hand, the pyridine group and the BINOL –OH at the 2′-position are proposed to have significant roles in activating the α,β-unsaturated carbonyl substrate that leads to high enantioselectivities. This hypothesis was further supported by computational studies on the conformation of **92** where the aforementioned intramolecular hydrogen bond was shown to take place (Figure 2.6).

The scope of reactions catalyzed by BINOL-based hydrogen bond donors was expanded to the formal DA reactions between dienamines and nitrosobenzene derivatives by Yamamoto and coworkers [57]. Even though TADDOL **16** was found to be able to catalyze the formal DA reaction of dienamine **101** and nitrosobenzene **31**, BINOL derivatives proved to be better hydrogen bonding catalysts for this transformation in terms of both yield and enantioselectivity (Scheme 2.24). A detailed screening of various BINOLs led to the identification of tris-*m*-xylylsilyl-substituted BINOL derivative **102** as the optimized catalyst. The formal DA reaction between morpholine dienamine **101** and various aromatic nitroso compunds proceeded in good to high yields (52%–90%), with complete regioselectivities and high enantioselectivities (80%–90% ee). However, substituents at the 4-position of dienamines were observed to have a profound effect on the reaction outcome. When 2-morpholino-4,4-diphenylcyclohexadiene (**103**) was used as the diene component, *N*-nitroso aldol adduct **104** was obtained as the sole product (27% yield, 61% ee). In order to shed light on the reaction mechanism, a series of control experiments were undertaken. Based on these mechanistic studies, it was proposed by the authors that the DA-type reactions involved sequential *N*-nitroso aldol and Michael reactions

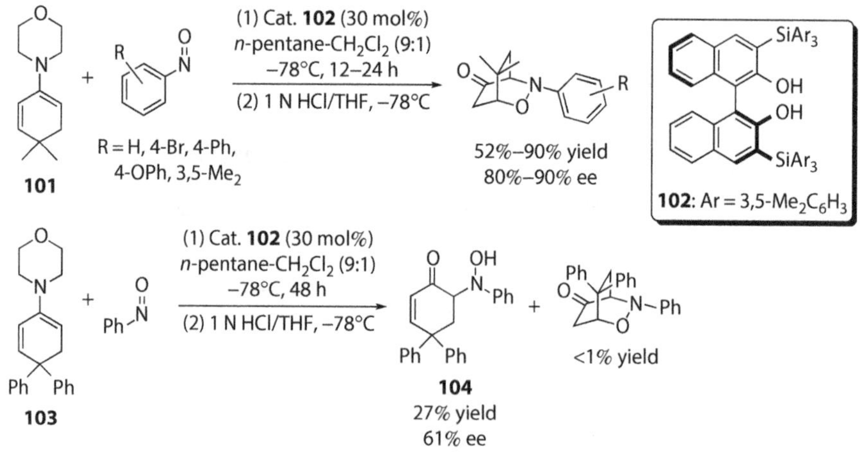

SCHEME 2.24 Enantioselective formal nitroso-DA reactions catalyzed by BINOL **102**.

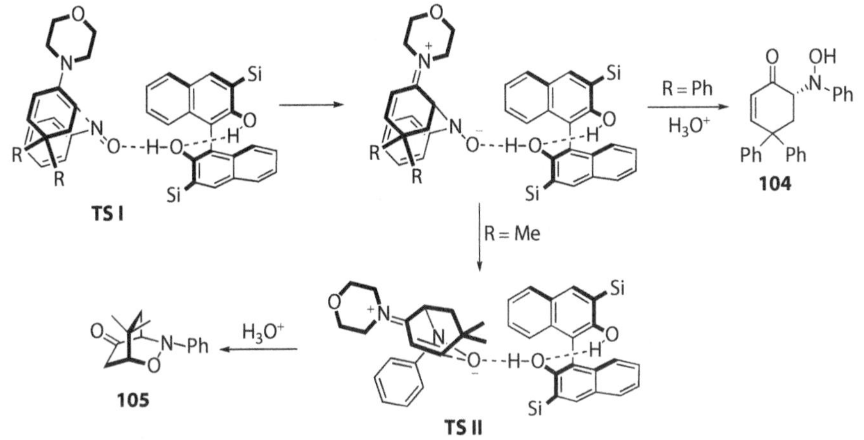

SCHEME 2.25 Transition states for the enantioselective nitroso-aldol and nitroso-DA reactions.

rather than following a concerted [4 + 2] pathway. Although the two methyl groups of dienamine **101** allowed the initial *N*-nitroso aldol adduct to proceed with a subsequent Michael addition leading to the formal DA products, sterically more demanding phenyl groups of dienamine **103** prevented the Michael reaction affording only the *N*-nitroso aldol product **104**.

The mechanism of the formal DA reaction proposed by Yamamoto and coworkers is shown in Scheme 2.25 [57]. First, nitrosobenzene derivative is expected to be activated by one of the hydroxyl groups of BINOL **102**, the conformation of which is locked by an intramolecular hydrogen bond. According to TS I described by the authors, the bulky tris-*m*-xylylsilyl substituents on BINOL catalyst **102** orient the approach of the dienamine substrate such that nucleophilic attack occurs from the *re* face during the *N*-nitroso aldol reaction. When the R groups are methyl as in the case of dienamine **101**, the *N*-nitroso aldol adduct finds the opportunity to undergo a conformational change so that the subsequent Michael addition takes place to afford the overall DA-type bicyclic product **105**. On the other hand, such a conformational change is energetically disfavored due to the steric clash between the tris-*m*-xylylsilyl substituents of the catalyst and the two phenyl groups for dienamine **103**, leading to the formation of *N*-nitroso aldol product **104** upon hydrolysis under acidic conditions.

2.5 SUMMARY

Catalysis of organic reactions using noncovalent interactions has emerged as a powerful strategy in the field of organocatalysis. Among various noncovalent interactions, hydrogen bonding has been the most frequently utilized interaction due to its strength, directionality, and predictability. In this respect, alcohol and phenol derivatives were among the first classes of hydrogen bond donors to be used as catalysts. The seminal work of Hine on biphenylenediols as dual hydrogen bond donors

[13–19] and the more recent studies of Rawal on TADDOLs as chiral hydrogen bonding catalysts [23,24] represent major breakthroughs in the field that paved the way for the catalysis of a broad range of transformations using alcohol- and phenol-based organocatalysts. The relative low acidity of alcohols and the use of high catalyst loadings can be considered as the major drawbacks associated with these catalytic systems. The recent introduction of fluorinated alcohols with enhanced acidities to the field of hydrogen bonding catalysis [41,43] has the potential to offer solutions to these limitations and is expected to lead to the development of more effective catalysts.

REFERENCES

1. Connon, S. J. *Chem. Eur. J.* **2006**, *12*, 5418–5427.
2. McGilvra, J. D., Gondi, V. B., Rawal, V. H. Asymmetric proton catalysis. In *Enantioselective Organocatalysis*, Dalko, P. I., (Ed.), Wiley-VCH: Weinheim, Germany, pp. 189–254, 2007.
3. Doyle, A. G., Jacobsen, E. N. *Chem. Rev.* **2007**, *107*, 5713–5743.
4. Yu, X., Wang, W. *Chem. Asian J.* **2008**, *3*, 516–532.
5. Connon, S. J. *Chem. Commun.* **2008**, 2499–2510.
6. Türkmen, Y. E., Zhu, Y., Rawal, V. H. Brønsted acids. In *Comprehensive Enantioselective Organocatalysis*, Dalko, P. I., (Ed.), Wiley-VCH: Weinheim, Germany, Vol. 2, pp. 239–288, 2013.
7. Chauhan, P., Mahajan, S., Kaya, U., Hack, D., Enders, D. *Adv. Synth. Catal.* **2015**, *357*, 253–281.
8. Akiyama, T. *Chem. Rev.* **2007**, *107*, 5744–5758.
9. Terada, M. *Synthesis* **2010**, 1929–1982.
10. Cheon, C. H., Yamamoto, H. *Chem. Commun.* **2011**, *47*, 3043–3056.
11. Rueping, M., Nachtsheim, B. J., Ieawsuwan, W., Atodiresei, I. *Angew. Chem. Int. Ed.* **2011**, *50*, 6706–6720.
12. Mahlau, M., List, B. *Angew. Chem. Int. Ed.* **2013**, *52*, 518–533.
13. Hine, J., Ahn, K., Gallucci, J. C., Linden, S.-M. *J. Am. Chem. Soc.* **1984**, *106*, 7980–7981.
14. Hine, J., Linden, S.-M., Kanagasabapathy, V. M. *J. Am. Chem. Soc.* **1985**, *107*, 1082–1083.
15. Hine, J., Hahn, S., Miles, D. E., Ahn, K. *J. Org. Chem.* **1985**, *50*, 5092–5096.
16. Hine, J., Linden, S.-M., Kanagasabapathy, V. M. *J. Org. Chem.* **1985**, *50*, 5096–5099.
17. Hine, J., Hahn, S., Miles, D. E. *J. Org. Chem.* **1986**, *51*, 577–584.
18. Hine, J., Ahn, K. *J. Org. Chem.* **1987**, *52*, 2083–2086.
19. Hine, J., Ahn, K. *J. Org. Chem.* **1987**, *52*, 2089–2091.
20. Emsley, J. *Chem. Soc. Rev.* **1980**, *9*, 91–124.
21. Kelly, T. R., Meghani, P., Ekkundi, V. S. *Tetrahedron Lett.* **1990**, *31*, 3381–3384.
22. Huang, Y., Rawal, V. H. *J. Am. Chem. Soc.* **2002**, *124*, 9662–9663.
23. Huang, Y., Unni, A. K., Thadani, A. N., Rawal, V. H. *Nature* **2003**, *424*, 146–146.
24. Thadani, A. N., Stankovic, A. R., Rawal, V. H. *Proc. Natl. Acad. Sci. U.S.A.* **2004**, *101*, 5846–5850.
25. Seebach, D., Beck, A. K., Heckel, A. *Angew. Chem. Int. Ed.* **2001**, *40*, 92–138.
26. Tanaka, K., Toda, F. *Chem. Rev.* **2000**, *100*, 1025–1074.
27. McGilvra, J. D., Unni, A. K., Modi, K., Rawal, V. H. *Angew. Chem. Int. Ed.* **2006**, *45*, 6130–6133.
28. Unni, A. K., Takenaka, N., Yamamoto, H., Rawal, V. H. *J. Am. Chem. Soc.* **2005**, *127*, 1336–1337.

29. Gondi, V. B., Gravel, M., Rawal, V. H. *Org. Lett.* **2005**, *7*, 5657–5660.
30. Gondi, V. B., Hagihara, K., Rawal, V. H. *Angew. Chem. Int. Ed.* **2009**, *48*, 776–779.
31. Gondi, V. B., Hagihara, K., Rawal, V. H. *Chem. Commun.* **2010**, *46*, 904–906.
32. Okino, T., Hoashi, Y., Takemoto, Y. *J. Am. Chem. Soc.* **2003**, *125*, 12672–12673.
33. Momiyama, N., Yamamoto, H. *J. Am. Chem. Soc.* **2005**, *127*, 1080–1081.
34. Akakura, M., Kawasaki, M., Yamamoto, H. *Eur. J. Org. Chem.* **2008**, 4245–4249.
35. Imahori, T., Yamaguchi, R., Kurihara, S. *Chem. Eur. J.* **2012**, *18*, 10802–10807.
36. Nakashima, D., Yamamoto, H. *J. Am. Chem. Soc.* **2006**, *128*, 9626–9627.
37. García-García, P., Lay, F., Rabalakos, C., List, B. *Angew. Chem. Int. Ed.* **2009**, *48*, 4363–4366.
38. Berkessel, A., Andreae, M. R. M., Schmickler, H., Lex, J. *Angew. Chem. Int. Ed.* **2002**, *41*, 4481–4484.
39. Berkessel, A., Adrio, J. A. *J. Am. Chem. Soc.* **2006**, *128*, 13412–13420.
40. Kim, W. H., Lee, J. H., Danishefsky, S. J. *J. Am. Chem. Soc.* **2009**, *131*, 12576–12578.
41. Coulembier, O., Sanders, D. P., Nelson, A., Hollenbeck, A. N., Horn, H. W., Rice, J. E., Fujiwara, M., Dubois, P., Hedrick, J. L. *Angew. Chem. Int. Ed.* **2009**, *48*, 5170–5173.
42. Gennen, S., Alves, M., Méreau, R., Tassaing, T., Gilbert, B., Detrembleur, C., Jerome, C., Grignard, B. *ChemSusChem* **2015**, *8*, 1845–1849.
43. Berkessel, A., Vormittag, S. S., Schlörer, N. E., Neudörfl, J.-M. *J. Org. Chem.* **2012**, *77*, 10145–10157.
44. Ruppert, I., Schlich, K., Volbach, W. *Tetrahedron Lett.* **1984**, *25*, 2195–2198.
45. Christ, P., Lindsay, A. G., Vormittag, S. S., Neudörfl, J.-M., Berkessel, A., O'Donoghue, A. C. *Chem. Eur. J.* **2011**, *17*, 8524–8528.
46. Berkessel, A., Adrio, J. A., Hüttenhain, D., Neudörfl, J.-M. *J. Am. Chem. Soc.* **2006**, *128*, 8421–8426.
47. Berkessel, A., Krämer, J., Mummy, F., Neudörfl, J.-M., Haag, R. *Angew. Chem. Int. Ed.* **2013**, *52*, 739–743.
48. Braddock, D. C., MacGilp, I. D., Perry, B. G. *Synlett* **2003**, 1121–1124.
49. Braddock, D. C., MacGilp, I. D., Perry, B. G. *Adv. Synth. Catal.* **2004**, *346*, 1117–1130.
50. Türkmen, Y. E., Rawal, V. H. *J. Org. Chem.* **2013**, *78*, 8340–8353.
51. Saied, O., Simard, M., Wuest, J. D. *Organometallics* **1996**, *15*, 2345–2349.
52. Saied, O., Simard, M., Wuest, J. D. *J. Org. Chem.* **1998**, *63*, 3756–3757.
53. Saied, O., Simard, M., Wuest, J. D. *Organometallics* **1998**, *17*, 1128–1133.
54. Yamada, Y. M. A., Ikegami, S. *Tetrahedron Lett.* **2000**, *41*, 2165–2169.
55. McDougal, N. T., Schaus, S. E. *J. Am. Chem. Soc.* **2003**, *125*, 12094–12095.
56. Matsui, K., Takizawa, S., Sasai, H. *J. Am. Chem. Soc.* **2005**, *127*, 3680–3681.
57. Momiyama, N., Yamamoto, Y., Yamamoto, H. *J. Am. Chem. Soc.* **2007**, *129*, 1190–1195.

3 Phosphoric Acid Catalysis

Jia-Hui Tay and Pavel Nagorny

CONTENTS

3.1 INTRODUCTION

Brønsted acids[1] have played a critical role in the development of organic chemistry, and, not surprisingly, the discussion of Brønsted acid-promoted transformations has become an integral part of any introductory organic chemistry textbook. Indeed, a wide range of organic transformations including fundamentally important C–C and C–X bond-forming reactions, cycloadditions, skeletal rearrangements, isomerizations, fragmentation, and eliminations involve acids. As a result of this, Brønsted acid-containing organic reagents such as *p*-toluenesulfonic, methanesulfonic, camphorsulfonic, or trifluoroacetic acids have become essential reagents for any organic laboratory regardless of its main focus.[2]

The first systematic study and application of phosphorous (V) acids in organocatalysis dates back to 1970s when Sir John Cornforth explored the use of phosphinic acids with five-membered dibenzo backbone as the catalysts for olefin hydration.[3] Recent advances in the field of asymmetric catalysis further highlighted the utility of organic acids for the preparation of chiral molecules and resulted in even stronger interest in Brønsted acid catalysis. Consequently, many chiral organic Brønsted acids have emerged as powerful and broadly applicable catalysts for a variety of asymmetric transformations. Since the seminal contributions by Akiyama[4] and Terada[5] in 2004, phosphoric acids and their derivatives have occupied a unique niche in these studies, and a large portion of the reports in this area is dedicated to exploring the design and reactivity of chiral and achiral phosphorous (V)-based acids.[6] Phosphorous (V)-based acids possess several important features that make them particularly useful for the field of asymmetric organocatalysis. These features include chemical stability to a wide range of reagents and conditions, the ease of preparation, ability to amend the privileged chiral C2-symmetric ligands, the tunability of the acidic properties, and the presence of a Lewis basic site that allows bifunctional catalysis. The acidity of phosphoric acids and derivatives is in the range that allows efficient activation of many different functional groups. At the same time, the conjugate bases of phosphorous (V)-based acids often possess significant Lewis basicity and nucleophilicity, which improves stereocontrol and greatly expands the scope of the reactions that could be promoted by phosphoric acids. It is the purpose of this chapter to provide an overview of the major mechanisms, by which phosphoric acids and derivatives catalyze various organic transformations and the factors determining their catalytic activity.

3.2 OVERVIEW OF BASIC STRUCTURAL FEATURES

A large number of phosphorus-based acids are available as the catalysts for various organic transformations. Such catalysts share some common design features such as tetrahedral geometry for the phosphorus-containing functionality and often have common substituents R_1 and R_2 attached to the phosphorus backbone. Although orthophosphoric acid $(HO)_3P{=}O$ is a tribasic acid, the introduction of one or two organic substituents to the hydroxyl groups can result in either di- or monobasic derivatives (Figure 3.1). Although the orthophosphoric acid monoesters are achiral, more frequently employed in organocatalysis diesters can possess a center of chirality if both substituents R_1 and R_2 are different. This phenomenon is typically not observed for orthophosphoric acid diesters due to the tautomerization that equilibrates the enantiomers, but configurationally stable chiral orthophosphoric acid derivatives can be obtained when such interconversion of enantiomers is not possible.

It should be noted that the less soluble in organic solvents and more conformationally flexible monoesters are rarely employed in asymmetric catalysis, and the majority of the recent applications of phosphorus-based Brønsted acids involve the use of orthophosphoric acid aryl diesters and their derivatives (Figure 3.2, acids **I–IV** and **IX–XIX**).[7–21] In addition, the examples of C–P bond-containing acids such as phosphinic (**V**)[3] and thiophosphonic acids (**VI**),[9] and

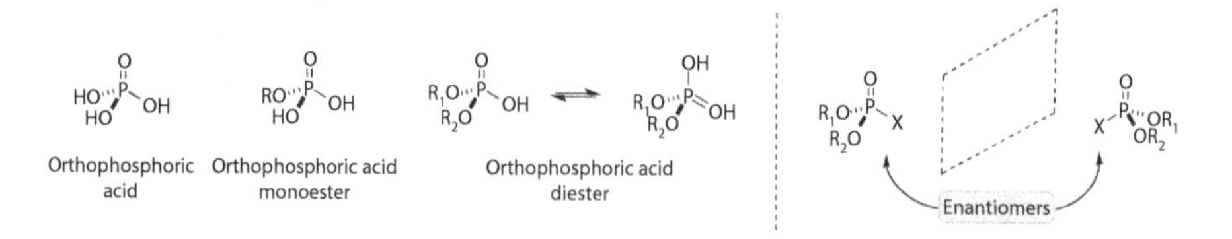

FIGURE 3.1 Orthophosphoric acid derivatives.

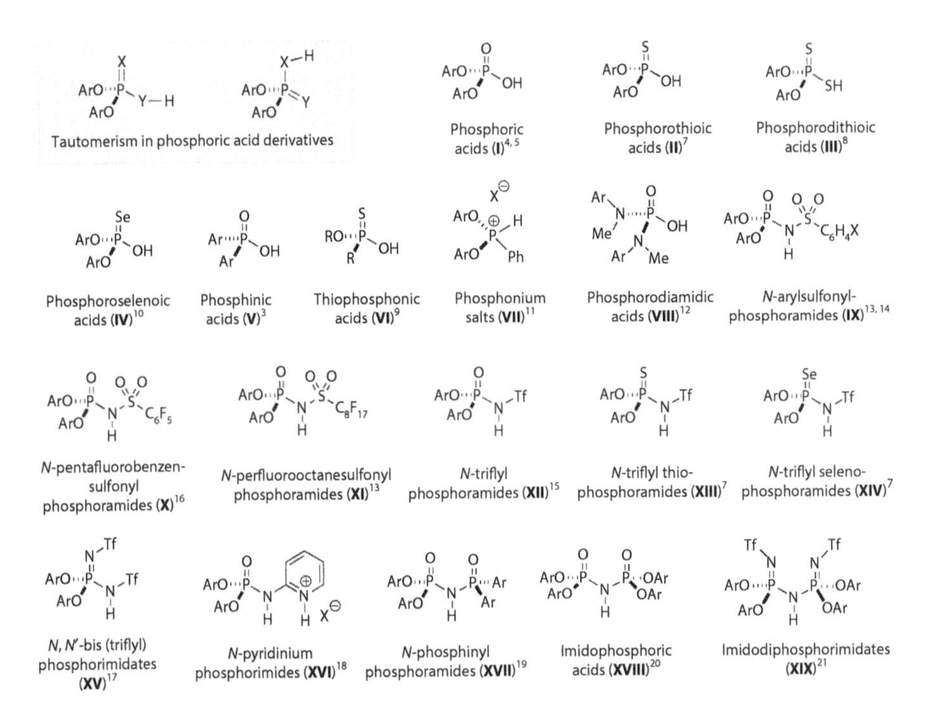

FIGURE 3.2 Examples of phosphorous-based acids employed in organocatalysis.

P–N bond-containing phosphorodiamidic acids (**VIII**)[12] are known. While being more stable to hydrolytic cleavage, the derivatives such as **V** and **VI** are more difficult to synthesize than O–P bond containing acids and hence their use has been limited to only few examples. In addition to –OAr groups, the X and HY substituents on phosphorus (V) can significantly affect the properties of phosphorus (V)-based Brønsted acids. Similar to orthophosphoric acid diesters **I** (X = O, HY = HO), equilibrium between two tautomeric forms of acids **II–XIX** (with the exception of **VII**) can exist.

Although one specific form may predominate in solution, the less dominant tautomeric form may serve as the actual active catalyst that promotes the reaction. For example, the recent computational investigation of *N*-triflyl phosphoramide (**XII**)[15]-catalyzed enantioselective Nazarov cyclizations suggests that although the tautomeric form of **XII** depicted in Figure 3.2, that is, $(ArO)_2P(=O)NHTf$ is more thermodynamically stable, the $(ArO)_2P(=NTf)OH$ tautomer is more likely to be the active catalyst for this reaction.[22] In addition to the extent of tautomerization, the nature of the X and HY substituents is of importance to some other properties such as solubility in organic solvents, acidity, Lewis basicity, coordinating properties of counterion, and steric hindrance and electrostatic field around the catalytic site. These parameters may play an important role for both achieving the catalytic transformation and for rendering this catalytic transformation enantio- or diastereoselective.

3.3 OVERVIEW OF THE KEY FACTORS DETERMINING THE ACIDITY OF PHOSPHORIC ACIDS AND THEIR DERIVATIVES

Although a variety of mechanisms by which acids **I–XIX** may catalyze organic transformations exist, the substrate activation by a direct protonation or formation of a hydrogen bond is among the most important mechanistic manifolds for phosphoric acid organocatalysis. Not surprisingly, Brønsted acidity of the catalyst plays an important role for its catalytic activity, and understanding the factors that enhance or attenuate the pK_a values of such acids is of great importance to phosphoric acid catalysis. Several computational[23] and experimental[24,25] studies have been focused on quantifying the acidity of phosphorus (V)-based Brønsted acids (*cf.* Figure 3.3),

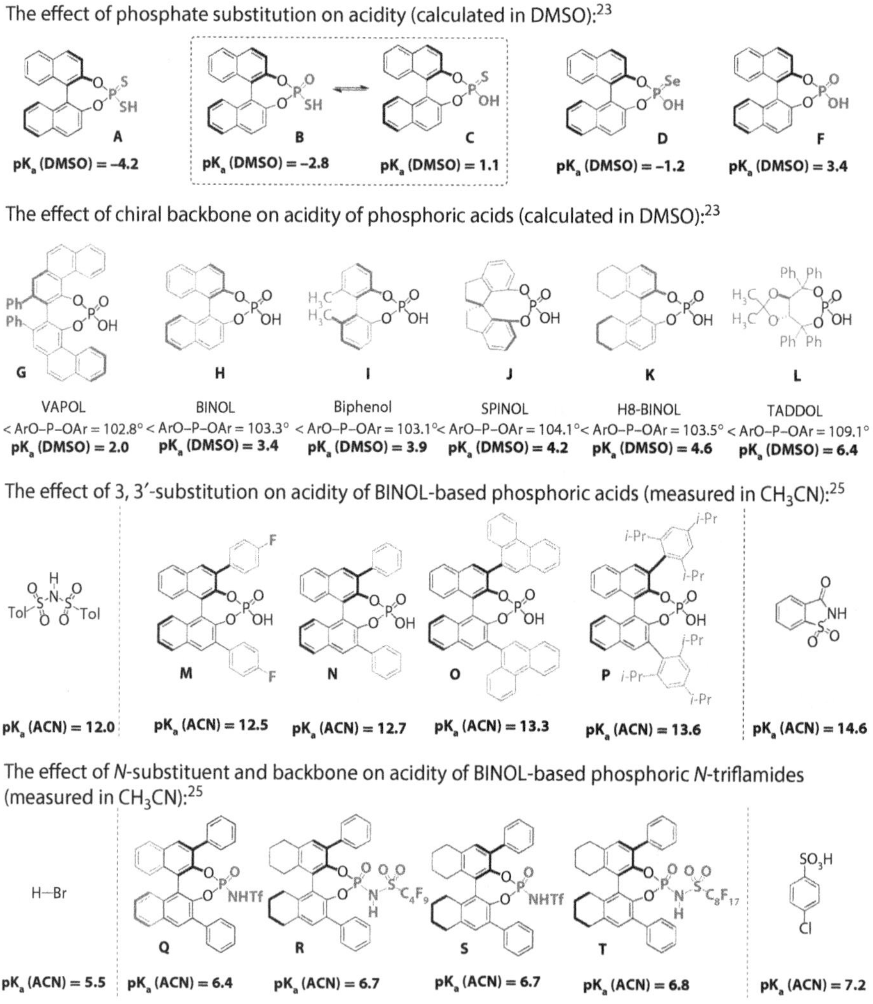

The effect of phosphate substitution on acidity (calculated in DMSO):[23]

A pKa (DMSO) = −4.2
B pKa (DMSO) = −2.8
C pKa (DMSO) = 1.1
D pKa (DMSO) = −1.2
F pKa (DMSO) = 3.4

The effect of chiral backbone on acidity of phosphoric acids (calculated in DMSO):[23]

G VAPOL < ArO–P–OAr = 102.8° pKa (DMSO) = 2.0
H BINOL < ArO–P–OAr = 103.3° pKa (DMSO) = 3.4
I Biphenol < ArO–P–OAr = 103.1° pKa (DMSO) = 3.9
J SPINOL < ArO–P–OAr = 104.1° pKa (DMSO) = 4.2
K H8-BINOL < ArO–P–OAr = 103.5° pKa (DMSO) = 4.6
L TADDOL < ArO–P–OAr = 109.1° pKa (DMSO) = 6.4

The effect of 3, 3′-substitution on acidity of BINOL-based phosphoric acids (measured in CH_3CN):[25]

pKa (ACN) = 12.0
M pKa (ACN) = 12.5
N pKa (ACN) = 12.7
O pKa (ACN) = 13.3
P pKa (ACN) = 13.6
pKa (ACN) = 14.6

The effect of *N*-substituent and backbone on acidity of BINOL-based phosphoric *N*-triflamides (measured in CH_3CN):[25]

H–Br pKa (ACN) = 5.5
Q pKa (ACN) = 6.4
R pKa (ACN) = 6.7
S pKa (ACN) = 6.7
T pKa (ACN) = 6.8
pKa (ACN) = 7.2

FIGURE 3.3 Acidity trends for phosphoric acids and their derivatives.

and both the nature of the X and YH group attached to phosphorus (V) and the chiral backbone were found to play an important role in determining the acidity of the catalyst.

The acidity of phosphoric acids falls into a relatively broad window with pK_a values ranging from 2 to 6[23] in dimethyl sulfoxide (DMSO) (12–14 in acetonitrile).[25] The acidity may be further enhanced if one or both oxygens of hydrogen phosphate are replaced with heteroatoms (S, Se, or NR). The higher acidity of sulfur and selenium-substituted phosphoric acids can be attributed to the increased polarizability of sulfur and selenium relative to oxygen, which results in more stable anionic conjugate bases. Thus, the pK_a (DMSO) trend observed for the phosphorous (V)-based acids **F, C,** and **D** (Figure 3.3) is consistent with the trends observed for the other heteroatom-substituted acids such as PhOH, PhSH, and PhSeH (pK_a (DMSO) are 18.0, 10.3, and 7.1, respectively). It should be noted that computations indicate that tautomeric phosphorothioic acid **B** and **C** would have significantly different acidities (pK_a(DMSO) = –2.8 and 1.1 correspondingly) with H-S tautomer **B** being more acidic than H-O tautomer **C**. These effects are further enhanced if both oxygens of hydrogen phosphate are replaced with heteroatoms, and phosphorodithioic acid **A** is significantly more acidic (pK_a(DMSO) = –4.2) than both **B** and **C**.

The nature of the chiral backbone plays an important role for the acidity of phosphoric acids and their derivatives as evident by examining the computed pK_a values in DMSO for acids **G–L** (Figure 3.3). Although such trends are often hard to rationalize, it is probably the combination of several factors (<ArO–P–OAr angle, the inductive stabilization of negative charge by chiral backbone, electrostatic stabilization of the negative charge by the neighboring groups, etc.) that determine the acidity of phosphoric acids. The <ArO–P–OAr angle value and inductive effects exhibited by backbone are certainly important parameters to consider. Based on the computed <ArO–P–OAr angles by the Nagorny group (DFT, B97-D, 6-31g**), the lower <ArO–P–OAr angle value seems to correlate with the higher acidity in DMSO. Thus, VAPOL-based acid **G** has the lowest <ArO–P–OAr angle = 102.8°, but the highest acidity in DMSO (pK_a = 2.0). Similarly the TADDOL-based acid **L** has the largest angle <ArO–P–OAr = 109.1°, and the lowest acidity in DMSO (pK_a = 6.4).

The inductive effects arising from the chiral backbone also play an important role, in particular for the acids with the similar <ArO–P–OAr angles (acids **H–K**). Higher degree of unsaturation in the backbone seems to result in increased acidity and, hence, BINOL-derived acid **H** (pK_a = 3.4) is more acidic than H8-BINOL-derived acid **K** (pK_a = 4.6). Similarly, the introduction of electron-withdrawing groups into the backbone typically results in an increased acidity (*vide infra*). These effects are particularly noticeable for the 3,3′-substituted BINOL-based acids **M–P**, pK_a values of which was recently measured in acetonitrile by Rueping and coworkers (Figure 3.3).[25] The acidity of **M–P** clearly correlates with the nature of the substitution at the 3,3′-positions. Thus, when compared to acid **N**, more electron-withdrawing 3,3′-groups (i.e., fluorinated benzene rings in **M**) result in enhanced acidity of phosphoric acids whereas more electron-donating 3,3-substituents (i.e., alkylated benzene rings in **P**) lower the acidity of hydrogen phosphate. In line with these results, a good linear correlation between the computed pK_a values and Hammett constants has

been observed by Li, Cheng, and coworkers[23] 3,3′-disubstituted phenyl BINOL-based chiral phosphoric acids (CPAs).

Introduced in 2006 by Yamamoto and Nakashima,[15] *N*-triflyl phosphoramides (**XII**) have gained significant popularity as more acidic alternatives to phosphoric acids. Recently, the Rueping group determined the pK_a values for several BINOL- and H8-BINOL-based *N*-sulfonyl phosphoramides (**Q–T**) in acetonitrile and correlated the acidity of these catalysts with some other types of Brønsted acids including phosphoric acids (*vide supra*). Indeed, *N*-triflyl phosphoramides (pK_a in acetonitrile = 6–7) were found to be stronger than *p*-TsOH (pK_a in acetonitrile = 8.5) but weaker than hydrogen bromide (pK_a in acetonitrile = 5.5) with BINOL-based *N*-triflyl phosphoramides being ~0.3 pK_a units more acidic than H8-BINOL-based *N*-triflyl phosphoramides. The acidity of phosphoric acids was found to be significantly lower (pK_a in acetonitrile = 12–14) and closer to the acidity of sulfonylimides (pK_a of Ts_2NH in acetonitrile = 12) and acylated sulfonylamides (pK_a of saccharine in acetonitrile = 14.6). Not surprisingly, the nature of the perfluoroalkylsulfonyl group (CF_3SO_2–, $C_4F_9SO_2$–, or $C_8F_{17}SO_2$–) of *N*-perfluoroalkylsulfonyl phosphoramides (i.e., H8-BINOL acids **R**, **S**, and **T**) played only minor role in affecting the acidity of these compounds ($\Delta pK_a \sim 0.1$). Although not investigated in the aforementioned studies, *N*-tosyl and *N*-mesyl phosphoramides are expected to be significantly less acidic than *N*-triflyl phosphoramides as the pK_a values for $CH_3SO_2NH_2$, $PhSO_2NH_2$, and $CF_3SO_2NH_2$ in DMSO are 17.5, 16.1, and 9.7, respectively.

Other strategies to enhance the acidity of phosphoric acids have been explored (Figure 3.4). It has been known since the 1960s that a combination of Lewis and Brønsted acids results in enhanced Brønsted acidity of the resulting complex, and magic acid ($HSO_3F{\bullet}SbF_5$) and fluoroantimonic acid ($HF{\bullet}SbF_5$) are two classical examples of this effect. Yamamoto and coworkers have pioneered the application of this phenomenon to asymmetric catalysis and introduced the term Lewis acid-assisted Brønsted acid catalysis (LBA) to describe it.[26] In addition, the Yamamoto group has explored other types of combined acid catalysis including Brønsted acid-assisted Brønsted acid catalysis (BBA). Both LBA and BBA concepts have been applied to enhance the acidity of CPAs (Figure 3.4). Thus, Luo and coworkers have discovered that addition of Mg(II) and In(III)-based Lewis acids could enhance the reactivity of BINOL-based phosphoric acids.[27] Although the Lewis acid was proposed to

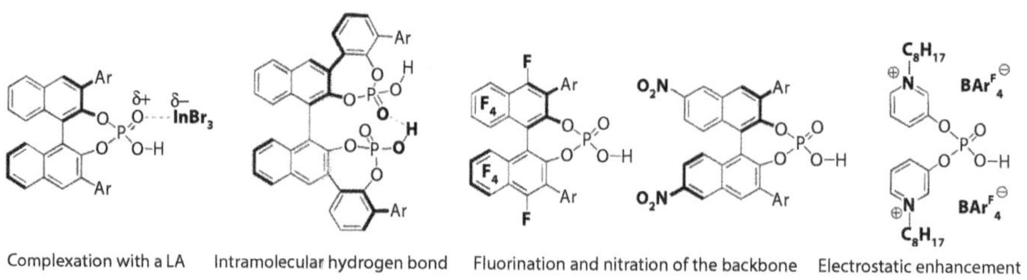

Complexation with a LA   Intramolecular hydrogen bond   Fluorination and nitration of the backbone   Electrostatic enhancement

FIGURE 3.4 Additional strategies to enhance the acidity of phosphoric acids.

coordinate the oxygen of phosphate and enhance its acidity through this complexation, in some instances it was suggested to play a more complex role and be engaged in substrate activation.

In 2011, the Terada group designed a new *bis*-phosphoric acid catalyst derived from (R)-3,3′-di(2-hydroxy-3-arylphenyl) binaphthol (Figure 3.4).[28] The authors obtained and analyzed these catalysts by X-ray and observed the formation of an intramolecular hydrogen bond between the OH group of one of the phosphates and P=O groups of the other. Based on this observation, the authors proposed BBA mode of catalysis with single-point activation. A related catalyst with carboxylic acid-activated phosphoric acid was later explored.

As it is evident from the pK_a trends summarized in Figure 3.3, introduction of electron-withdrawing groups into 3,3′-aryl substituents of BINOL backbone enhances the acidity of phosphorous (V)-based acids. Alternatively, extensive fluorination,[29] nitration of BINOL backbone,[30] or oxidation of the benzylic positions of the H8-BINOL backbone[31] at positions other than 3,3′ resulted in more active phosphoric acids with higher acidity (Figure 3.4). Kass and coworkers have explored an alternative strategy that was based on introducing positively charged nitrogen with noncoordinating counterion into the aromatic ring of the phenol bound to phosphoric acid (Figure 3.4).[32] Such electrostatically enhanced phosphoric acids were found to be significantly more acidic and could promote reactions ~2000–2500 times faster than diphenylphosphoric acid.

3.4 OVERVIEW OF THE BASIC REACTION MECHANISMS INVOLVING PHOSPHORIC ACIDS AND THEIR DERIVATIVES

Among all the Brønsted acids derived from phosphorous, CPAs with pentavalent phosphorous (V) center are arguably the most studied class of compounds and their applications are widespread in synthesis and catalysis. The rigid and easily tunable chiral backbones of CPAs make them excellent catalysts for asymmetric catalysis. Most interestingly, CPAs with both Brønsted acidic site and Lewis basic site can participate in various unique modes of activation in catalysis. Although their specific role in a given transformation may not always be obvious, CPAs are often found to be superior in catalyzing a plethora of organic reactions and leading to the formation of products with high yield and selectivity. Due to their structural similarities, the concepts outlined in the following can be extended well beyond CPA catalysis to describe other phosphorous (V) acids (i.e., **II–XIX**) as these acids also feature Brønsted acidic site and Lewis basic site.

These observations have inspired numerous studies aimed at gaining better mechanistic understanding of catalysis involving CPAs, and several general mechanistic manifolds have been identified to date (Figure 3.5). In addition to using CPAs as Brønsted acid or hydrogen bond donor (HBD) catalysts, numerous applications of CPAs in counterion-directed or covalent intermediate-based catalysis have been uncovered. In the following sections we will discuss and highlight some of the insights and implications derived from these studies on each proposed mode of activation by CPAs.

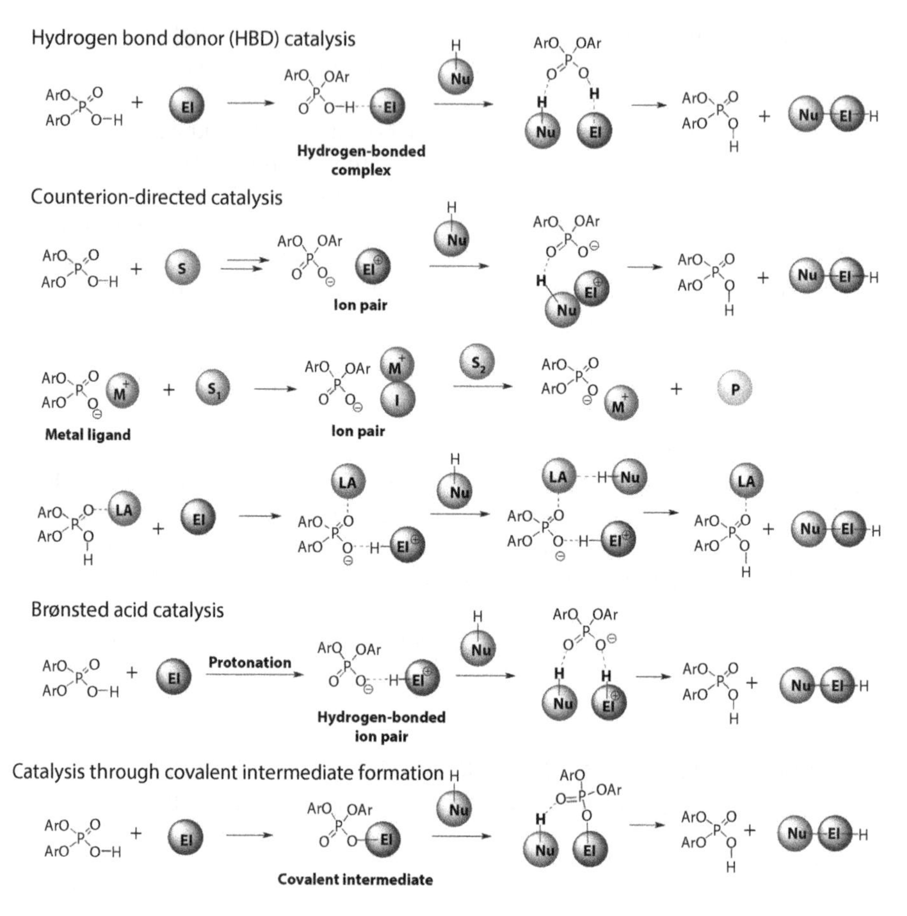

FIGURE 3.5 **(See color insert.)** General reaction mechanisms observed for phosphoric acid catalysis.

3.4.1 Bifunctional Activation through Hydrogen Bonding

The Lewis basic phosphoryl oxygen of a CPA can participate in hydrogen bonding with an acidic proton of the reacting substrate and can provide stabilization to the intermediary complex or transition state. On the other hand, the acidic proton on a CPA can activate basic electrophiles through either protonation or hydrogen bonding. Both pathways effectively enhance the electrophilicity or reactivity of electrophiles by lowering their LUMO energies. From a practical standpoint, any mechanistic information is useful for the development and optimization of the CPA-catalyzed reactions, which can generally proceed through either a hydrogen-bonding complex with the substrates or an ion pair through protonation. These intermediates have significant differences, and different factors may affect their properties in a different way. For example, hydrogen bonds are inherently directional and their strengths are highly dependent on the electronic properties of interacting substrates.

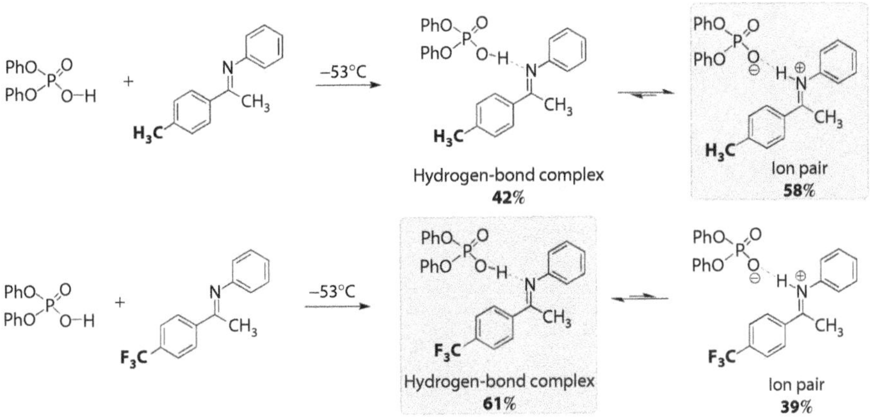

FIGURE 3.6 Determining the extend of proton transfer versus hydrogen bonding for the reaction of imines and diphenylphosphoric acid by Rueping and coworkers. (From Fleischmann, M. et al., *Angew. Chem. Int. Ed.*, 50, 6364–6369, 2011. With permission.)

On the other hand, electrostatic interaction between ion pair is less directional, and they can exist as contact ion pair, solvent-separated ion pair, or anything in between, depending on the polarity of the medium. Often, the distinctions between the two pathways are very subtle, and researchers have mostly relied on computational studies, and sometimes nuclear magnetic resonance (NMR) experiments, to unveil the mechanistic action of the CPA in a transformation. For instance, the Rueping group recently investigated the protonation of imines with diphenylphosphoric acid by NMR (Figure 3.6).[33] Although a full protonation of imine by a CPA is typically assumed, a more complex picture was observed in this study. Not only significant amounts of hydrogen-bond complex were detected along with the ion pair but also the relative amounts of the two were shown to be highly dependent on the nature of the electronic properties of imine and temperature.

In their pioneering report on CPA-catalyzed Mannich reaction,[4] Akiyama and coworkers proposed a mechanism involving an iminium ion pair as the reactive intermediate that resulted from protonation of the aldimine nitrogen by the CPA. Later, through DFT calculations, a more favorable reaction pathway proceeding through a dicoordinated intermediate was identified (Figure 3.7).[34] In this proposal, the phenolic hydroxyl group of the aldimine forms a hydrogen bond with the Lewis basic phosphoryl oxygen on the CPA **1** while the nitrogen of aldimine is hydrogen bonded to the CPA. The authors rationalized that the two-point contact between the CPA and the aldimine rigidified the transition state geometry, which resulted in high facial selectivity for the nucleophilic attack. This was verified experimentally by replacing the phenolic hydroxyl of aldimine with methyl ether or hydrogen and observing drastic decrease in enantioselectivity for both cases. Finally, the computational studies indicated that this reaction proceeds through an ionic (rather than a neutral) transition state where the hydrogen of phosphate is being transferred to imine to form an ion pair.

Entry	R	Yield (%)	ee (%)
1	2-OH	98	89
2	4-OH	28	20
3	2-OCH$_3$	56	2
4	H	76	39

FIGURE 3.7 Akiyama's enantioselective Mannich reaction. (From Yamanaka, M. et al., *J. Am. Chem. Soc.*, 129, 6756–6764, 2007. With permission.)

In 2012, List[19] and Nagorny[35] independently reported the development of asymmetric chiral phosphoric acid-catalyzed spiroketalization of cyclic enol ethers. Although the List group employed spatially confined chiral imidophosphoric acids as the catalysts, the Nagorny group discovered that TRIP catalyst **P** could catalyze both the enantioselective and diastereoselective spiroketalizations with excellent stereocontrol (Figure 3.8).

Nagorny, Zimmerman, and coworkers followed up the aforementioned report by conducting detailed mechanistic studies,[36] and they concluded that the reaction is likely proceeding through a single-step asynchronous mechanism. Their calculations based on growing string method (GSM)[37] pointed to a favorable concerted pathway

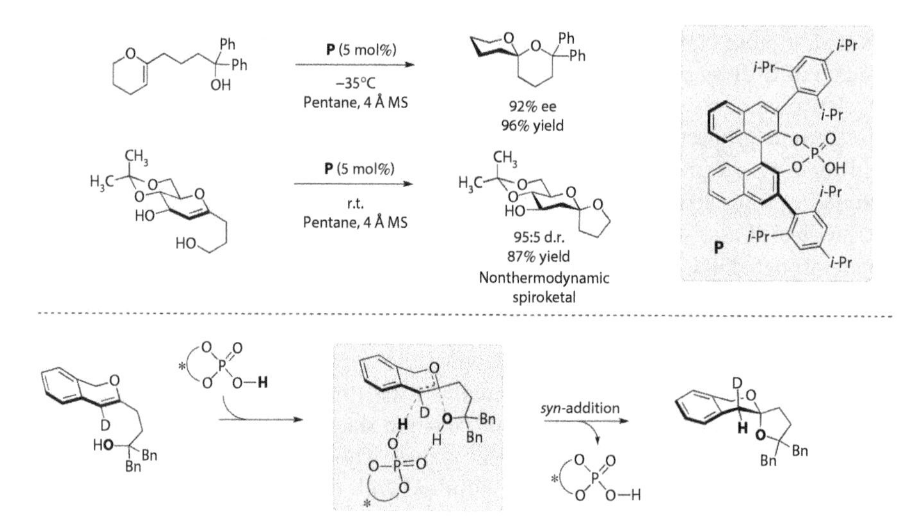

FIGURE 3.8 Enantioselective spiroketalization by Nagorny and coworkers. (From Sun, Z. et al., *J. Am. Chem. Soc.*, 134, 8074–8077, 2012; Khomutnyk, Y.Y. et al., *J. Am. Chem. Soc.*, 138, 444–456, 2016. With permission.)

for the cyclization, in which the protonation of the alkene and re-protonation of the catalyst by the hydroxy proton happened in a single step. By using deuterium-labeled substrates and CPA, an exclusive *syn* delivery of proton and oxygen nucleophile across the alkene was observed for several substrates. The Hammett analysis of this reaction provided a ρ value of -2.9, which suggested that there was a positive charge build-up in the transition state of the rate-determining step. The value was, however, much lower than what was reported for similar system investigated by Tan and coworkers that underwent S_N1-like mechanism ($\rho = -5.1$). In addition, the dynamic simulation studies and kinetic isotope effect measurement were inconsistent with the formation of an ionic intermediate but more consistent with the concerted albeit asynchronous mechanism. Based on all the mechanistic observations, it was concluded that the stepwise mechanism or the formation of long-lived ion-pair intermediate was unlikely under the reaction condition.

In 2016, Sun and coworkers reported enantioselective CPA-catalyzed oxetane desymmetrization through ring opening using HCl generated *in-situ* from chlorotrimethoxysilane and wet molecular sieves (Figure 3.9).[38] This work was later followed up by Houk and coworkers,[39] who investigated this transformation computationally to clarify the reaction mechanism and the origins of stereocontrol.

A free energy profile of the reaction was computed, and it revealed the formation of a catalyst-substrate complex via hydrogen bonding. The complex, which is 6.9 kcal/mol more stable than the separated reactants, then underwent an S_N2 reaction through a concerted transition state that led to the ring-opened product. Importantly, the bifunctional CPA catalyst enforced the cyclic transition state and assisted the proton shuffling from the nucleophile to the electrophile. These studies also suggested that the steric repulsions between the substrate and 3,3′-substituents of catalyst **2** resulted in significant strain due to the catalyst's rigid spirobiindane backbone. In the disfavored transition state, the twisting of the backbone forces the catalyst to adopt an unfavored distorted conformation, which was avoided in the favored transition state. This finding was in an agreement with the experimental observations by Sun and coworkers, where catalysts with a more flexible BINOL backbone exhibited lower enantioselectivity.

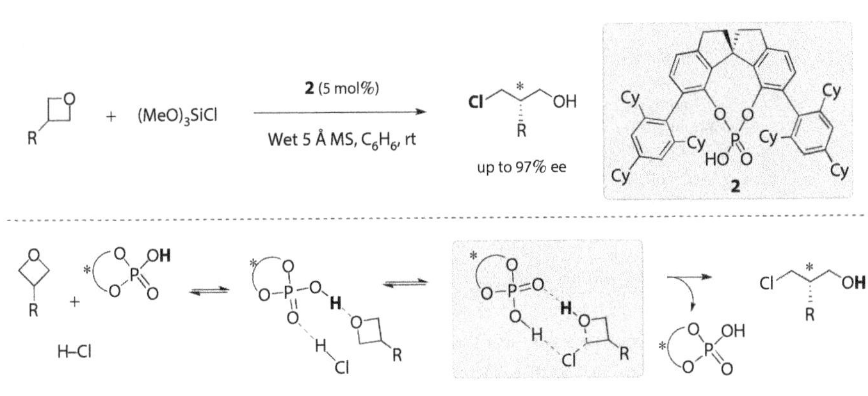

FIGURE 3.9 Enantioselective desymmetrization of oxetanes by Sun and coworkers. (From Yang, W. et al., *Angew. Chem. Int. Ed.*, 55, 6954–6958, 2016. With permission.)

The wealth of experimental data accumulated since the original reports of Akiyama and Terada on enantioselective Mannich reactions suggest that the substituent at the 3,3′-position of BINOL backbone plays a crucial role in determining the degree of asymmetric induction.[4,5] In most cases, the observed selectivity might be rationalized by considering the steric interactions between the catalyst and substrates. The selectivity outcome, however, often does not simply correlate with the size of the substituents. In some cases, the use of a catalyst with excessively bulky groups at the 3,3′-positions of BINOL backbone can actually lead to a significant lower reaction rate or reversal in selectivity. To better understand the effect of various factors determining the catalytic activity of BINOL-based CPAs, Goodman developed quantitative parameters to describe structural features of the catalysts that were important for stereoinduction (Figure 3.10).[40] He utilized A-value (a) and rotation barrier (b) parameters to quantify the proximal sterics effects and ligand AREA(θ) (c) to quantify sterics distant from the phosphorous center. Most interestingly, Goodman's case studies on Hantzsch ester hydrogenation of imines showed that both distal and proximal sterics played important but distinct roles in stereoinduction.

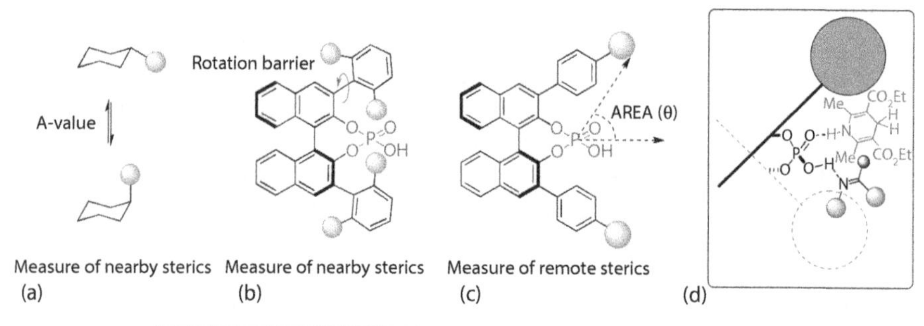

3,3′ group	Computed A-value (a)	Rotation barrier (b)	AREA (θ) (c)
H	0.00	0.00	107
$CHPh_2$	1.76	1.13	47
$SiPh_3$	4.85	1.35	29
Ph	4.39	2.05	70
1-naphth	4.07	13.63	62
2-naphth	4.35	2.13	49
9-anthyryl	14.40	28.31	61
9-phenanthryl	4.09	14.45	48
$4\text{-}PhC_6H_4$	4.35	2.01	50
$4\text{-}(t\text{-}Bu)C_6H_4$	4.37	2.02	49
$3,5\text{-}(Ph)_2C_6H_3$	4.21	2.21	36
$3,5\text{-}(CF_3)_2C_6H_3$	4.03	2.02	62
$2,4,6\text{-}(Me)_3C_6H_2$	15.97	21.58	61
$2,4,6\text{-}(i\text{-}Pr)_3C_6H_2$	26.33	28.40	51

FIGURE 3.10 Stereoselective model for the addition to imines by Goodman and coworkers. He utilized A-value (a) and rotation barrier (b) parameters to quantify the proximal steric effects and ligand AREA (theta) (c) to quantify sterics distant from the phosphorous center and (d) predictive model. (From Reid, J.P. and Goodman, J.M., *J. Am. Chem. Soc.*, 138, 7910–7917, 2016. With permission.)

The calculations revealed that the distal sterics controlled the presence of E/Z isomer of imine, whereas the proximal bulk dictated the relative orientation of imine to the catalyst. Nonetheless, steric bulk is certainly not the only tunable parameter for achieving high selectivity. When comparing the results from Akiyama's and Terada's independent studies on Mannich-type reaction, it is clear that the electronic properties of the catalyst have significant influence on the stereoselectivity outcome of the reaction too (Figure 3.11). In Terada's report, the ee of Mannich product increases as the size of aromatic substituents at the 3,3′-position of BINOL-based phosphoric acid increases (Figure 3.11a). On the other hand, Akiyama and coworkers discovered that the catalyst with an electron-deficient 4-nitro-phenyl group at the same position afforded the best enantioselectivity for their Mannich-type reaction (Figure 3.11b). The bifunctional capability of CPAs is well explored in many other polar organic reactions, such as Strecker reaction,[41] Baeyer–Villiger oxidation,[42] and Michael addition,[43] just to name a few. For each of the examples, CPAs are found to play an important role in simultaneously activating both the electrophiles and nucleophiles and in organizing the geometry of the substrates through hydrogen-bonding, which helps to accelerate the reaction and to maximize the stereoinduction.

The rapid development of imine reactions catalyzed by BINOL-based CPAs eventually inspired Goodman to develop a theoretical calculation-based model, which correctly predicted the stereoselectivity for more than 40 reactions reported at the time (Figure 3.12).[44] According to Goodman, each imine reaction catalyzed by bifunctional CPAs would undergo one of the four transition state models (i.e., **IE**, **IIE**, **IZ**, **IIZ**) that are depicted in Figure 3.12. The transition states can be classified

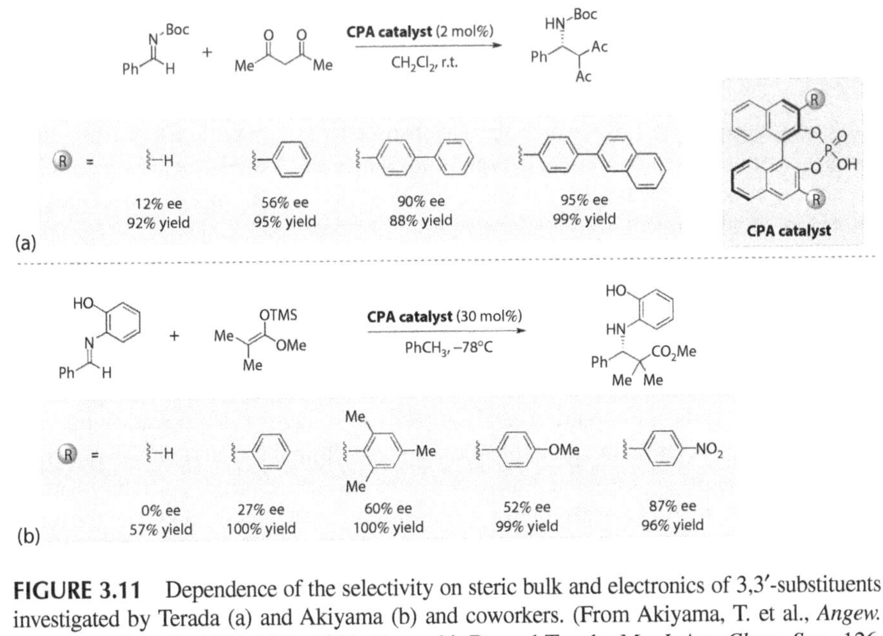

FIGURE 3.11 Dependence of the selectivity on steric bulk and electronics of 3,3′-substituents investigated by Terada (a) and Akiyama (b) and coworkers. (From Akiyama, T. et al., *Angew. Chem. Int. Ed.*, 43, 1566–1568, 2004; Uraguchi, D., and Terada, M., *J. Am. Chem. Soc.*, 126, 5356–5357, 2004. With permission.)

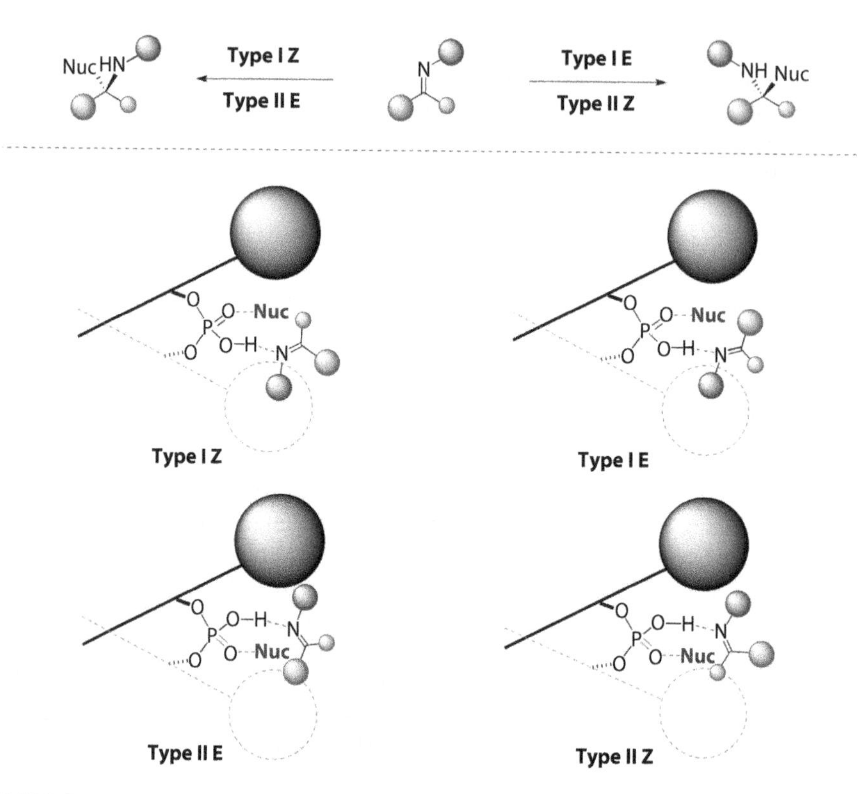

FIGURE 3.12 **(See color insert.)** Stereochemical model for chiral phosphoric acid-catalyzed addition to imines by Goodman and coworkers. (Based on Simón, L. and Goodman, J.M., *J. Org. Chem.*, 76, 1775–1788, 2011. With permission.)

into two groups: Type I, for which the *N*-substituents are pointing away from catalyst's bulk at the 3,3′-positions, and Type II, for which the *N*-substituents are pointing toward the bulk. For each type, the imine can exist in either *E*– or *Z*– conformations. Using Goodman's model, a simple analysis on the relative steric size of nucleophiles and substituents on imine and the *E*/*Z* configuration of imines can be performed to predict the stereochemistry of the product for a given enantiomer of the catalyst.

3.4.2 Counterion Catalysis

Counterion catalysis or ion-pairing catalysis takes advantage of electrostatic attraction between ionic species as the main handle for controlling the course of the reaction. Given that a majority of organic transformations proceed through partially charged transition states, there is a great interest in developing strategies to modulate the reactivity of discrete ionic species, especially in the field of asymmetric catalysis. The idea of conferring stereoinduction using chiral counterions dates back to the 1980s, when chiral cation-binding polyether catalysts[45] and then quaternary ammonium cations derived from cinchona alkaloids[46] were developed into phase transfer catalysts (PTCs) for enantioselective Michael reactions and alkylations. However, it was not

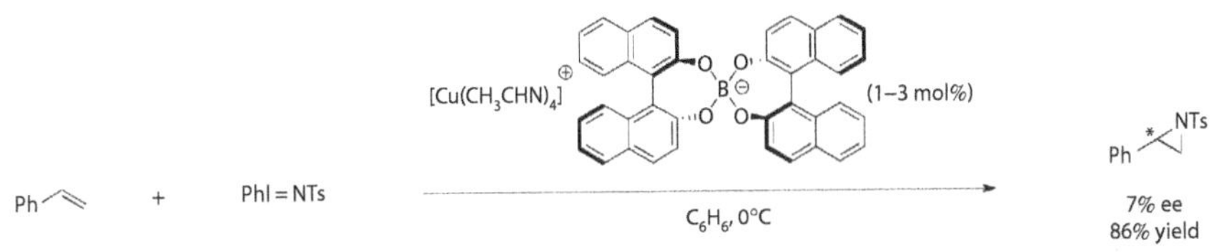

[$Cu(CH_3CHN)_4$]$^{\oplus}$ (1–3 mol%)

Ph⁄⁄ + PhI=NTs → (C_6H_6, 0°C) Ph–aziridine–NTs, 7% ee, 86% yield

FIGURE 3.13 Early example of an enantioselective chiral anion-controlled reaction by Arndtsen and coworkers. (From Llewellyn, D.B. et al., *Org. Lett.*, 2, 4165–4168, 2000. With permission.)

until 2000 when chiral anions were employed by Arndtsen and coworkers to induce chirality in aziridination reactions.[47] The Arndtsen group observed a significant effect of achiral counterions on the enantioselectivity of aziridination reaction catalyzed by chiral bis(oxazoline)-copper complex in benzene solvent but the counterion-depended effects were not observed in more polar acetonitrile solvent. The results motivated them to synthesize cationic copper(I) catalysts with a chiral borate anion derived from BINOL and apply them to aziridination (Figure 3.13) and cyclopropanation reactions. They were able to validate their hypothesis that the chiral borate may serve as a counterion and the source of chirality in the reaction by observing that (1) the enantioselectivity of the catalyzed reactions correlated with the dielectric constant of solvent and (2) the crystal structure of the copper complex revealed that the borate did not coordinate to copper center as a ligand.

The same chiral borate was later used by Nelson and coworkers in their study of the effect of counterions on the ring-opening of prochiral aziridinium ions.[48] Despite the low enantioselectivity observed for this reaction, these reports served as an important prove of concept, which opened the door for subsequent investigations on the use of chiral anion as organocatalyst in asymmetric catalysis.

With the introduction of BINOL-based CPA catalysis by Akiyama and Terada, it was not long before the chiral phosphates were found to be highly effective in counterion catalysis. The easily tunable backbone of BINOL-based phosphate facilitated the process of catalyst screening and overcame the hurdle of low stereoselectivity exhibited by the previous approaches. The great efficacy of CPAs as counterion catalyst was demonstrated by List and coworkers when they reported a highly enantioselective transfer hydrogenation of α,β-unsaturated β,β-disubstituted aldehyde using Hantzsch ester as the reductant (Figure 3.14).[49] By combining morpholine and sterically hindered TRIP CPA (catalyst **P**), ammonium phosphate **3** that served as the actual catalyst for this reaction was generated and used to achieve enantioselectivities as high as 99:1 e.r. The reaction was proposed to proceed through the formation of morpholine-derived fully substituted iminium ion with phosphate counterion. Since the putative fully substituted iminium intermediate has no handle for hydrogen-bonding activation, the authors proposed that the stereoselectivity was imparted by chiral counter anion.

In general, ion pair with chiral phosphate anion can be generated in several ways: (1) protonation of substrate, (2) dehydration of substrate, or (3) anion exchange with phosphate salt. As discussed in Section 3.4.1, it is hard to pinpoint if a substrate with a basic lone pair was activated by a CPA through hydrogen bonding or protonation (*vide supra*). Formation of tight or contact ion pair (cf. Figure 3.15), which is favored

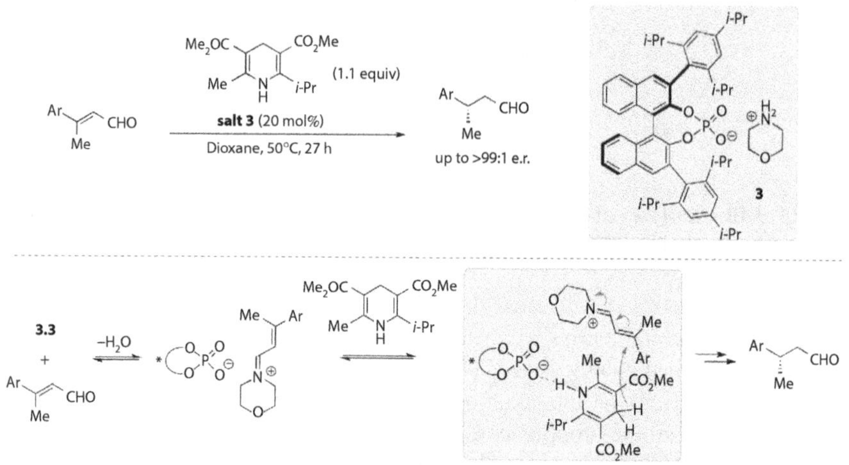

FIGURE 3.14 Transfer hydrogenation of iminium phosphate ion pair by List and coworkers. (From Mayer, S. and List, B., *Angew. Chem. Int. Ed.*, 45, 4193–4195, 2006. With permission.)

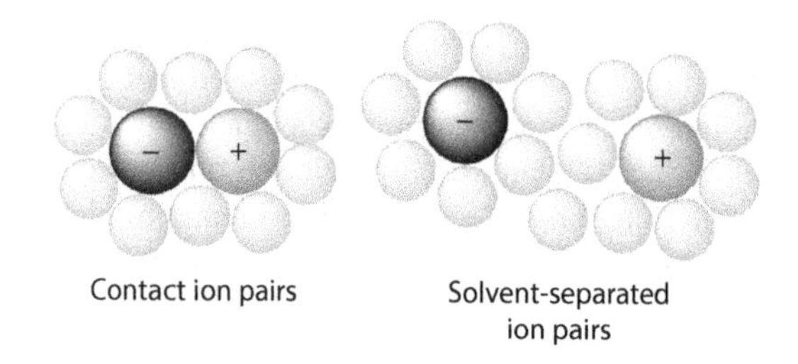

FIGURE 3.15 Contact and solvent-separated ion pairs.

in nonpolar medium, is ideal for maximum asymmetric induction because it allows stronger interactions between an ionic substrate and the chiral elements of its counter ion. Not surprisingly, most reports generally describe higher selectivity in nonpolar solvent.

In 2011, Huang and coworkers developed a method for *N*-alkylation of indoles catalyzed by CPAs (Figure 3.16).[50] They proposed a cyclic *N*-acyliminium ion, generated through the protonation of α,β-unsaturated lactam, as the reactive electrophile. To support their proposal, the authors mixed benzylated lactam with thiosphosphoric acid **4** in toluene and monitored the reaction by *in-situ* IR. As the reaction progressed, a decrease in absorption that corresponds to the lactam's double bond stretch was observed along with the increase in a new absorption peak, which was assigned to enol ether double bond stretch. When the sample from the reaction was subjected to high-resolution mass spectrometry (HRMS) analysis, a peak with the mass consistent with the proposed iminium ion pair was observed. Furthermore, when catalyst **4** with acidic deuterium was added to the lactam, followed by 3-methyl indole, the *N*-alkylated product has deuterium atom incorporated at multiple carbons

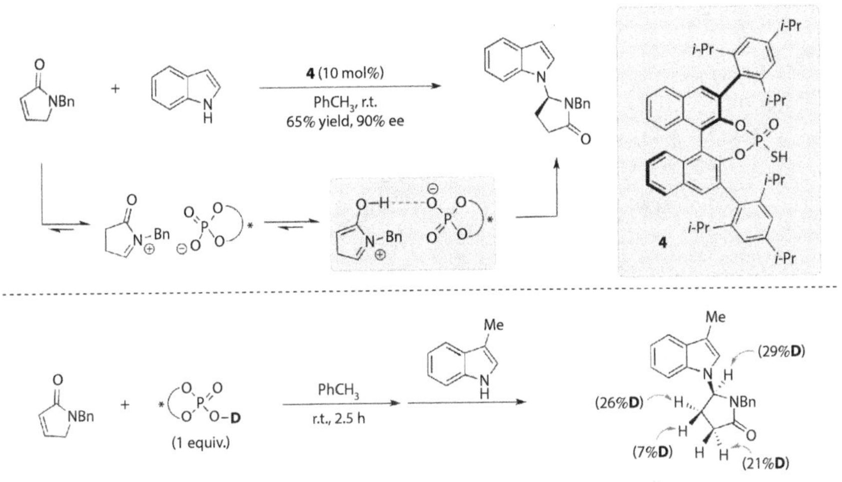

FIGURE 3.16 N-alkylation of indoles by Huang and coworkers. (From Xie, Y. et al., *Angew. Chem. Int. Ed.*, 50, 5682–5686, 2011. With permission.)

in the lactam ring. These results support the formation of iminium ion through the protonation by **4** and point out that the protonation is reversible.

In a related study, Rueping and coworkers demonstrated that similar cyclic *N*-acyliminium ions could be generated through dehydration of β-hydroxy lactams (Figure 3.17).[51] In this case, the authors observed that *N*-triflyl phosphoramide **5** was highly effective in facilitating the formation of a reactive iminium ion pair and its reaction with indole at the C3-position.

Exploring the idea of utilizing chiral phosphates as PTCs, Toste and coworkers described the first example of catalytic chiral anion PTC in 2008 (Figure 3.18).[52] Starting with racemic β-chloro tertiary amine, *meso*-aziridinium ion was generated *in situ* in the presence of CPA **P** and solid Ag_2CO_3 in toluene. The lipophilic chiral phosphate transferred silver cation from the surface of silver carbonate into the bulk solution where it abstracted chloride and triggered the ring closure to form chiral azridinium ion pair. The formation of insoluble AgCl salt drove the formation of aziridinium-phosphate ion pair, which then underwent enantioselective ring opening

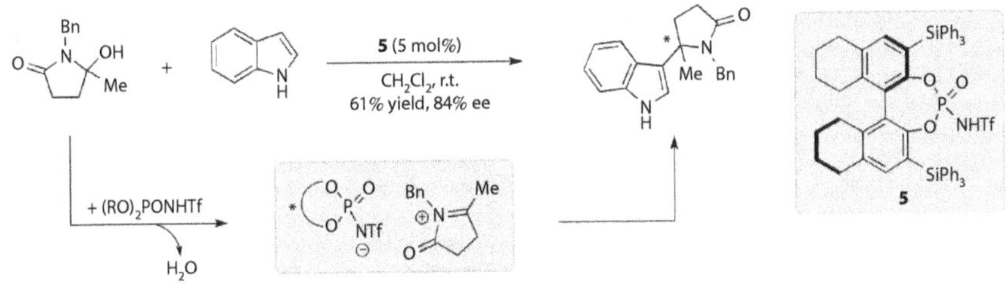

FIGURE 3.17 Alkylation of indoles by Rueping and coworkers. (From Rueping, M. and Nachtsheim, B.J., *Synlett*, 1, 119–122, 2010. With permission.)

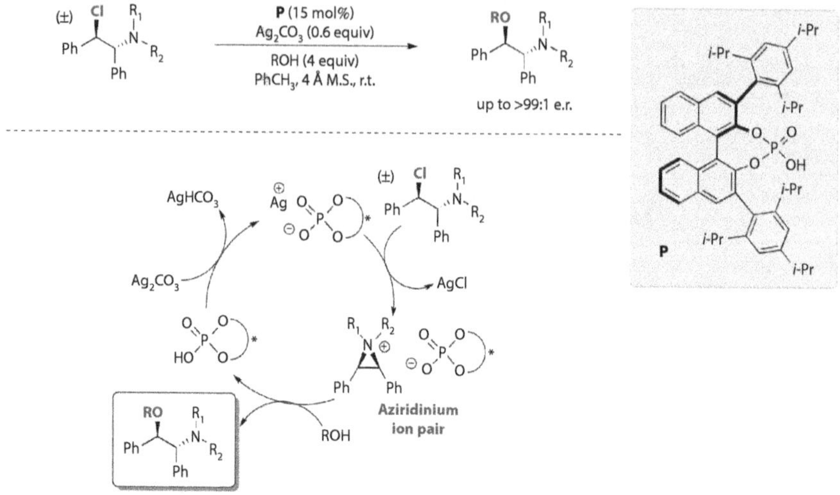

FIGURE 3.18 Use of CPA P as a PTC for the enantioselective formation of β-aminoalcohols by Toste and coworkers. (From Hamilton, G.L. et al., *J. Am. Chem. Soc.*, 130, 14984–14986, 2008. With permission.)

through the nucleophilic attack by alcohol. The proposed roles of chiral phosphate and the intermediacy of aziridinium ion were supported by several observations: (1) only double inversion products were isolated, (2) the reaction did not proceed in the absence of either CPA **P** or Ag_2CO_3, and (3) low enantioselectivity (56%) was obtained when more soluble AgOTf was used. In other words, silver cation was essential for the reaction to proceed, and there was no silver cation in the bulk solution without the presence of CPA as the phase transfer agent. The authors also discovered that the silver phosphate preformed from CPA **P** was equally effective in catalyzing this reaction.

Using a similar strategy, Toste and coworkers were able to generate a chiral fluorination reagent in nonpolar medium and used it for enantioselective fluorocyclization (Figure 3.19).[53] Selectfluor, which is typically insoluble in nonpolar

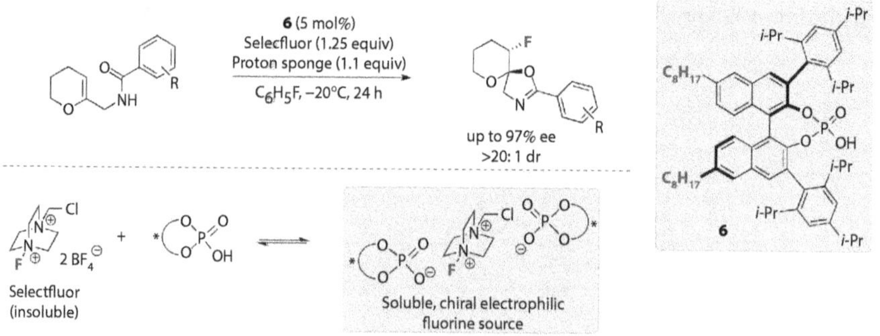

FIGURE 3.19 Use of CPA as the PTC catalysts for enantioselective fluorocyclization. (From Rauniyar, V. et al., *Science*, 334, 1681–1684, 2011. With permission.)

solvents, was used as an electrophilic fluorine source to activate the π-bond of enol ether. In the presence of catalytic CPA **6**, Selectfluor, and base, the enol ether underwent antiselective fluorination/cyclization to form spiro-fused oxazoline with high diastereo- and enantioselectivity. Consistent with their proposal that CPAs may act as effective PCTs, they found that CPA **6** with lipophilic alkyl chain attached to the 6,6′-positions of the BINOL backbone afforded better enantioselectivity. In addition, nonlinear relationship between the ee of the catalyst and the ee of the fluorocyclization product was observed, suggesting that both tetrafluoroborate counteranions of Selectfluor were exchanged with chiral phosphates before it reacted with the enol ether substrate. Toste's reports demonstrated the versatility of CPAs as counterion catalysts that could be paired with cationic reagents to facilitate asymmetric reactions.

3.4.3 Lewis Acid-Assisted Activation

In the quest to further expand the application of phosphorous-derived Brønsted acids in catalysis, external Lewis acids were found to be highly effective in augmenting the reactivity of CPAs (*vide supra*).[27] Given the amphoteric nature of phosphorous acid, additional Lewis acid can either exchange with the acidic proton or coordinate to the Lewis basic phosphoryl oxygen (Figure 3.20). In the first case, Lewis acid enhances the basicity of phosphoryl oxygen for the activation of a protic nucleophile, whereas it itself serves as a binding site for the activation of a basic electrophile. When Lewis acid binds to the phosphoryl oxygen, it effectively enhances the acidity of phosphoric acid through induction. Such complexes may act as both hydrogen bond-donors or Brønsted acids (Figure 3.5) depending on the substrate being activated. In addition, activation by Lewis acid is also frequently invoked in explaining the outcome of these reactions.[27] Due to the aforementioned ambiguities in determining the actual reaction mechanisms, these catalysts will be considered in this section along with phosphate/Lewis acidic counterion catalysis.

In 2008, Ishihara and coworkers published the first of the series of seminal studies describing the use of metal salt of CPAs in catalysis.[54] Before this report, Kagan and coworkers extensively investigated the use of chiral lithium binaptholate as a Lewis base to catalyze the addition of trimethylsilyl cyanide (TMSCN) to aldehydes.[55] Their method, however, was not applicable to the addition of TMSCN to ketone to generate tertiary cyanohydrin.

Ishihara and coworkers postulated that a stronger Lewis base was necessary to activate TMSCN toward addition to less reactive ketones (Figure 3.21). Based on

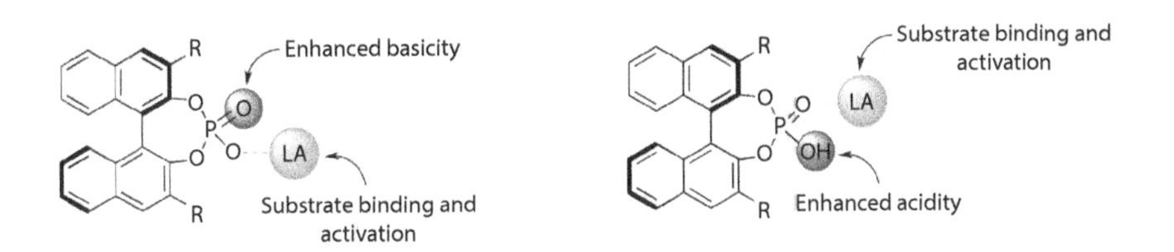

FIGURE 3.20 Enhancement of basicity or acidity of CPAs by Lewis acids.

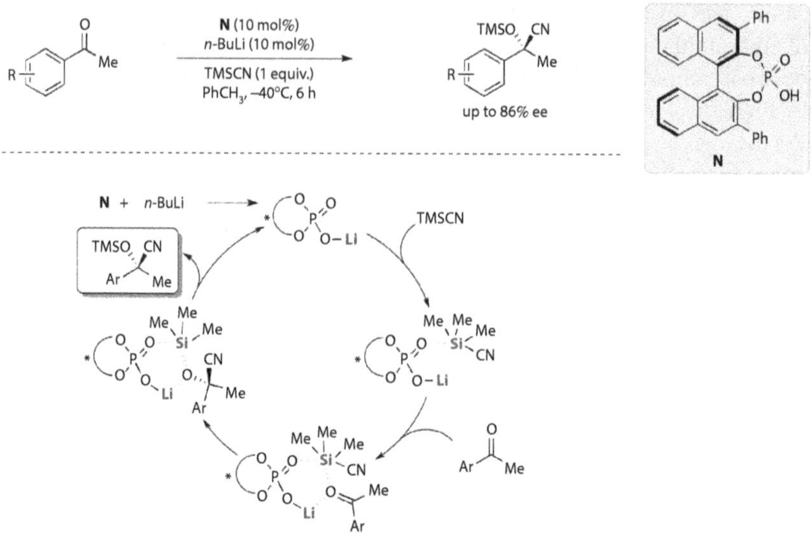

FIGURE 3.21 The use of CPA salts as Lewis base catalysts by Ishihara and coworkers. (From Hatano, M. et al., *Adv. Synth. Catal.*, 350, 1776–1780, 2008. With permission.)

this hypothesis, they proposed to use a chiral lithium phosphate as the catalyst and rationalized that Lewis acidic lithium ion should increase the basicity of P=O moiety, which could in turn activate TMSCN more effectively. In addition, they also envisioned that the ketone moiety would be doubly activated by both silicon and lithium through the formation of a six-membered chelation ring. After catalyst and conditions screening, lithium phosphate generated *in situ* by deprotonating CPA **N** with *n*-BuLi was found to be active in catalyzing the addition of TMSCN to various aromatic ketones with moderate to good enantioselectivity.

Subsequently, Ishihara and coworkers reported some interesting observations regarding the variation in performance of CPA catalysts for the Mannich reaction depending on the method of their purifications (Figure 3.22).[56] When they used **H[X]** purified by flash chromatography on silica gel to catalyze the addition of acetylacetone to aldimines, they were able to obtain results comparable to Terada's report.[5] However, when they tried the reaction under the same condition, except the catalyst was prewashed with HCl to generate free phosphoric acid, they observed a significant drop in enanatioselectivity, and the reaction favored the formation of the opposite enantiomer. They speculated that the CPA could be contaminated with metal impurities during the purification on silica gel, a phenomenon that was also observed by the List group.[57] Ishihara and coworkers went on to prepare individual alkali or alkali-earth metal phosphate salts and tested their catalytic performances. Although Li(I), Na(I), Mg(II), Ca(II), and Sr(II) salts all afforded excellent yield for the imine addition product, Ca(II) salt provided the best enantioselectivity. Based on HRMS and ^{31}P NMR analysis, they were able to verify **Ca[X]$_2$** as the sole catalyst species in the solution, ruling out the presence of free acid **H[X]**. They proposed a plausible transition state model to explain the selectivity observed, in which the

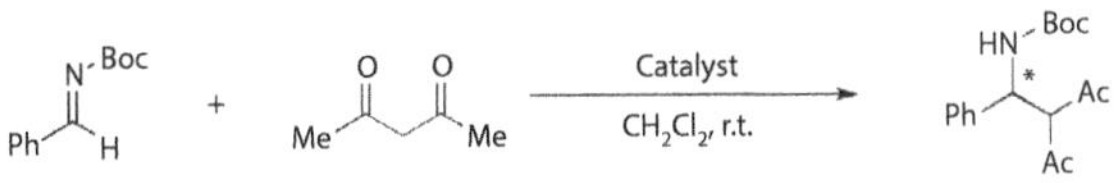

Entry	Catalyst (mol%)	Yield (%)	ee (%)(config.)
1	H[X] purified on silica gel (2)	86	92 (**R**)
2	H[X] washed with HCl (2)	88	27 (**S**)
3	Li[X] (5)	99	11 (**S**)
4	Na[X] (5)	88	9 (**S**)
5	$Mg[X]_2$ (2.5)	99	43 (**R**)
6	$Sr[X]_2$ (2.5)	99	59 (**R**)
7	**$Ca[X]_2$ (2.5)**	99	92 (**R**)

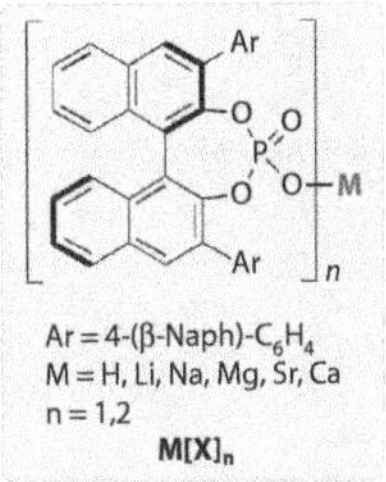

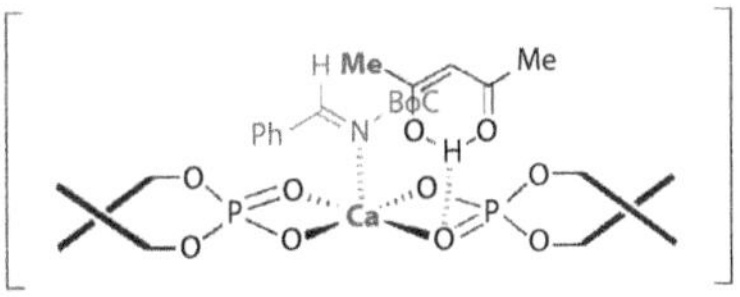

FIGURE 3.22 The study of phosphate counterion effect on enantioselectivity of Mannich reaction by Ishihara and coworkers. (From Hatano, M. et al., *Angew. Chem. Int. Ed.*, 49, 3823–3826, 2010. With permission.)

aldimine was activated by the Lewis acidic Ca(II) center, and the enol form of acetylacetone was activated by Lewis basic P=O moiety. Similar Ca(II) bis(phosphate) catalyst was also explored by Reuping and coworkers for the asymmetric addition of pyrone to aromatic aldimines.[58]

In subsequent studies, Ishihara and coworkers demonstrated that boron-based Lewis acids could also be used to effectively enhance the catalytic activity of CPAs in a Diels–Alder reaction (Figure 3.23).[59] Through optimization, they discovered that the combination of BBr_3 and CPA exhibited excellent catalytic activity for the Diels–Alder reaction of α-substituted acrolein with various dienenophiles, providing cycloaddition products with high diastereo- and enantioselectivity. Without any additional Lewis acid, CPA alone showed low reactivity and little stereocontrol. Aimed at identifying the active catalytic species, ^{31}P NMR analysis of the equimolar mixture of CPA **7** and BBr_3 in dichloromethane at −78°C was employed. This mixture provided a major ^{31}P NMR peak at −6.0 ppm, which was significantly different from the original peak for CPA **7** observed at 2.4 ppm. When the solution of the catalyst obtained earlier was subjected to reaction with cyclopentadiene and methacrolein, the authors were able to reproduce essentially the same yield and stereoselectivity which they obtained during optimization. In contrast, when the same catalyst solution was prepared at room temperature, multiple ^{31}P NMR signals were observed in +5 to −25 ppm range, possibly due to the formation of oligomeric CPA-BBr_3 complexes with concomitant loss of HBr.

Using methyl-1-cyclohexene as the acid scavenger, HBr formation was detected upon mixing of CPA **7** and BBr_3 at room temperature but not at −78°C. Not surprisingly, the adventitious HBr promoted background reaction and led to low ee. Based

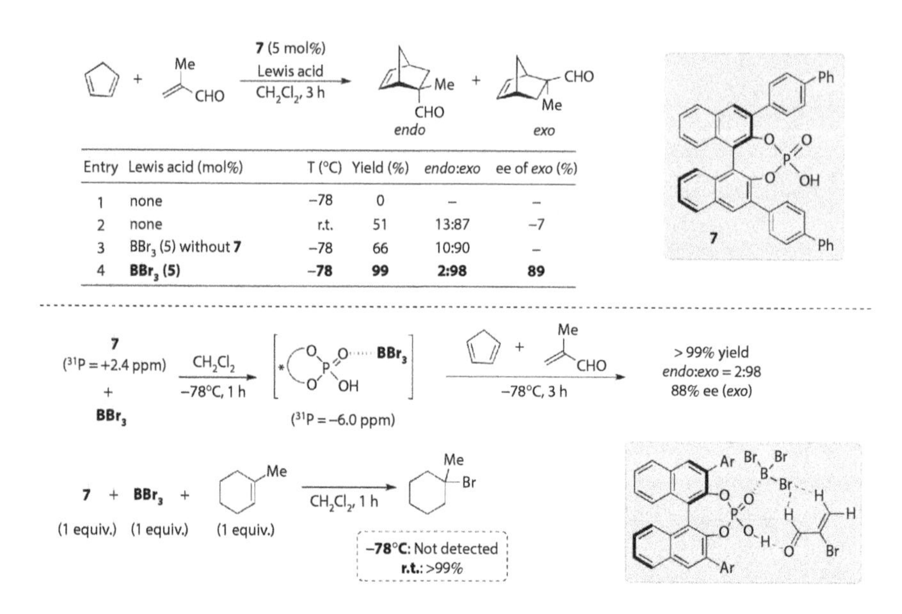

Entry	Lewis acid (mol%)	T (°C)	Yield (%)	endo:exo	ee of exo (%)
1	none	−78	0	–	–
2	none	r.t.	51	13:87	−7
3	BBr_3 (5) without **7**	−78	66	10:90	–
4	**BBr_3 (5)**	**−78**	**99**	**2:98**	**89**

FIGURE 3.23 Enantioselective [2 + 4] cycloaddition catalyzed by Lewis acid•CPA complex by Ishihara and coworkers. (From Hatano, M. et al., *J. Am. Chem. Soc.*, 137, 13472–13475, 2015. With permission.)

on further theoretical calculations, the complex with BBr_3 coordinated to the Lewis basic phosphoryl oxygen was proposed as the active catalytic species. When the BBr_3•**7** complex with α-bromo acrolein was subjected to geometry optimization, the hydrogen bond between carbonyl oxygen and the acidic proton of CPA **7** and two halogen bonds between Br and C-H bonds of α-bromo acrolein were observed. This suggests that BBr_3 not only enhances the acidity of **7** but also plays an important role in controlling the dienophile orientation.

A similar strategy was employed by Tsogoeva and coworkers in their search for a new catalyst for an enantioselective [3 + 2] cycloaddition of hydrazones to olefins (Figure 3.24).[60] The authors discovered that combining CPA **8** and silicon-based Lewis acid, Ph_2SiCl_2, resulted in greatly enhanced reactivity and stereocontrol. Although Ph_2SiCl_2 was inactive in catalyzing the reaction, CPA **8** alone showed significantly lower reactivity and stereoselectivity. Based on NMR analysis, the authors obtained strong support for the *in situ* formation of chiral silane species in solution. The ^{31}P NMR of a 2:1 mixture of CPA **8** and Ph_2SiCl_2 contained a major peak at 0.918 ppm, which corresponds to free CPA **8**, and a new peak at −12.011 ppm, which was assigned to the silylated phosphate species. Furthermore, the authors observed only one ^{29}Si signal at −33.864 ppm, which was shifted by nearly 40 ppm upfield from the signal of Ph_2SiCl_2 at +6.815 ppm. Since the 2:1 ratio of **8** to Ph_2SiCl_2 was found to be the optimal for the [3 + 2] cycloaddition, the authors proposed two transition state models, where hydrazone could coordinate to Si center in either a monodentate or a bidentate fashion, and the second molecule of CPA helped to stabilize the transition state through hydrogen bonding.

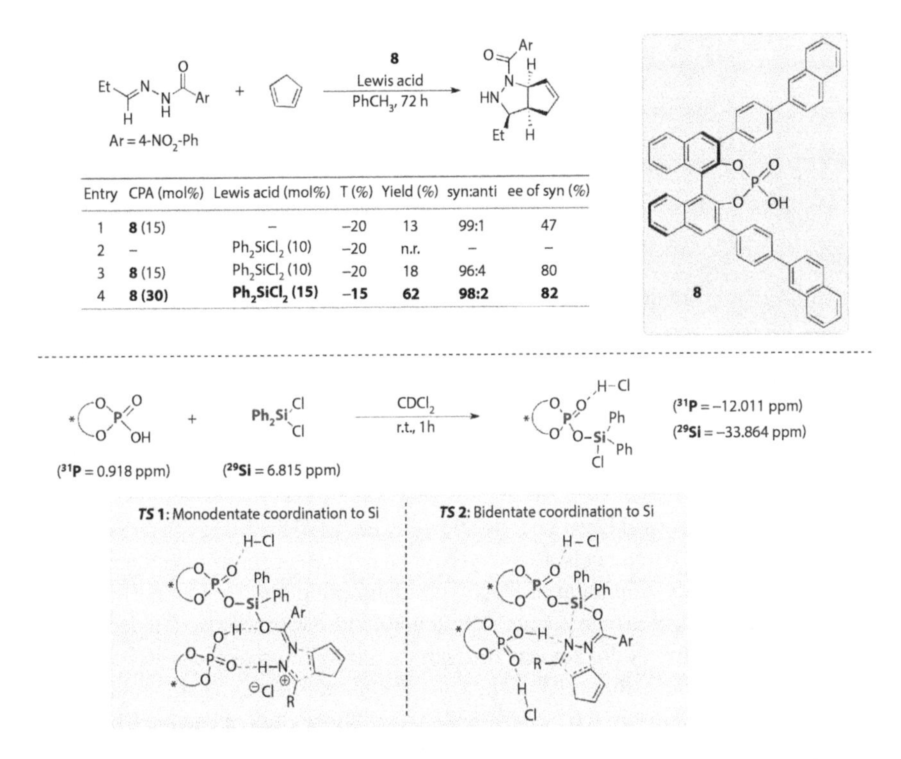

Entry	CPA (mol%)	Lewis acid (mol%)	T (%)	Yield (%)	syn:anti	ee of syn (%)
1	**8** (15)	–	–20	13	99:1	47
2	–	Ph_2SiCl_2 (10)	–20	n.r.	–	–
3	**8** (15)	Ph_2SiCl_2 (10)	–20	18	96:4	80
4	**8 (30)**	**Ph_2SiCl_2 (15)**	**–15**	**62**	**98:2**	**82**

FIGURE 3.24 Application of Lewis acid/CPA catalysis to dipolar cycloaddition by Tsogoeva and coworkers. (From Serdyuk, O.V. et al., *Adv. Synth. Catal.*, 354, 3115–3121, 2012. With permission.)

3.4.4 Brønsted Acid Catalysis and Its Application to Enantioselective Protonation

Brønsted acid catalysis represents an important subtype of reaction mechanisms, characterized by complete transfer of phosphorous acid to substrate. The resulting ion pair may have significant hydrogen bonding between the phosphate anion and the protonated substrate (Section 3.4.1). In case of the protonated iminium ions, such ion pairs may also coexist with the corresponding neutral hydrogen bond complexes (cf. Figure 3.6). However, cases with irreversible proton transfer in particular, via protonation of π-bonds, are also known. Asymmetric protonation of a π-bond represents one of the simplest ways to access chiral compounds; however, some challenges have to be considered in implementing this simple transformation into life. First of all, the rapid reaction rate of protonation makes it difficult to exert high level to stereocontrol through the interaction between the prochiral intermediate and chiral catalyst or reagent. Furthermore, the newly formed stereocenter can potentially undergo undesired racemization. The development of various CPAs with diverse range of pKa enabled the studies of phosphorous (V)-based acids as a chiral surrogate for proton atom, and this strategy was found to be useful in several organic transformations.

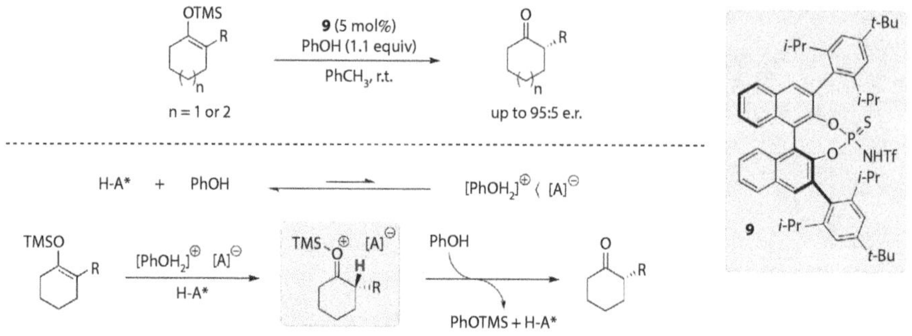

FIGURE 3.25 Enantioselective protonation of silyl enol ether with *N*-triflylthiophosphoramides by Yamamoto and coworkers. (From Cheon, C.H. and Yamamoto, H.A., *J. Am. Chem. Soc.*, 130, 9246–9247, 2008. With permission.)

In 2008, Yamamoto and Rueping independently reported the first applications of chiral phosphorous (V)-based acids as catalysts in enantioselective protonation reactions. By combining a catalytic amount of *N*-triflyl thiophosphoramide **9** with a stoichiometric phenol as an achiral proton source, Yamamoto and coworkers were able to generate chiral α-substituted cyclic ketones through the protonation of their silyl enol ether precursors (Figure 3.25).[7] Remarkably, in some instances such reactions could proceed at catalyst loadings as low as 0.05 mol%, without significant loss in enantioselectivity.

In the absence of an achiral proton source, however, no reaction was observed even with stoichiometric amount of **9**. Based on this result, they suggested that phenol was first protonated by **9**, and then either another molecule of the *N*-triflyl thiophosphoramide or the resultant phenolic oxocarbenium ion was responsible for the protonation of the silyl enol ether. This step was followed by desilylation of the resultant ionic intermediate with phenol to form chiral ketone and to regenerate the catalyst **9**. However, it remained unclear whether the protonation or the desilylation was the rate (and ee)-determining step for the reaction.

Using Hantzsch ester as the reductant and CPA **10** as the catalyst, Rueping and coworkers were able to achieve formation of optically active 3-susbtituted tetrahydroquinolines through transfer hydrogenation of quinolines (Figure 3.26).[61]

A probable mechanism involved protonation of quinolone nitrogen with phosphoric acid resulting in an ion pair. This ion pair undergoes a 1,4-hydride reduction with the Hantszch ester to produce enamine. This enamine intermediate undergoes subsequent enantioselective protonation at the C3-position with **10**, which establishes the stereocenter and results in an iminium ion. Finally, the iminium intermediate undergoes 1,2-reduction with the Hantzsch ester to furnish the product. It should be noted that the CPA serves as a Brønsted acid catalyst for both 1,4-reduction and enamine protonation steps. Other transformations where phosphorous (V) acids play a dual role and protonate both a heteroatom and a π-bond are also known. Thus, Rueping and coworkers went on to develop an asymmetric Nazarov cyclization reaction depicted in Figure 3.27.[62] Although this transformation could be promoted by CPAs, the use of more acidic *N*-triflyl phosphoramide **11** resulted in improved reactivity. Catalyst **11** first acted as a general Brønsted acid and activated

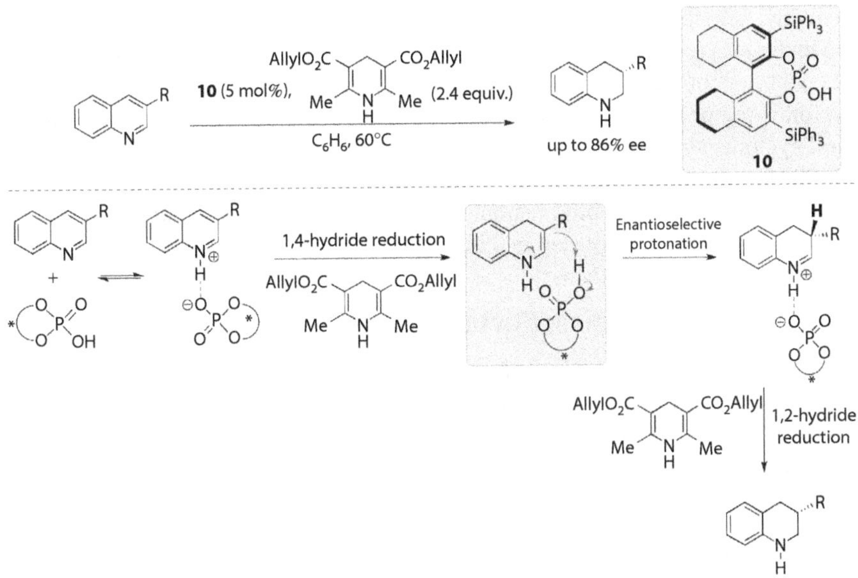

FIGURE 3.26 Enantioselective reduction of quinolines proceeding through enantioselective protonation by Rueping and coworkers. (From Rueping, M. et al., *Adv. Synth. Catal.*, 350, 1001–1006, 2008. With permission.)

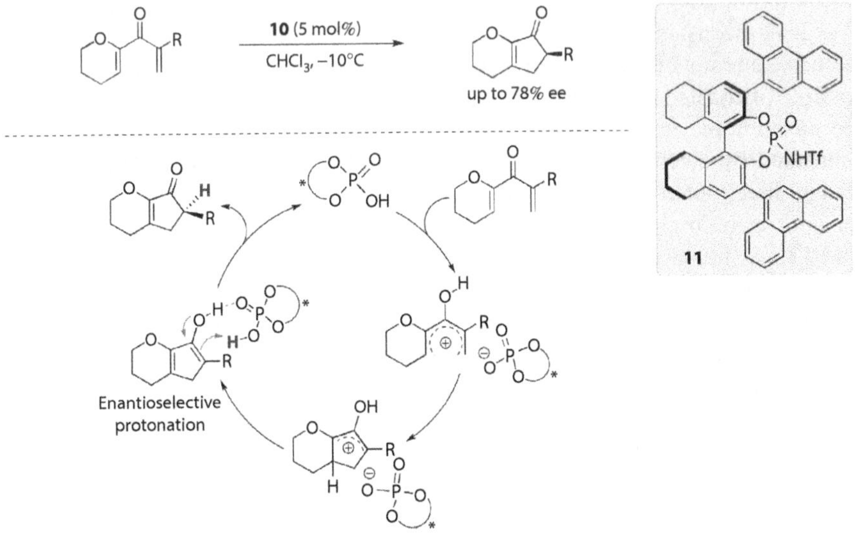

FIGURE 3.27 Enantioselective Nazarov cyclization by Rueping and coworkers. (From Rueping, M. and Leawsuwan, W., *Adv. Synth. Catal.*, 351, 78–84, 2009. With permission.)

the divinyl ketone by protonating the carbonyl oxygen. This protonated intermediate then undergoes a 4π-electrocylization to generate a cyclic oxyallyl cation that undergoes deprotonation to provide cycopentadiene intermediate and regenerates acid **11**. The following protonation of this intermediate by acid **11** results in the formation of chiral cyclopentenone. Since 4π-electrocyclization does not result in the formation of β-stereocenters, and the cyclopentadiene intermediate formed prior to the second protonation is achiral, the protonation of enol-containing cyclopentadiene by **11** represents the enantioselective step in this sequence.

3.4.5 Activation through Covalent Phosphate Intermediate

Covalent catalysis involves the formation of transient covalent bond between substrate and catalyst and subsequent breakage of the bond to regenerate the catalyst. In theory, the rigid and highly directional covalent linkage should enhance the interactions between substrate and its surrounding chiral elements, and hence, it helps to maximize the asymmetric induction from the catalyst. One of the major challenges in developing covalent catalysis is to identify a catalyst that is nucleophilic enough to form the covalent bond with substrate, but also possesses properties of a good leaving group to facilitate the turnover of the catalyst. The formation of covalently linked with CPA intermediates had not been invoked to rationalize the CPA-catalyzed reactions before 2011 until the seminal study by the Toste group demonstrates the feasibility of such intermediates.[63] Since then, several other reports have described related covalent intermediates, and a growing number of such transformations indicate that this is a rather general class of reactions promoted by CPAs.[64–67]

In 2011, Toste and coworkers discovered that chiral dithiophosphoric acids are excellent catalysts that promoted the formation of chiral pyrrolidines via asymmetric intramolecular hydroamination reaction (Figure 3.28).[63] Although the use of organic Brønsted acids to promote addition to unactivated alkene was known at the time, attempts to develop an asymmetric variant of this reaction were unsuccessful.[68] Brønsted acids can generally activate alkenes through protonation, and the enantiotopic discrimination of the resulting carbocation can potentially be controlled by electrostatic interaction with a chiral counteranion. However, the lack of directionality and rigidity of such interactions were presumably the reasons for poor enantioselectivities that are previously reported. Through optimization, acid **12** was found to be a particularly effective catalyst for highly enanotioselective hydroamination of diene or allene with tethered amine as nucleophile. Toste reasoned that chiral dithiophosphoric acids are more acidic and nucleophilic than their oxygenated counterparts due to the presence of more polarizable sulfur atom. Subsequent mechanistic and computational studies indicated that this reaction might proceed through a covalent intermediate. Indeed, such a covalent intermediate could be observed by time-of-flight mass spectrometry (TOF-MS) analysis of the reaction mixture. Furthermore, deuterium labeling studies suggested that the reaction underwent *syn*– addition of phosphoric acid across the alkene, followed by a *syn*– S_N2' substitution to displace the phosphate catalyst and to form the cylic amine product. It is likely that hydrogen bonding between phosphoryl sulfur and N-H helped to direct the stereoselective substitution too.

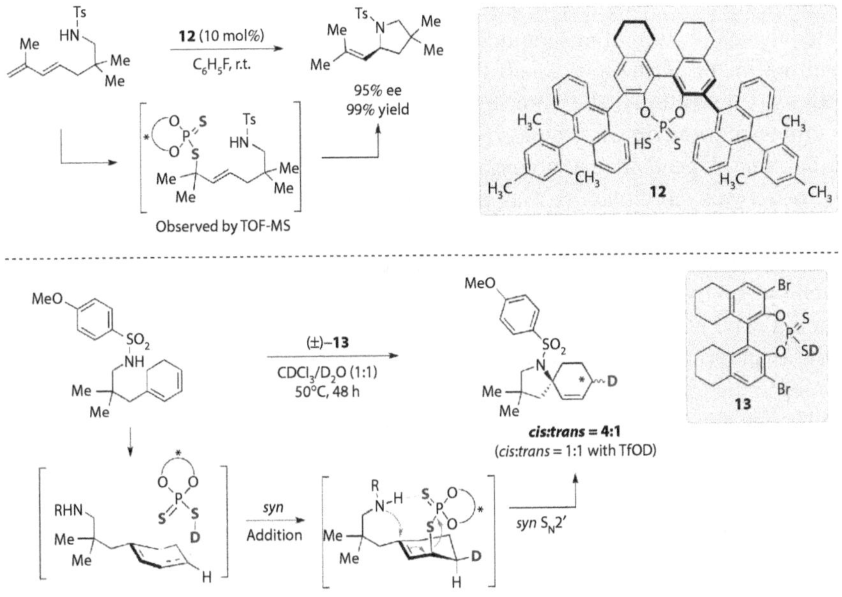

FIGURE 3.28 Enantioselective hydroamination of dienes proceeding through covalent dithiophosphate by Toste and coworkers. (Shapiro, N.D. et al., *Nature*, 470, 245–249, 2011. With permission.)

A similar type of mechanistic proposal, which involved the formation of a covalent phosphate intermediate followed by a S_N2' substitution, was invoked by Nagorny and coworkers for their enantioselective synthesis of piperidines (Figure 3.29).[64] It was initially envisioned that protonation of α,β-unsaturated acetal by a CPA would generate a chiral oxocarbenium ion pair, which could then undergo enantioselective

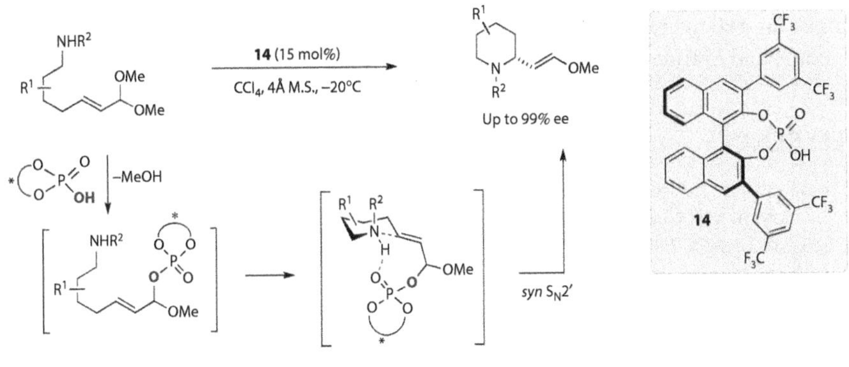

FIGURE 3.29 Enantioselective S_N2'-like cyclization leading to formation of chiral piperidines by Nagorny and coworkers. (From Sun, Z. et al., *Angew. Chem. Int. Ed.*, 53, 11194–11198, 2014. With permission.)

intramolecular aza-Michael reaction. Indeed, catalytic amount of CPA **14** promoted the cyclization of unsaturated acetals tethered with protected primary amine resulting in the formation of substituted chiral piperidines with great enantioselectivity. Interestingly, the enol ether-containing product underwent subsequent kinetic resolution via CPA-catalyzed acetalization with methanol. This resulted in further time-dependent enantioenrichment of the material, as the minor enantiomer of product was more reactive than the major enantiomer. In contrast to their initial hypothesis, the results of computational studies ruled out the formation of unsaturated oxocarbenium ion or the concerted mechanism, in which a bifunctional CPA synchronized the ring closure and the departure of leaving group through hydrogen bonding. The energy barriers of those pathways were too high to be considered feasible under the actual reaction condition. Instead, the full system simulations pointed to the formation of an energetically favorable mixed phosphorous acid acetal. The intermediate then underwent a lower barrier, concerted, but asynchronous ring closure and displacement of the phosphate catalyst to form the final product.

3.5 CONCLUSION

In conclusion, phosphorous (V)-based acids are versatile catalysts that can effectively promote a variety of different organic transformations, and the number and scope of such transformations have been rapidly growing in the past years. A great amount of information has been gathered to gain better understanding of the structural features of these acids that are particularly important to catalysis, and as a result, various strategies to enhance phosphorous (V) acid reactivity have emerged over the years. Due to their directionality, symmetry, high degree of tunability, and bifunctional nature of activation, BINOL-based phosphoric acids have been extensively used in asymmetric catalysis. Several different mechanistic manifolds for CPA catalysis have been uncovered and included HBD, Brønsted acid, counter anion, metal counterion, and covalent intermediate catalysis. Often, it is hard to pinpoint the precise reaction mechanism, by which a specific transformation is catalyzed and additional experimental and computational studies are required to identify feasible mechanistic option. At the same time, certain types of reactions such as nucelophilic addition to imines are relatively well understood, and general predictive models for selectivity outcome and optimal catalyst selection are available.

REFERENCES

1. Brønsted, J. N. *Chem. Rev.* **1928**, *5*, 231–338.
2. Akiyama, T. *Chem. Rev.* **2007**, *107*, 5744–5758.
3. Cornforth, J. *J. Chem. Soc. Perkin Trans.* **1996**, *1*, 2889–2893.
4. Akiyama, T., Itoh, J., Yokota, K., Fuchibe, K. *Angew. Chem. Int. Ed.* **2004**, *43*, 1566–1568.
5. Uraguchi, D., Terada, M. *J. Am. Chem. Soc.* **2004**, *126*, 5356–5357.
6. Parmar, D., Sugiono, E., Raja, S., Rueping, M. *Chem. Rev.* **2014**, *114*, 9047–9153.
7. Cheon, C. H., Yamamoto, H. A. *J. Am. Chem. Soc.* **2008**, *130*, 9246–9247.
8. Pousse, G., Devineau, A., Dalla, V., Humphreys, L., Lasne, M.-C., Rouden, J., Blanchet, J. *Tetrahedron* **2009**, *65*, 10617–10622.

9. Ferry, A., Stemper, J., Marinetti, A., Voituriez, A., Guinchard, X. *Eur. J. Org. Chem.* **2014**, *1*, 188–193.
10. Zhou, F., Yamamoto, H. *Angew. Chem. Int. Ed.* **2016**, *55*, 8970–8974.
11. Sakuma, M., Sakakura, A., Ishihara, K. *Org. Lett.* **2013**, *15*, 2838–2841.
12. Nelson, H. M., Patel, J. S., Shunatona, H. P., Toste, F. D. *Chem. Sci.* **2015**, *6*, 170–173.
13. Desai, A. A., Wulff, W. D. *Synthesis* **2010**, *21*, 3670–3680.
14. Rueping, M., Nachsheim, B. J., Koenigs, R. M., Leawsuwan, W. *Chem. Eur. J.* **2010**, *16*, 13116–13126.
15. Nakashima, D., Yamamoto, H. *J. Am. Chem. Soc.* **2006**, *128*, 9626–9627.
16. Borovika, A., Nagorny, P. *Tetrahedron* **2013**, *69*, 5719–5725.
17. Kaib, P. S., List, B. *Synlett* **2016**, *27*, 156–158.
18. Nishikawa, Y., Nakano, S., Tahira, Y., Terazawa, K., Yamazaki, K., Kitamura, C., Hara, O. *Org. Lett.* **2016**, *18*, 2004–2007.
19. Vellalath, S., Corich, I., List, B. *Angew. Chem. Int. Ed.* **2010**, *49*, 9749–9752.
20. Coric, I., List, B. *Nature* **2012**, *483*, 315–319.
21. Lee, S., Kaib, P. S. J., List, B. *Synlett* **2017**, *28*, A–C.
22. Lovie-Toon, J. P., Tram, C. M., Flynn, B. L., Krenske, E. H. *ACS Catal.* **2017**, *7*, 3466–3476.
23. Yang, C., Xue, X.-S., Jin, J.-L., Li, X., Cheng, J.-P. *J. Org. Chem.* **2013**, *78*, 7076–7085.
24. Christ, P., Lindsay, A. G., Vormittag, S. S., Neudorfl, J.-M., Berkessel, A., O'Donoghue, A. C. *Chem. Eur. J.* **2011**, *17*, 8524–8528.
25. Kaupmees, K., Tolstoluzhsky, N., Raja, S., Rueping, M., Leito, I. *Angew. Chem. Int. Ed.* **2013**, *52*, 11569–11572.
26. Yamamoto, H., Futatsugi, K. *Angew. Chem. Int. Ed.* **2005**, *44*, 1924–1942.
27. Lv, J., Luo, S. *Chem. Commun.* **2013**, *49*, 847–858.
28. Momivama, N., Konno, T., Furiya, Y., Iwamoto, T., Terada, M. *J. Am. Chem. Soc.* **2011**, *133*, 19294–19297.
29. Momivama, N., Okamoto, H., Kikuchi, J., Korenaga, T., Terada, M. *ACS Catal.* **2016**, *6*, 1198–1204.
30. Harada, S., Kuwano, S., Yamaoka, Y., Yamada, K., Takasu, K. *Angew. Chem. Int. Ed.* **2013**, *52*, 10227–10230.
31. Wang, Y., Liu, W., Ren, W., Shi, Y. *Org. Lett.* **2015**, *17*, 4976–4979.
32. Ma, J., Kass, S. R. *Org. Lett.* **2016**, *18*, 5812–5815.
33. Fleischmann, M., Drettwan, D., Sugiono, E., Rueping, M., Gschwind, R. M. *Angew. Chem. Int. Ed.* **2011**, *50*, 6364–6369.
34. Yamanaka, M., Itoh, J., Fuchibe, K., Akiyama, T. *J. Am. Chem. Soc.* **2007**, *129*, 6756–6764.
35. Sun, Z., Winschel, G. A., Borovika, A., Nagorny, P. *J. Am. Chem. Soc.* **2012**, *134*, 8074–8077.
36. Khomutnyk, Y. Y., Arguelles, A. J., Winschel, G. A., Sun, Z., Zimmerman, P. M., Nagorny, P. *J. Am. Chem. Soc.* **2016**, *138*, 444–456.
37. Zimmerman, P. M. *J. Chem. Theory Comput.* **2013**, *9*, 3043–3050.
38. Yang, W., Wang, Z., Sun, J. *Angew. Chem. Int. Ed.* **2016**, *55*, 6954–6958.
39. Champagne, P. A., Houk, K. N. *J. Am. Chem. Soc.* **2016**, *138*, 12356–12359.
40. Reid, J. P., Goodman, J. M. *J. Am. Chem. Soc.* **2016**, *138*, 7910–7917.
41. Simón, L., Goodman, J. M. *J. Am. Chem. Soc.* **2009**, *131*, 4070–4077.
42. Xu, S., Wang, Z., Li, Y., Zhang, X., Wang, H., Ding, K. *Chem. Eur. J.* **2010**, *16*, 3021–3035.
43. Akiyama, T., Katoh, T., Mori, K. *Angew. Chem. Int. Ed.* **2009**, *48*, 4226–4228.
44. Simón, L., Goodman, J. M. *J. Org. Chem.* **2011**, *76*, 1775–1788.
45. Cram, D. J., Sogah, G. D. Y. *J. Chem. Soc. Chem. Commun.* **1981**, *13*, 625–628.
46. Dolling, U. H., Davis, P., Grabowski, E. J. J. *J. Am. Chem. Soc.* **1984**, *106*, 446–447.

47. Llewellyn, D. B., Adamson, D., Arndtsen, B. A. *Org. Lett.* **2000**, *2*, 4165–4168.
48. Carter, C., Fletcher, S., Nelson, A. *Tetrahedron: Asymmetry* **2003**, *14*, 1995–2004.
49. Mayer, S., List, B. *Angew. Chem. Int. Ed.* **2006**, *45*, 4193–4195.
50. Xie, Y., Zhao, Y., Qian, B., Yang, L., Xia, C., Huang, H. *Angew. Chem. Int. Ed.* **2011**, *50*, 5682–5686.
51. Rueping, M., Nachtsheim, B. J. *Synlett* **2010**, *1*, 119–122.
52. Hamilton, G. L., Kanai, T., Toste, F. D. *J. Am. Chem. Soc.* **2008**, *130*, 14984–14986.
53. Rauniyar, V., Lackner, A. D., Hamilton, G. L., Toste, F. D. *Science* **2011**, *334*, 1681–1684.
54. Hatano, M., Ikeno, T., Matsumura, T., Torii, S., Ishihara, K. *Adv. Synth. Catal.* **2008**, *350*, 1776–1780.
55. Holmes, I. P., Kagan, H. B. *Tetrahedron Lett.* **2000**, *41*, 7453–7456.
56. Hatano, M., Moriyama, K., Maki, T., Ishihara, K. *Angew. Chem. Int. Ed.* **2010**, *49*, 3823–3826.
57. Klussmann, M., Ratjen, L., Hoffmann, S., Wakchaure, V., Goddard, R., List, B. *Synlett* **2010**, *14*, 2189–2192.
58. Rueping, M., Bootwicha, T., Sugiono, E. *Synlett* **2011**, *3*, 323–326.
59. Hatano, M., Goto, Y., Izumiseki, A., Akakura, M., Ishihara, K. *J. Am. Chem. Soc.* **2015**, *137*, 13472–13475.
60. Serdyuk, O. V., Zamfir, A., Hampel, F., Tsogoeva, S. B. *Adv. Synth. Catal.* **2012**, *354*, 3115–3121.
61. Rueping, M., Theissmann, T., Raja, S., Bats, J. W. *Adv. Synth. Catal.* **2008**, *350*, 1001–1006.
62. Rueping, M., Leawsuwan, W. *Adv. Synth. Catal.* **2009**, *351*, 78–84.
63. Shapiro, N. D., Rauniyar, V., Hamilton, G. L., Wu, J., Toste, F. D. *Nature* **2011**, *470*, 245–249.
64. Sun, Z., Winschel, G. A., Zimmerman, P., Nagorny, P. *Angew. Chem. Int. Ed.* **2014**, *53*, 11194–11198.
65. Liu, L., Leutzsch, M., Zheng, Y., Alachraf, M. W., Thiel, W., List, B. *J. Am. Chem. Soc.* **2015**, *137*, 13268–13271.
66. Lv, J., Zhang, Q., Zhang, X., Luo, S. *J. Am. Chem. Soc.* **2015**, *137*, 15576–15583.
67. Kuroda, Y., Harada, S., Oonishi, A., Kiyama, H., Yamaoka, Y., Yamada, K., Takasu, K. *Angew. Chem. Int. Ed.* **2016**, *55*, 13137.
68. Ackermann, L., Althammer, A. *Synlett* **2008**, *7*, 995–998.

4 Halogen Bond Catalysis
An Emerging Paradigm in Organocatalysis

Choon-Hong Tan and Choon Wee Kee

CONTENTS

4.1 BACKGROUND

4.1.1 Definition

According to the definition proposed by IUPAC, "A halogen bond occurs when there is evidence of a net attractive interaction between an electrophilic region associated with a halogen atom in a molecular entity and a nucleophilic region in another, or the same, molecular entity."[1] Halogen bond is depicted by "···" in the halogen-bonded complex R-X···Y; X is a halogen, and Y is an electron donor or

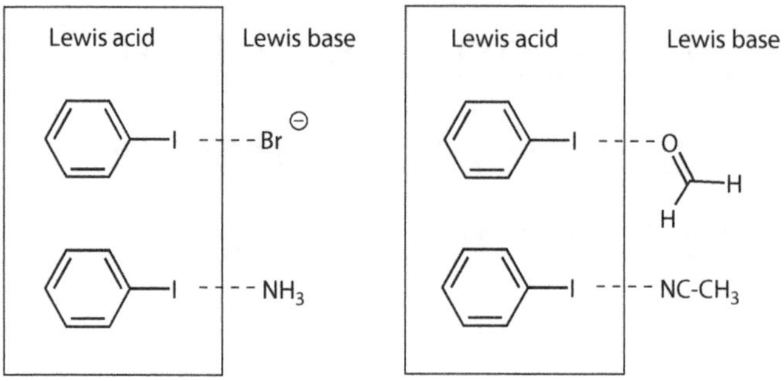

FIGURE 4.1 Examples of halogen-bonded complexes.

Lewis base (Figure 4.1). The basic features of halogen bond are similar to that of hydrogen bond and they are: (a) interatomic distance between X and Y is usually less than the sum of the *van der Waals* radii, (b) analysis of the electron density topology usually shows a bond path and a (3,–1) bond critical point, (c) nuclear magnetic resonance (NMR) chemical shifts are usually affected by formation of halogen bond, (d) the IR and Raman scattering frequencies of both R-X and Y are affected by halogen bond formation; new vibration modes associated with X···Y are also observed, and (e) UV–vis bands of the halogen bond donor usually shift to shorter wavelengths.[1]

4.1.2 Sigma Hole Model

C-X bonds are generally thought to be polarized (the halogen that is more electronegative bears a partial negative charge), therefore it appears that interaction with a Lewis base (an electron-rich species) would be repulsive, at least from an electrostatic point of view. The preceding description is, however, an over-simplification of the electron distribution of halogen in a compound. For instance, from the electrostatic potential map of bromobenzene derived from density functional theory (DFT) calculation (Figure 4.2), it is evident that the electron distribution around the bromine atom is anisotropic. This is a general phenomenon for many halogens present in organic molecules. A red ring of negative electrostatic potential could be observed perpendicular to the planes of bromobenzene. This peripheral red ring converges to a region of positive electrostatic potential, which is termed the *sigma hole*.[2] The sigma hole is confined to a small region centred along the C-Br axis. The sigma hole model is consistent with both the high directionality and increasing halogen bond's strength with electron-withdrawing substituents as advocated by Politzer and coworkers.

Halogen bonding has been recognized as an important noncovalent interaction in supramolecular chemistry[3] and biology.[4] The wide implication of halogen bonding

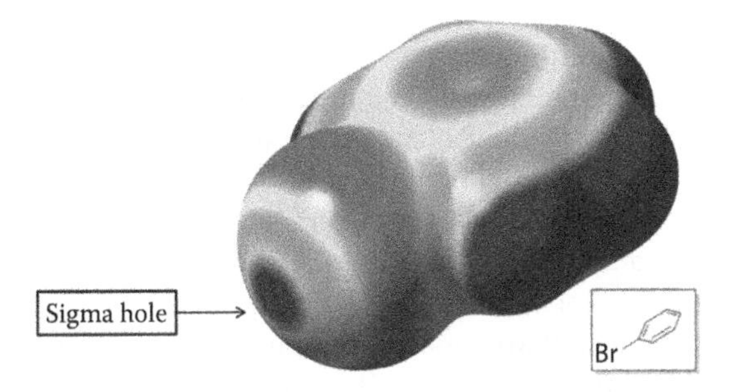

FIGURE 4.2 **(See color insert.)** Electrostatic potential of bromobenzene; blue region: repulsive interaction with respect to (w.r.t) a positive test charge and red region: attractive interaction w.r.t a negative test charge.

in both chemistry and biology has motivated the development of various experimental techniques to study halogen bonding, both in gas phase[5] and in solution.[6] Experimental techniques that allow the determination of thermodynamics of halogen bonding in solution are helpful to understand the role of halogen bonding in supramolecular chemistry and biology.[7]

4.2 HALOGEN BONDING IN ORGANIC REACTIONS

4.2.1 Bromination Reaction

The involvement of halogen bonding in reactions can be traced back to one of the classic reactions in organic chemistry—the addition of bromine to alkene. It was observed that upon mixing Br_2 with alkenes, a new transient UV–vis absorption band at 270 nm appeared.[8] This was ascribed to the 1:1 π-complex (Scheme 4.1).[9] Subsequently, a 2:1 π-complex was reported with an absorption band at 310 nm.[10] The interaction between the π-bond of alkene and Br_2 could be regarded as a type of halogen bond in which the C=C acts as a Lewis base and Br_2 acts as the Lewis acid.[11] These halogen-bonded complexes are regarded as key intermediates in the electrophilic addition of bromine alkene (Scheme 4.1).

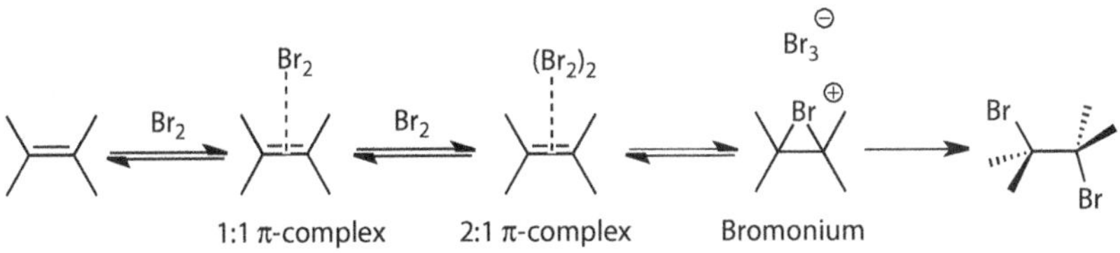

Br_2 Br_2 $(Br_2)_2$ Br_2 $Br_3^{\ominus}$ $Br^{\oplus}$ Br Br

1:1 π-complex 2:1 π-complex Bromonium

SCHEME 4.1 Mechanism for the electrophilic addition of bromine to alkene.

4.2.2 Ritter Reaction and Aniline Dimerization

Huber and coworkers reported that stoichiometric amount of iodo-imidazolium salts **2a** and **2b** is able to promote the Ritter reaction of benzyhydryl bromide with acetonitrile (Scheme 4.2).[12] The authors presented ^{13}C NMR studies to support their proposal that halogen bonds are important in promoting the reactions. They observed that after mixing **1** and **2b**, the ^{13}C signal for the iodine-bearing carbon of **2b** shifted downfield in the presence of **1** (102 ppm–110 ppm).[12]

Minataka, Takeda, and coworkers reported the use of stoichiometric amount of *t*-BuOI to perform an oxidative dimerization of aniline, thereby forming a diazo compound (Scheme 4.3).[13] Aniline bearing electron-donating group (EDG) is proposed to have stronger interaction with *t*-BuOI *via* halogen bond. This stronger halogen bond is exploited to achieve heterodimerization between anilines with EDG and anilines with electron-withdrawing group (EWG). Halogen bonding between *t*-BuOI and aniline is crucial for the formation of substituted *N, N*-diiodoaniline **4**, which is a key intermediate for the dimerization. For additional reviews on halogen bonding in organic synthesis, the readers are encouraged to refer to the works of Taylor[7] and Huber.[14]

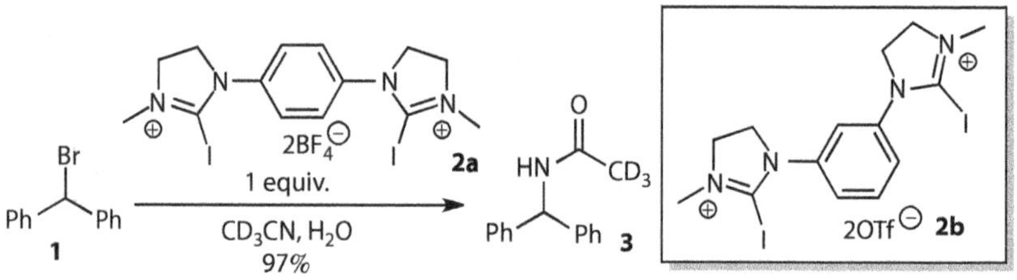

SCHEME 4.2 Ritter reaction promoted with iodoimidazolium salt.

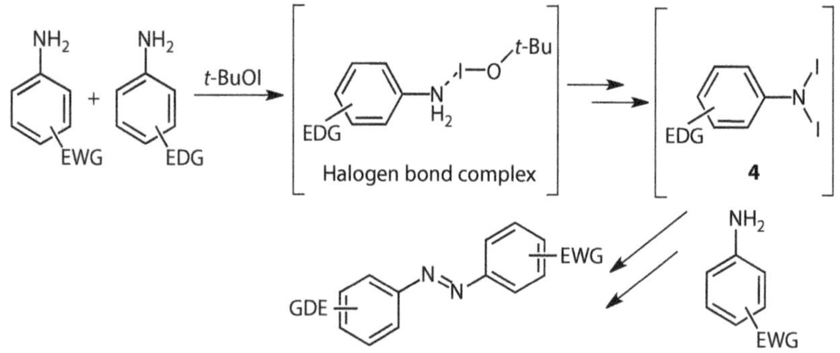

SCHEME 4.3 Oxidative dimerization of aniline.

4.3 HALOGEN BONDING IN CATALYSIS

4.3.1 Introduction

Halogen bond is similar to the hydrogen bond in its directionality and strength, however, in contrast to hydrogen bonding, it is rarely used in the design of organocatalysts.[15,16] The possibility of hidden proton catalysis in Lewis acid-catalyzed Diels–Alder reaction was highlighted by Sammakia and Latham[17] and Oestreich.[18] Halogen bond catalysis is in principle a subset of Lewis acid catalysis, therefore, the possibility of Brønsted acid catalysis should be considered. One possible way to provide evidence for the presence of hidden proton catalysis in Lewis acid catalysis is to add a proton scavenger such as 2,6-di-*tert*-butylpyridine[17] or $(Mes)_3P$.[18] The protonated 2,6-di-*tert*-butylpyridine and $(Mes)_3P$ could be conveniently characterized by ^{1}H and ^{31}P NMR spectroscopy, respectively, and thus could provide positive evidence for the generation of Brønsted acid in the reaction system. Kinetic experiments could then be performed to provide further evidences to relate the proton formation to catalytic activity. This has practical implications in the design of asymmetric catalyst, as a competing achiral Brønsted acid catalyst in the same reaction system is detrimental to enantioselectivity.

We would provide examples of halogen bond being implicated in organocatalysis with discussion on evidences presented by the authors to corroborate the involvement of halogen bonds.

4.3.2 Transfer Hydrogenation with Hantzsch Ester

Bolm and coworkers reported that iodoperfluoroalkane could promote the reduction of quinoline with Hantzsch ester as the reducing agent and proposed that halogen bond is essential (Scheme 4.4).[19] However, attempts to characterize the proposed halogen bond between $CF_3(CF_2)_7I$ and the substrate were inconclusive from ^{13}C NMR (downfield shift of 0.01–0.06 ppm was observed). ^{19}F NMR gave a 0.06 ppm downfield shift for CF_3 and CF_2 moieties. The authors noted that $CF_3(CF_2)_7I$ decomposes in the presence of Hantzsch ester, when the substrate is not present.

Tan and coworkers reported a similar reaction with cationic halogen bond donors (Table 4.1).[20] Bis-imidazolium **C1·2OTf** was prepared and found to provide superior performance in terms of reaction rate. In the presence of 2 mol% **C1·2OTf**, the reaction was complete in 4.5 h and an excellent yield of 99% was obtained. They expanded the scope to include pyridine derivatives and imines.

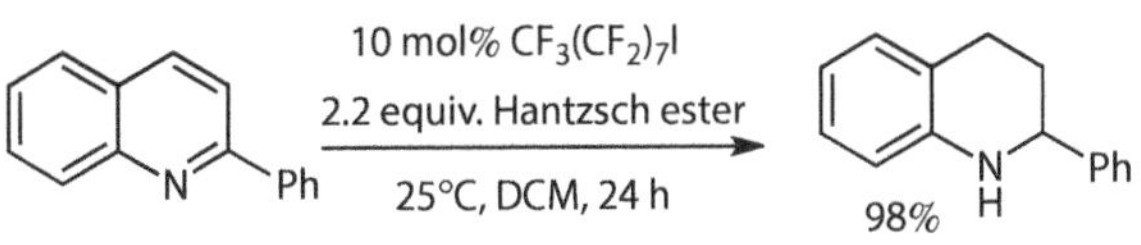

10 mol% $CF_3(CF_2)_7I$
2.2 equiv. Hantzsch ester
25°C, DCM, 24 h
N Ph → N H Ph 98%

SCHEME 4.4 Iodoperfluoroalkane-catalyzed reduction of quinoline.

TABLE 4.1
Bis-Imidazolium-Catalyzed Quinoline Reduction

10 mol% catalyst R (Linker) R
2.2 equiv. Hantzsch ester
rt, DCM
5 → 6
R = C1.2OTf, C2.2OTf, C3, C4.2OTf

Entry	Catalyst	Time (h)	Yield[a] (%)
1	**C1·2OTf**	1	99
2	**C2·2OTf**	24	68
3	**C3**	24	45
4	**C4·2OTf**	19	78
5[b]	**C1·2OTf**	2.5	99
6[c]	**C1·2OTf**	4.5	99

[a] Determined from GC-MS.
[b] 20 mol% of K_2CO_3 added.
[c] 2 mol% catalyst.

In order to eliminate the possibility of hidden proton catalysis, K_2CO_3 was added (Table 4.1, entry 5). The rate was decreased and 2.5 h is required instead of 1 h when the reaction was performed in the absence of K_2CO_3 (entry 1). This suggests that more definitive evidences for the involvement of halogen bonding are required. The authors attempted to support the formation of halogen bond between **C1·2OTf** and quinoline *via* ^{13}C NMR. They noted that upon addition of quinoline **5** to **C1·2OTf**, the ^{13}C peak of the carbon attached to iodine shifted upfield from 136.7 ppm to 122.2 ppm. In addition, the authors observed no concentration-dependent chemical shift; instead, the intensity of the peak at 122.2 ppm increases with increasing concentration of quinoline. This is in contrast to the downfield shift effect of on the sp^2 carbon-bearing iodine observed by Goroff,[21] Huber,[22] and Takeda.[23] The result of Goroff and coworkers was also supported by theoretical calculations.[24] More in-depth studies would be required to establish the involvement of halogen bonding in transfer hydrogenation that are claimed to be promoted/catalyzed by halogen bonds.

Isothermal titration calorimetry (ITC) titration was performed to determine binding constant between **C1·2OTf/C4·2OTf** and bromide. It was found that the binding constant for **C4·2OTf** is larger than **C1·2OTf**, however the catalytic activity of **C1·2OTf** is higher than **C4·2OTf** (3.3×10^4 vs. 4.6×10^5). If halogen bond catalysis is assumed, given the higher nucleophilicity of the amine product, the stronger binding affinity of **C4·2OTf** is likely to result in product deactivation of the catalyst and thus resulted in the lower catalytic activity of **C4·2OTf** relative to **C1·2OTf**.

4.3.3 Addition to Oxocarbenium

Huber and coworkers develop the use of iodofluorinated arenes as halogen-bonding catalysts.[25] These neutral multidentate compounds were shown to catalyze the reaction of 1-chloroisochroman with ketene silyl acetals (Table 4.2). Exploiting the anion (chloride or bromide)-binding capability of **C5** calculated from preliminary

TABLE 4.2
Addition to Oxocarbenium Catalyzed by Iodofluorinated Arenes

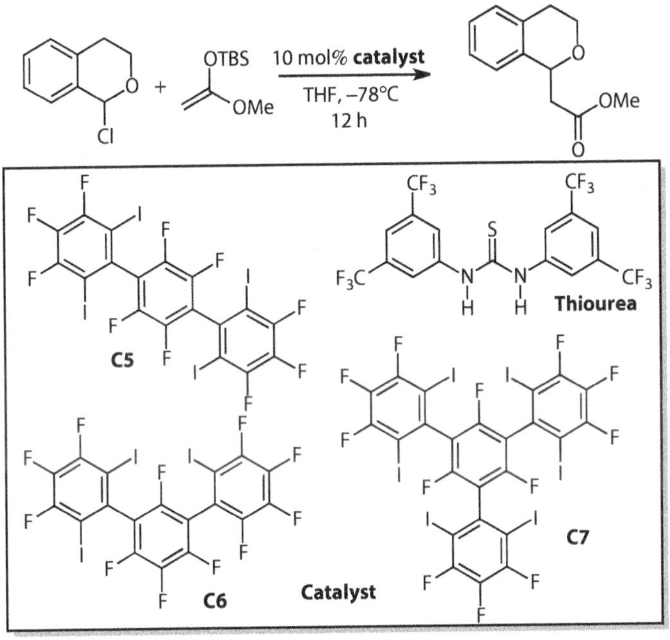

Entry	Catalyst	Yield (%)	ΔH°(bromide)[a]	N[b]
1	–	≤5	–	–
2	TfOH	27	–	
3	**Thiourea**	12	−4.6	0.8
3	**C5**	37	−7.9	1.0
4	**C6**	14	−6.6	1.2
5	**C7**	81	–	–

Source: Kniep, F. et al., *Angew. Chem. Int. Ed.*, 52, 27, 7028–7032, 2013.

[a] Determined from ITC titration with tetra-n-butylammonium bromide (TBAB).

[b] Stoichiometry of binding.

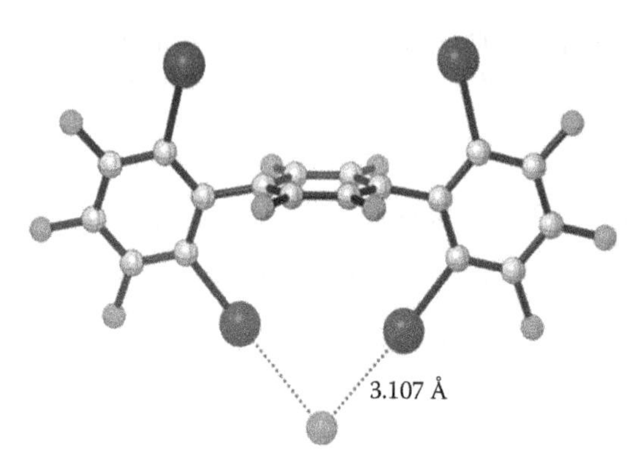

FIGURE 4.3 **(See color insert.)** XRD-derived structures of C5 chloride complex.

DFT calculations, the authors synthesized and tested a variety of related compounds related to **C5**. Control experiments indicate that negligible reaction occurs in the absence of catalyst. Triflic acid, a possible acid contaminant, was found to be slightly less effective than **C5** (entry 2–3). The C_3 symmetrical catalyst **C7** was found to be superior in catalytic performance (entry 5). The enantioselective version of this reaction was reported by Jacobsen and coworkers, which utilized hydrogen bond *via* thiourea.[26]

The authors determined the thermodynamics of binding chloride and bromide (in the form of quaternary ammonium salt) *via* ITC titrations at 30°C in tetrahydrofuran (THF). **C5** was found to have the highest ΔH° followed by **C6**, whereas **thiourea** was found to have the lowest. The stoichiometry of binding is approximately one for all three catalysts, thus implying a bidentate binding mode to bromide. This is supported by the XRD-derived structure of **C5** and chloride (Figure 4.3).

4.3.4 Diels–Alder Reaction

Huber and coworkers reported that dicationic halogen bond donor **C8** is capable of functioning as an organocatalyst, which activates carbonyl *via* halogen bonding (Table 4.3) for Diels–Alder reaction.[22] In the absence of catalyst, 24% yield of product and a selectivity of 5:1 (endo:exo) were obtained (Table 4.3, entry 1). Catalyst **C8** with OTf counteranion was found to be an ineffective catalyst; however, changing the counteranion to a noncoordinating anion BAr^F_4 resulted in a significant increase in both rate and selectivity toward the *endo*-product (entry 3). This is ascribed to the coordination of OTf to the catalyst and thus deactivating it. This hypothesis is supported by the addition of extraneous OTf in the form of NBu_4OTf (entry 4). Crystal structure showing halogen bond between iodine of a related cationic halogen bond donor and OTf was previously reported by Huber (Figure 4.4).[12]

From ^{13}C NMR, the authors observed that when 8 equivalents of cyclohexanone were added to **C8.BAr$^F_4$**, the carbon attached to the iodine experienced a shift of 2 ppm downfield (97.2 ppm–99.2 ppm). The binding constant between **C8.BAr$^F_4$** and cyclohexanone was determined to be about $4M^{-1}$ *via* ^{13}C NMR titration. The authors noted that this Diels–Alder reaction could be catalyzed by trace amounts of acid, thus it is important to rule out hidden proton catalysis. This is critical for further

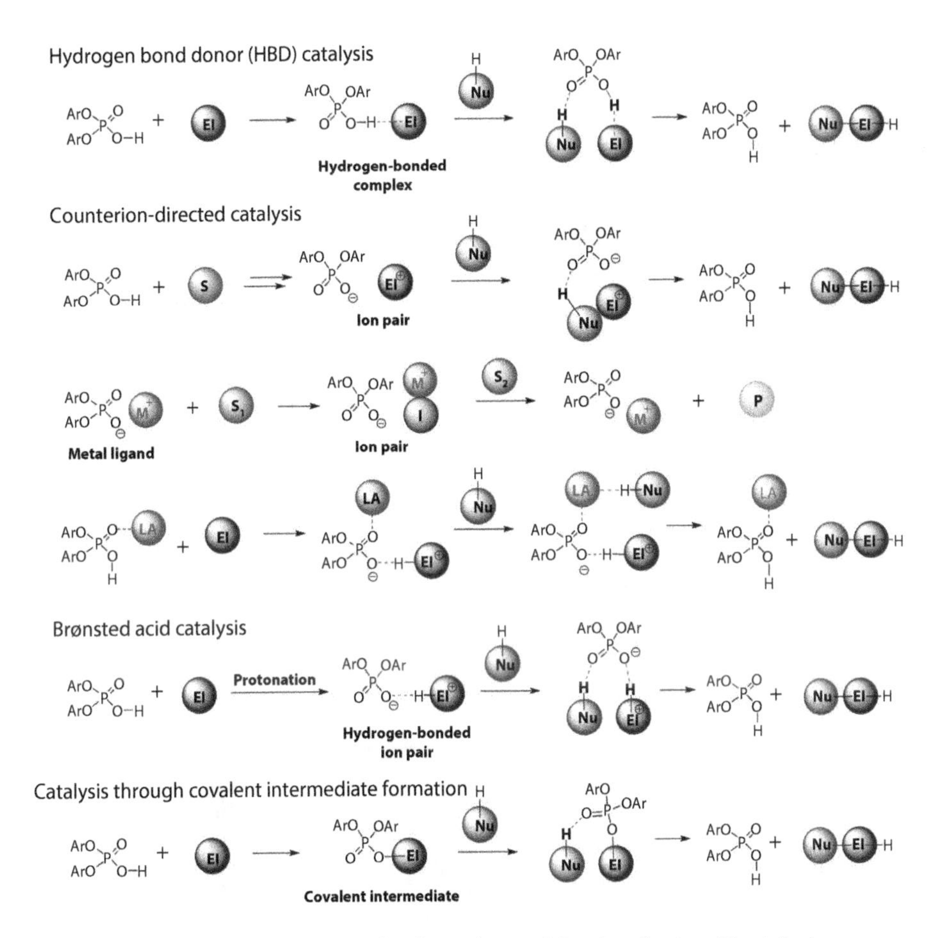

FIGURE 3.5 General reaction mechanisms observed for phosphoric acid catalysis.

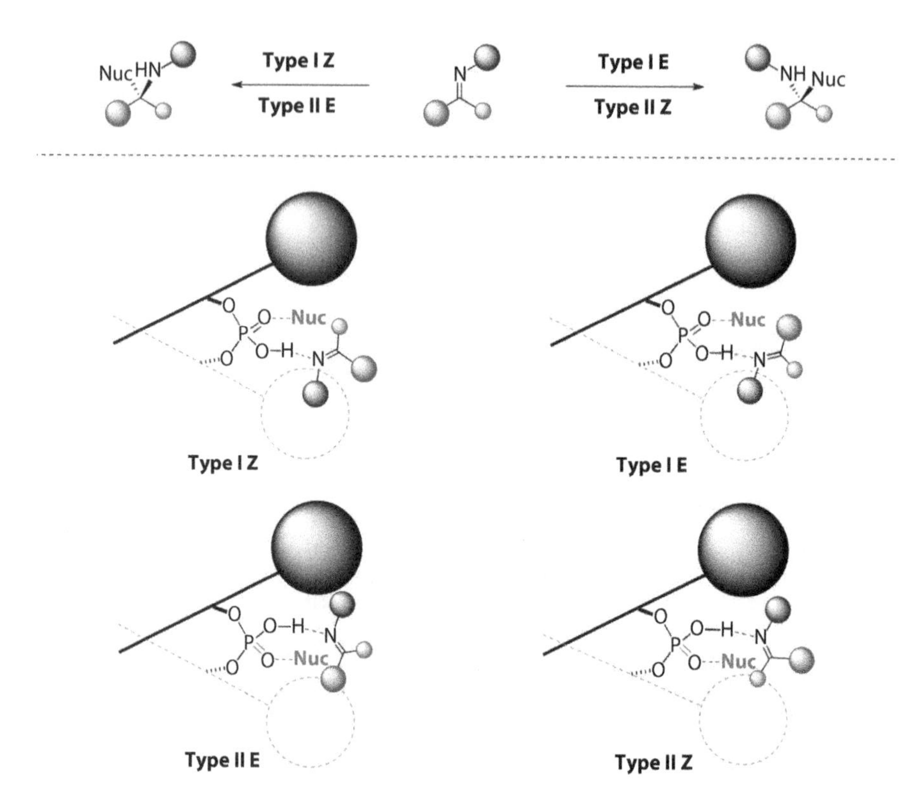

FIGURE 3.12 Stereochemical model for chiral phosphoric acid-catalyzed addition to imines by Goodman and coworkers. (Based on Simón, L. and Goodman, J.M., *J. Org. Chem.*, 76, 1775–1788, 2011. With permission.)

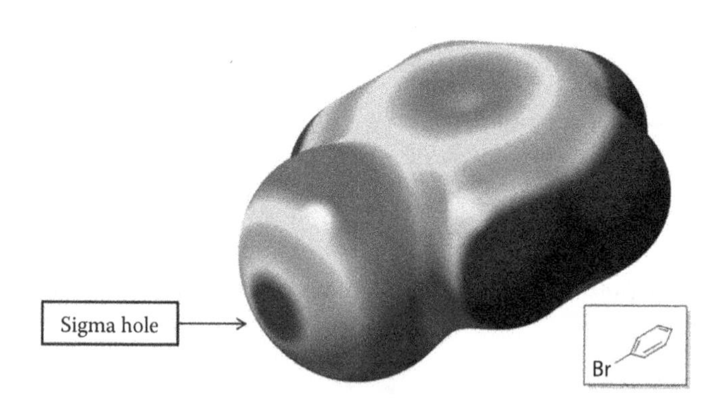

FIGURE 4.2 Electrostatic potential of bromobenzene; blue region: repulsive interaction with respect to (w.r.t) a positive test charge and red region: attractive interaction w.r.t a negative test charge.

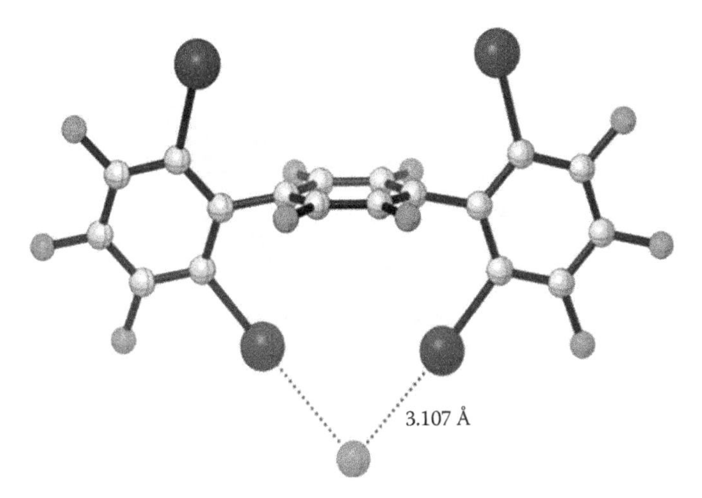

FIGURE 4.3 XRD-derived structures of C5 chloride complex.

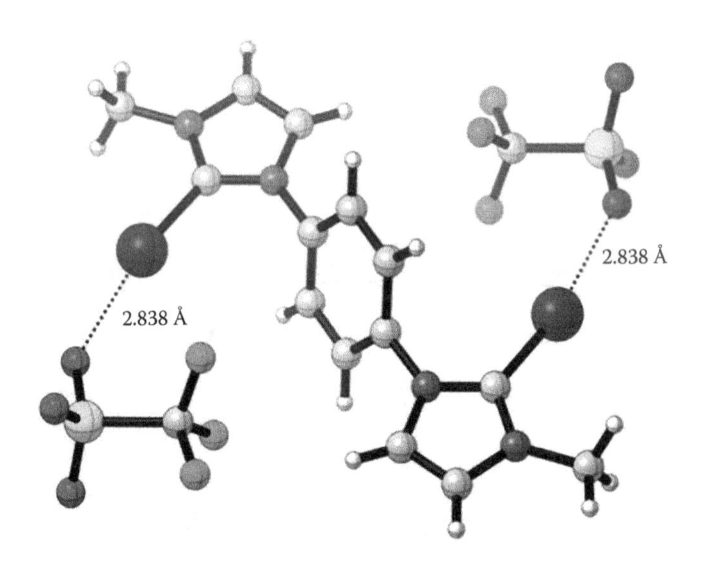

FIGURE 4.4 XRD-derived structure of 2a·2Otf (Scheme 4.2).

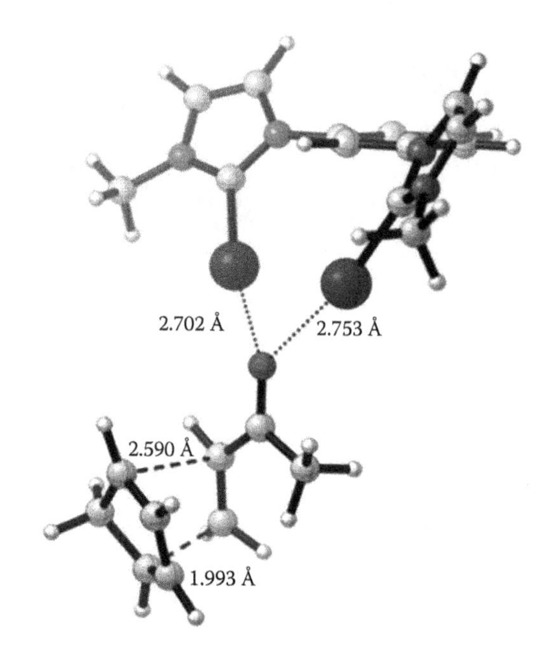

FIGURE 4.5 Halogen-bonded TS for Diels–Alder reaction at M06-2X/TZVPP.

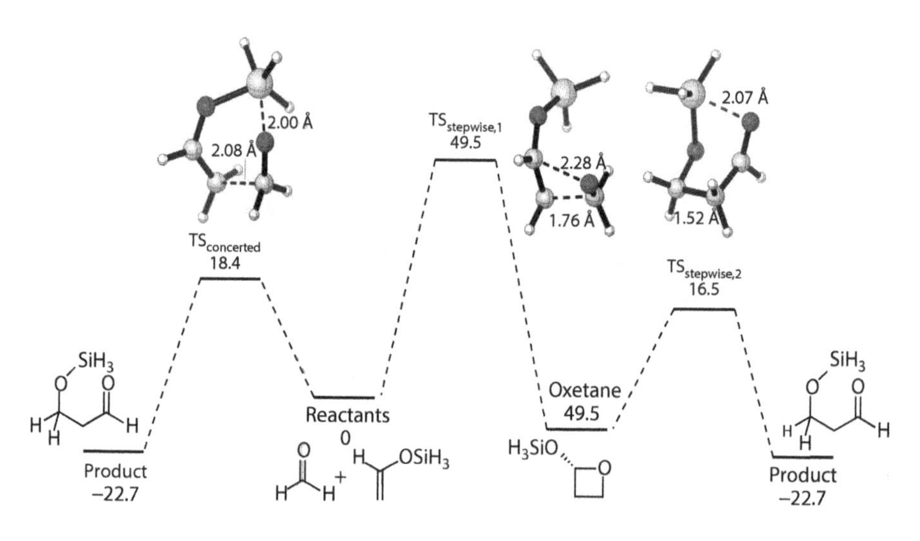

FIGURE 4.7 Uncatalyzed Mukaiyama aldol reaction profile calculated at G3(MP2).

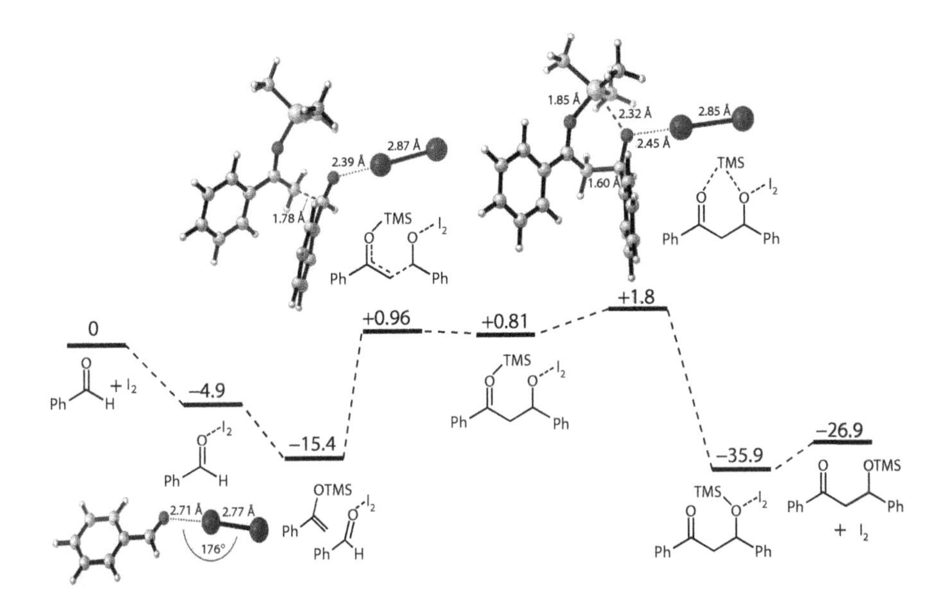

SCHEME 4.8 Iodine-catalyzed Mukaiyama Aldol reaction of benzaldehyde calculated using MP2/6-311+G(d, p)//B3LYP/6-31G(d) (relative energies, in kcal/mol, refer sum electronic energy and zero-point correction).

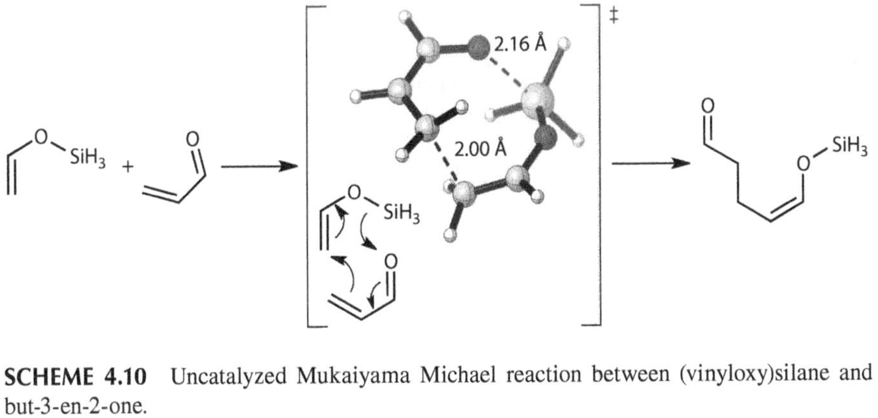

2.16 Å
2.00 Å
‡
O
SiH_3

SCHEME 4.10 Uncatalyzed Mukaiyama Michael reaction between (vinyloxy)silane and but-3-en-2-one.

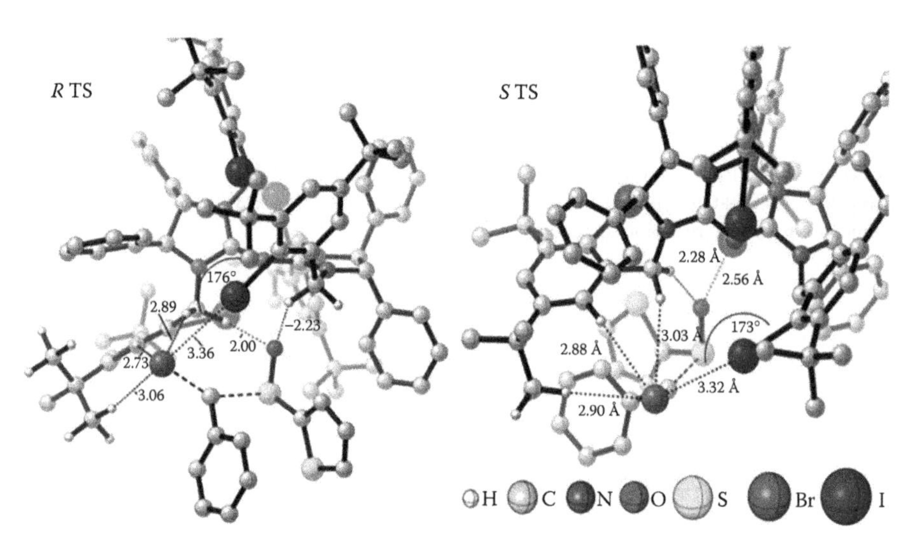

FIGURE 4.10 *R*- and *S*-TS with the lowest Gibbs free energy.

TABLE 4.3
Diels–Alder Reaction Catalyzed by Dicationic Halogen Bond Donor

20 mol% **C8**, CD_2Cl_2, r.t, 6 h

C8 (n-Oct)

Entry	Catalyst/Counteranion	Additive	Yield	Endo/Exo
1	–	–	24	5
2	**C8**.OTf	–	24	–
3[a]	**C8**.BAr^F_4	–	63	10
4[a]	**C8**.BAr^F_4	2.5 equiv. NBu_4OTf	24	–
5[a]	$NMe_4BAr^F_4$	–	24	–

Source: Jungbauer, S. H. et al., *Chem. Commun.*, 50, 47, 6281–6284, 2014.
[a] BAr^F_4 is B(3,5-ditrifluoromethylphenyl)$_4$ anion.

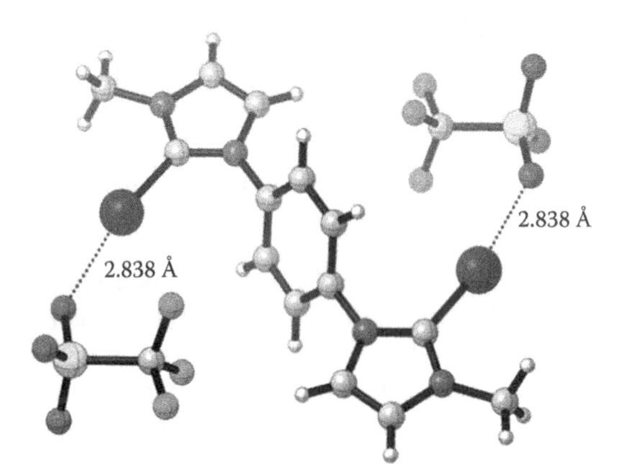

FIGURE 4.4 **(See color insert.)** XRD-derived structure of 2a·2Otf (Scheme 4.2).

development of the asymmetric catalytic variants. The authors presented a series of arguments based on indirect evidences to rule out impurities as the source of increased reactivity and to establish the importance of halogen bond as the origin of catalytic activity. DFT calculations predicted that the barrier of the Diels–Alder reaction was reduced by 3kcal/mol, when a model halogen-bonding catalyst is present (Figure 4.5).

Takeda, Minakata, and coworkers reported an Aza-Diels–Alder reaction catalyzed with iodo-imidazolium **C9**. Halogen bonding is proposed to play a key role and the addition of chloride in the form of n-Bu_4NCl inhibits the reaction (Scheme 4.5).[23] This was used as an indirect evidence to support the involvement of halogen bonding

2.702 Å
2.753 Å
2.590 Å
1.993 Å

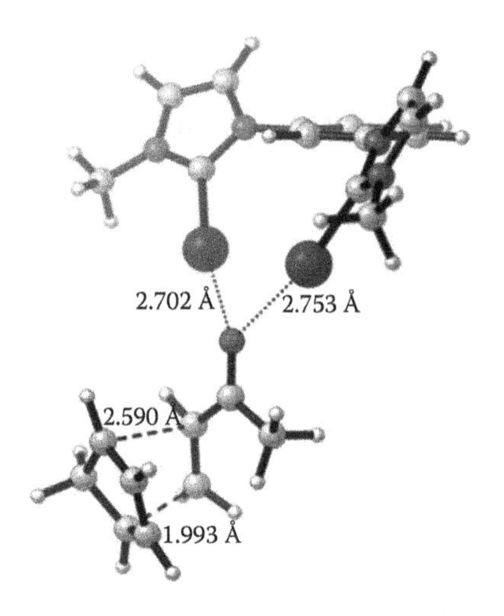

FIGURE 4.5 **(See color insert.)** Halogen-bonded TS for Diels–Alder reaction at M06-2X/TZVPP.

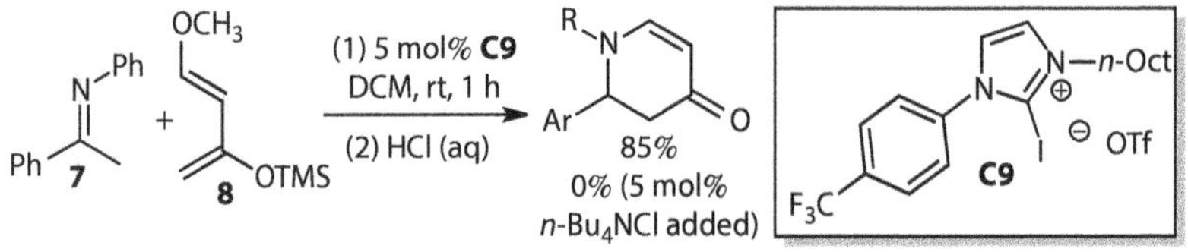

OCH3
Ph
N
+
Ph
7
8
OTMS
(1) 5 mol% C9
DCM, rt, 1 h
(2) HCl (aq)
R
N
Ar
O
85%
0% (5 mol%
n-Bu4NCl added)
N—n-Oct
N
OTf
I
C9
F3C

SCHEME 4.5 Aza-Diels–Alder reaction catalyzed by iodo-imidazolium.

in the reaction. Interestingly, in the Diels–Alder reaction reported by Huber and coworkers,[22] triflate anion was found to be detrimental to the reaction (Table 4.3, entry 2–3) but not in the case of this Aza-Diels–Alder reaction.

^{1}H NMR titration of catalyst **C9** with imine **7** in CD_2Cl_2 indicates that a complex is formed and a Job plot indicates a complex with a 1:1 stoichiometry. Further evidence for the formation of a halogen bond complex between **C9** and imine **7** is provided by ^{13}C NMR; a downfield shift of C_d (Figure 4.6) is consistent with previous study on the effect of halogen bond formation on ^{13}C NMR.[24]

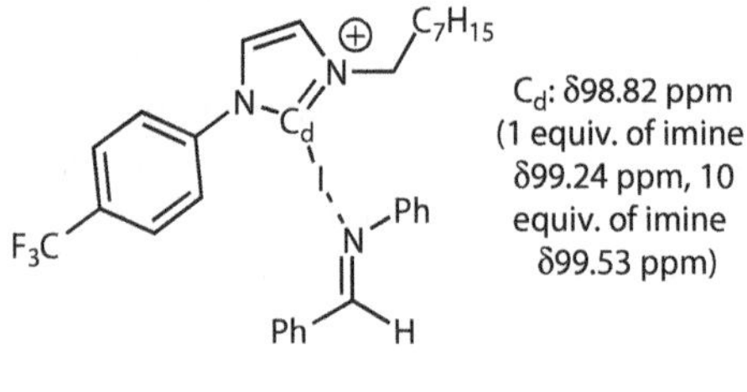

C7H15
N
N
Cd
I
N
Ph
Ph
H
F3C
Cd: δ98.82 ppm
(1 equiv. of imine
δ99.24 ppm, 10
equiv. of imine
δ99.53 ppm)

FIGURE 4.6 Halogen-bonded complex between C9 and imine 7.

TABLE 4.4
Qualitative Initial Rate and Binding Constants for Various Catalysts

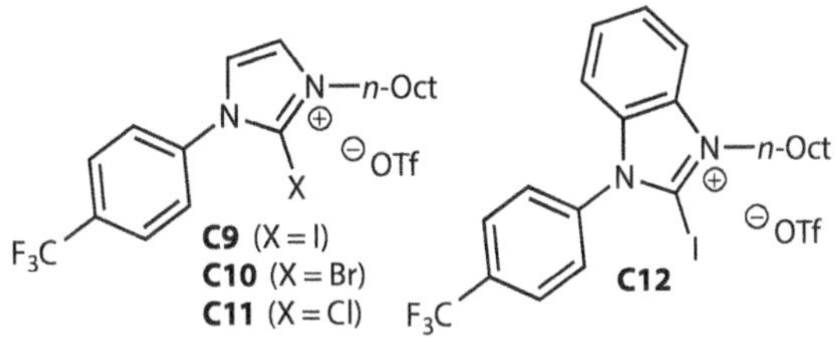

C9 (X = I)
C10 (X = Br)
C11 (X = Cl)
C12

Catalyst	Initial Rate (1-fastest, 4-slowest)	Binding Constants[a]/M^{-1}
C11	3	0.97
C10	2	1.19
C9	4	0.93
C12	1	1.59

[a] Binding constant between catalysts and imine.

They observed that the initial rate of reaction is in the order of Br>Cl>I (Table 4.4). This does not follow the general trend of halogen bond energy, where I>Br>Cl. The most active catalyst was found to be **C12**, which has a benzo-imidazole core. The initial rate follows the trend in binding constants qualitatively (Table 4.4).

4.3.5 Mukaiyama Aldol Reaction

Mukaiyama Aldol reaction is usually catalyzed using an organometallic Lewis acid.[27,28] The uncatalyzed Mukaiyama Aldol reaction is proposed to be concerted based on both experimental[29] and theoretical evidences.[30] The cross-over experiments by Denmark and coworkers provided strong support that the uncatalyzed reaction occurs *via* a concerted pathway (Scheme 4.6).

Wong studied the uncatalyzed Mukaiyama Aldol between formaldehyde and (vinyloxy)silane using highly accurate G3(MP2) method (Figure 4.7).[31] They found

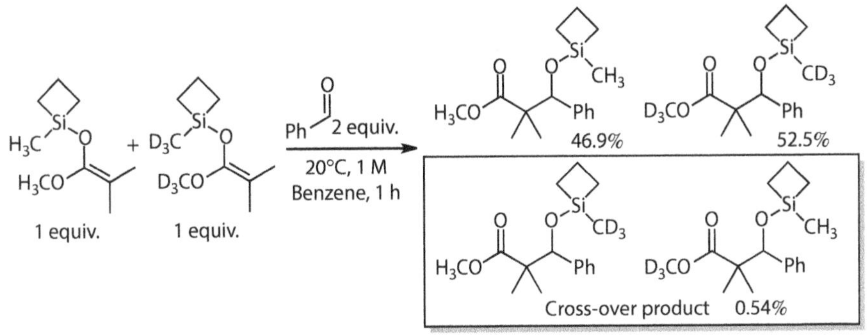

H_3C H_3CO 1 equiv. + D_3C D_3CO 1 equiv.; PhCHO 2 equiv.; 20°C, 1 M Benzene, 1 h; H_3CO … CH_3 Ph 46.9%; D_3CO … CD_3 Ph 52.5%; H_3CO … CD_3 Ph; D_3CO … CH_3 Ph; Cross-over product 0.54%

SCHEME 4.6 Cross-over experiments by Denmark and coworkers.

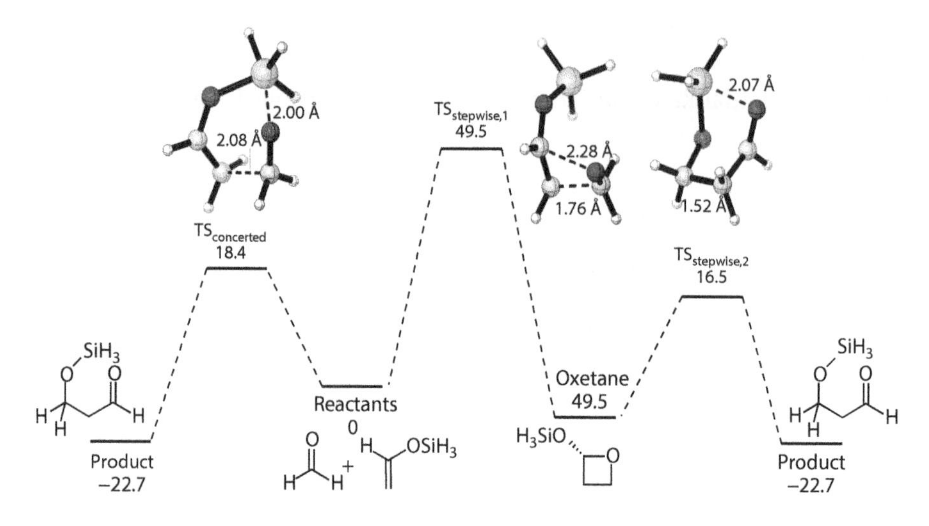

FIGURE 4.7 **(See color insert.)** Uncatalyzed Mukaiyama aldol reaction profile calculated at G3(MP2).

that the stepwise pathway, which proceeds *via* an oxetane intermediate, has an activation barrier of 49.5 kcal/mol for the first step and 16.5 kcal/mol for the subsequent step.[30] The concerted mechanism has a barrier of 18.4 kcal/mol. They concluded that the concerted pathway is strongly favored.

Wong also reported a computational study on iodine-catalyzed Mukaiyama Aldol reaction between formaldehyde and (vinyloxy)silane.[32] The concerted pathway, in which iodine is halogen bonded to formaldehyde, is found to have the lowest activation barrier of 12.4 kcal/mol (Scheme 4.7).

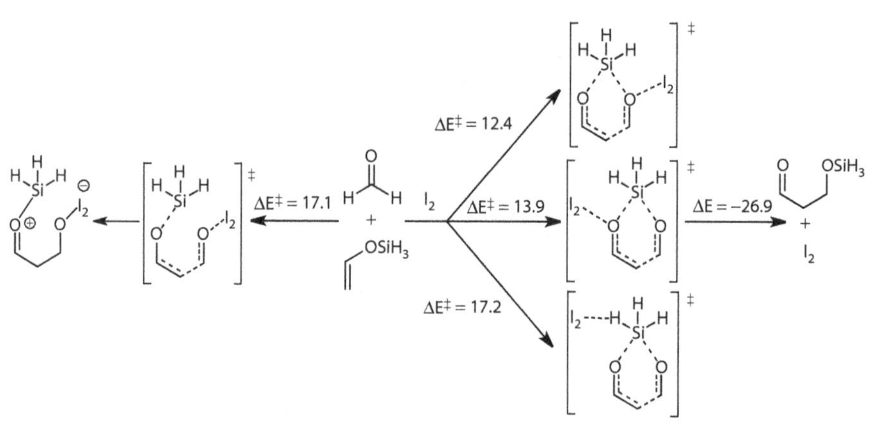

SCHEME 4.7 Pathways proposed by Wong based on DFT calculations (energies in kcal/mol).

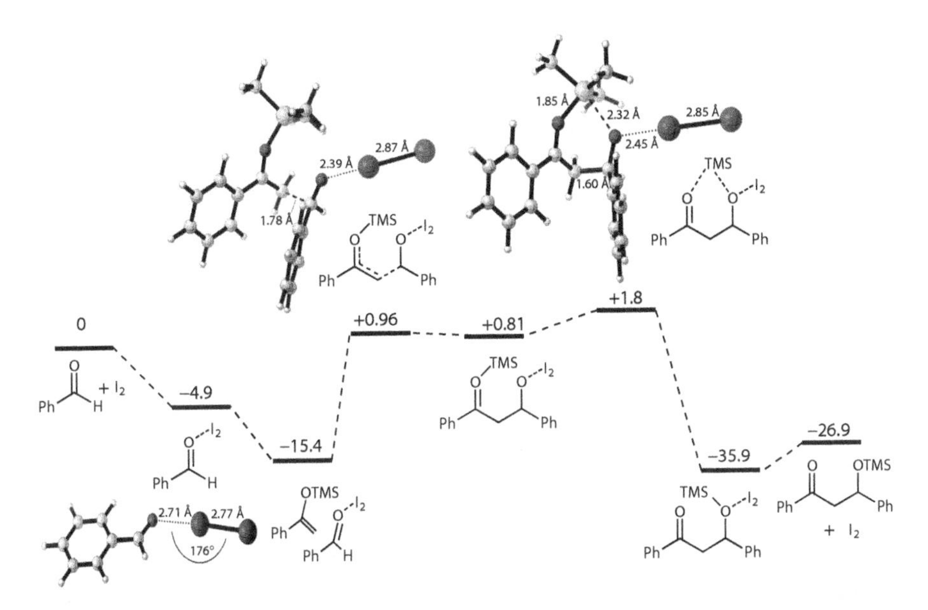

SCHEME 4.8 **(See color insert.)** Iodine-catalyzed Mukaiyama Aldol reaction of benzaldehyde calculated using MP2/6-311+G(d, p)//B3LYP/6-31G(d) (relative energies, in kcal/mol, refer sum electronic energy and zero-point correction).

When the iodine-catalyzed Mukaiyama Aldol reaction of benzaldehye reported by Phukan[33] was studied by Wong and Wang, they were unable to locate concerted transition structures (TS) that corresponds to the concerted pathway.[32] Calculations for the stepwise pathway were thus performed (Scheme 4.8). The halogen-bonded complex of benzaldehyde and iodine is 4.9 kcal/mol more stable than the separated reactants. The step with the highest barrier is the migration of trimethyl silyl (TMS) group from the ketone functional group to the iodine-bonded alcohol group. The activation barrier is 1.8 kcal/mol, which is very low and is consistent with the fast reaction rate that was observed experimentally (87% yield in 1.5 h).[33]

Phukan and Deuri then reported iodine-catalyzed Mukaiyama Michael reaction of methyl vinyl ketone (MVK) (Scheme 4.9).[34] As a model, they have located a TS for the uncatalyzed Mukaiyama Michael reaction between (vinyloxy)silane and but-3-en-2-one (Scheme 4.10). This pathway has a high activation energy of 26.7 kcal/mol. Stepwise pathway could not be located by the authors at B3LYP/6-31G(d).

For the iodine-catalyzed reaction (Scheme 4.9), the authors were only able to locate the stepwise mechanism. Two different pathways were proposed (Scheme 4.11). In path 1, 10-membered TS resulted in the formation of a hypoiodite and Me_3SiI. The activation barrier with respect to free reactants and catalyst is 8.1 kcal/mol. Subsequent exchange of iodine for TMS gave the product. In path 2, a Michael addition occurs followed by an intramolecular TMS transfer. The intramolecular

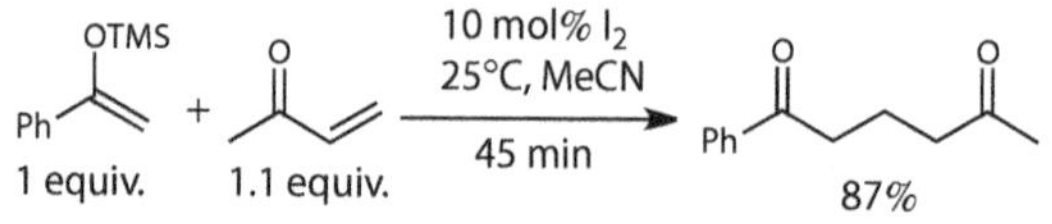

SCHEME 4.9 Mukaiyama Michael reaction of methyl vinyl ketone (MVK).

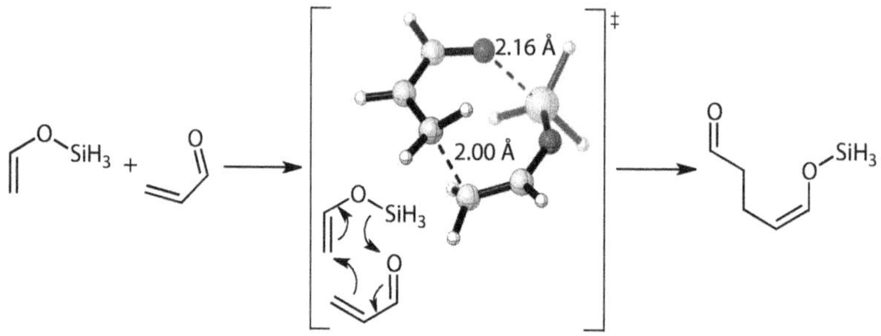

SCHEME 4.10 (See color insert.) Uncatalyzed Mukaiyama Michael reaction between (vinyloxy)silane and but-3-en-2-one.

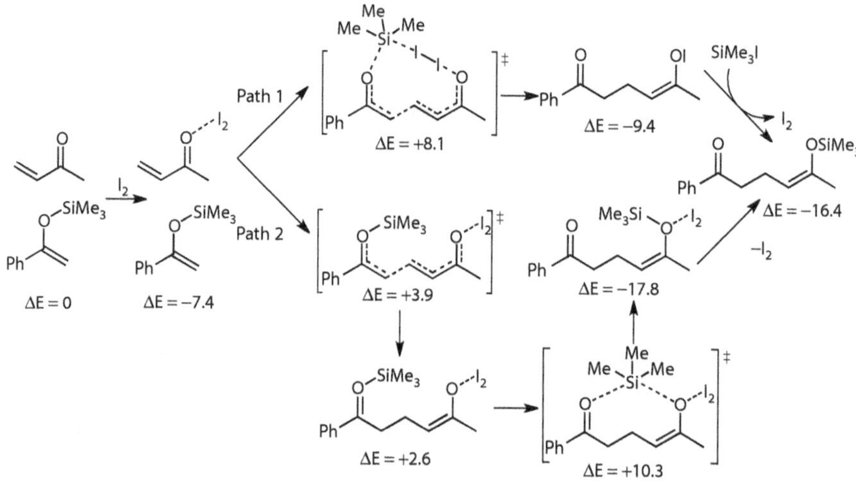

SCHEME 4.11 Pathways proposed by Phukan and Deuri based on DFT calculations (all ΔE are with respect to free reactants and iodine).

TMS transfer TS has the highest energy barrier of 10.3 kcal/mol. The author proposed that path 2 is disfavored due to the higher barrier of the intramolecular TMS transfer.

The authors applied parameters from conceptual DFT[35] to rationalize the reactivity of MVK upon formation of halogen bond with I_2. The formation of halogen

TABLE 4.5
Electrophilicity Index of Various Carbonyls and Their Iodine Complexes

Acrolein MVK Cyclohexenone

Acrolein-I_2 MVK-I_2 Cyclohexenone-I_2

Compound	ω
Acrolein	1.84
Acrolein-I_2	2.70
MVK	1.54
MVK-I_2	2.54
Cyclohexenone	1.40
Cyclohexenone-I_2	2.32

bond complexes resulted in an increase in the electrophilicity index (ω)[36] and this is interpreted by the authors as activation of the carbonyl *via* halogen bond (Table 4.5). Finally, we note that Taylor commented that it is unclear if reactions catalyzed by I_2 are due to halogen bond, as the formation of hydrogen iodide/trimethylsilyl iodide (HI/TMSI) in the reaction could contribute to the observed reactivity.[7]

4.3.6 Halolactonization

Catalytic halofunctionalization reaction is an important class of organic reactions where halogen bonding could play an important role. Although the involvement of halogen bonding was not explicitly stated by authors working on catalytic asymmetric halofunctionalization, interaction between Lewis base and halonium, which could be considered as halogen bond, is an important factor to consider to attain enantioselective halofunctionalization.[37] Halogen bonds between *N*-bromosuccinimide (NBS) and a Lewis basic component of the chiral catalyst have been proposed by Shi,[38] Denmark,[39] and Yeung[40] (Figure 4.8).

Ishihara reported an enantioselective iodolactonization catalyzed by nucleophilic catalysts (Scheme 4.12).[41] It was proposed that *N*-chlorophthalimide (NCP) and the catalyst **C13** activate I_2 or I-Cl cooperatively (Scheme 4.13). The halogen-bonded complex between I-X and NCP was characterized by Raman spectroscopy. Raman bands at 114 cm^{-1} and 163 cm^{-1} were postulated to be the I-I-I (Cl) bending by the authors. Similar to the model by Denmark (Figure 4.8), the pretransition state of this reaction was proposed to involve a halogen bond between the alkene π-electrons and sigma hole of iodine.

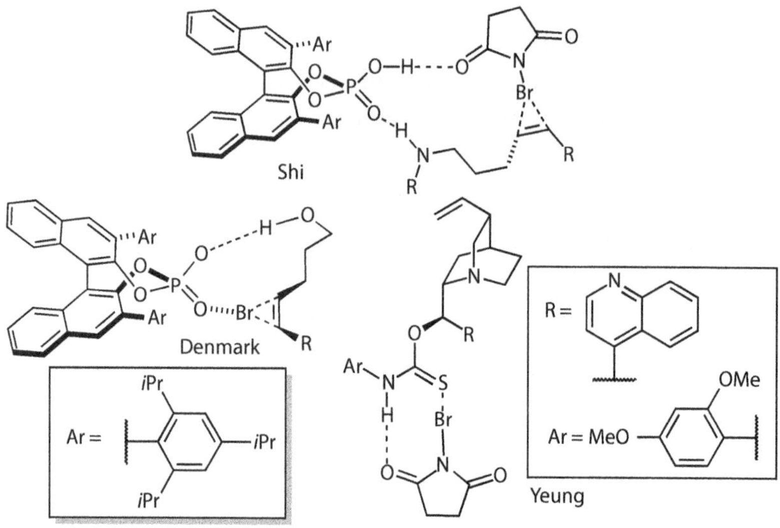

FIGURE 4.8 Halogen bond between a Lewis base catalyst and halonium proposed by various groups.

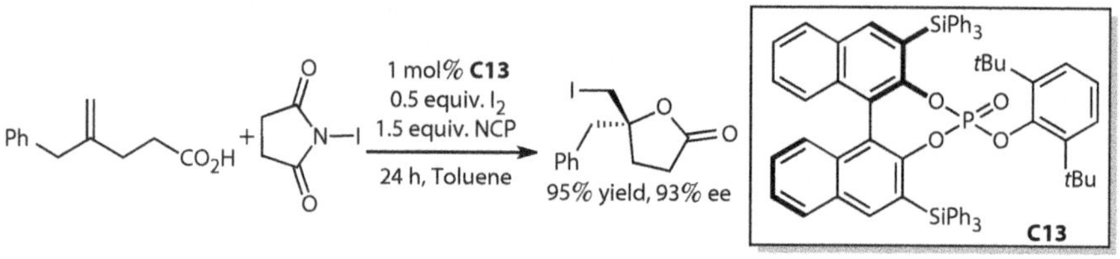

SCHEME 4.12 Enantioselective iodolactonization.

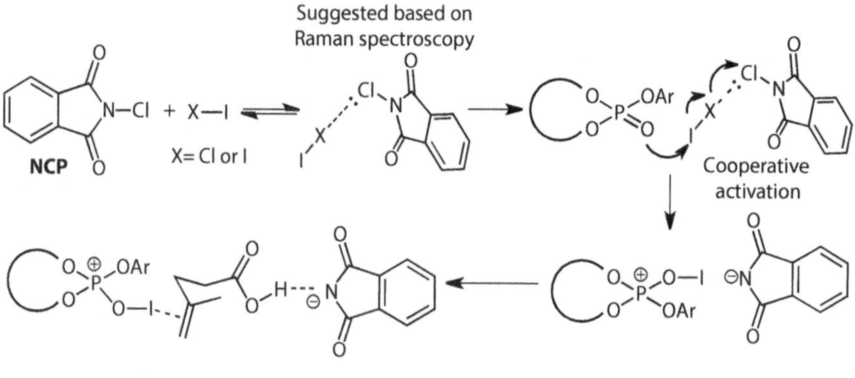

SCHEME 4.13 Postulated cooperative activation of I-X (X=Cl or I).

4.3.7 Sulfenate Alkylation

Tan, Kee, and coworkers reported that halogenated pentanidium-based phase transfer catalysts are capable of stabilizing substrates in TS *via* halogen bonds and nonclassical hydrogen bonds (NCHB[42]).[43] In their report, sulfenate anions[44] were generated *in situ* and react with electrophiles present in a highly enantioselective manner (Scheme 4.14). Experimentally, the introduction of halogen to pentanidiums resulted in significant improvements in both yield and enantioselectivity (Table 4.6). This is particularly pronounced when a less active electrophile such as benzyl chloride was used. The non-halogenated catalyst **C17** results in exclusive formation of the achiral by-product **11**.

Calculations were performed with the ONIOM methods. Several conformations for *R*- and *S*-TS were located for the iodo-catalyst **C14**$_{\mathbf{noanion}}$ (Figure 4.9).[45]

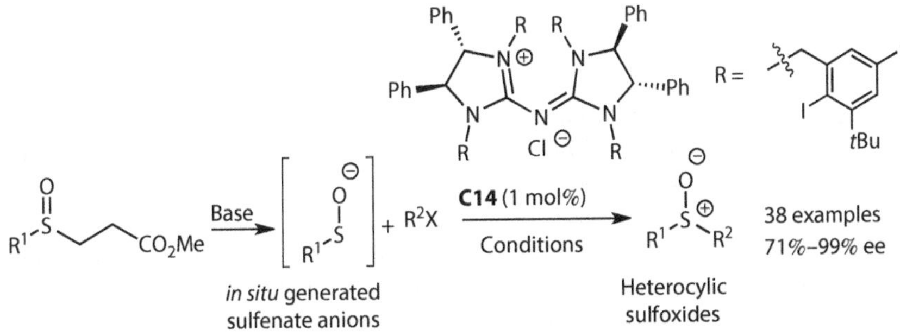

Base; in situ generated sulfenate anions; + R^2X; **C14** (1 mol%), Conditions; Heterocylic sulfoxides; 38 examples, 71%–99% ee; R = ... tBu, I, tBu

SCHEME 4.14 Pentanidium-catalyzed enantioselective alkylation of sulfonate.

TABLE 4.6
Effects on Structural Variations of Catalysts on Yield and ee-Value

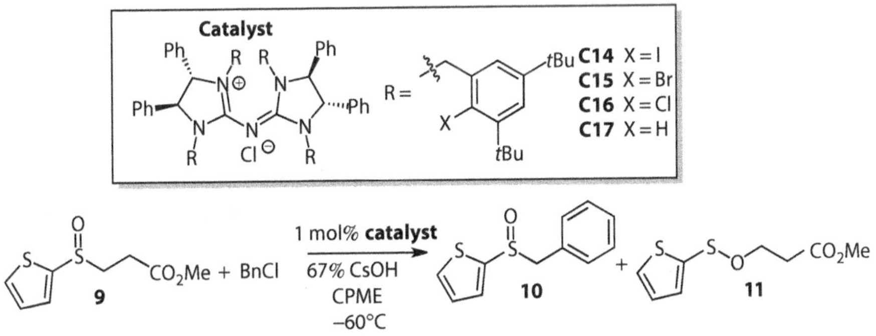

Catalyst: R = ...; **C14** X = I; **C15** X = Br; **C16** X = Cl; **C17** X = H

9 + BnCl → 1 mol% **catalyst**, 67% CsOH, CPME, −60°C → **10** + **11**

Entry	Catalyst	ee (%)	Yield of 10	Yield of 11
1	C17	–	0	27
2	C16	83	3	6
3	C15	91	15	0
4	C14	90	29	0

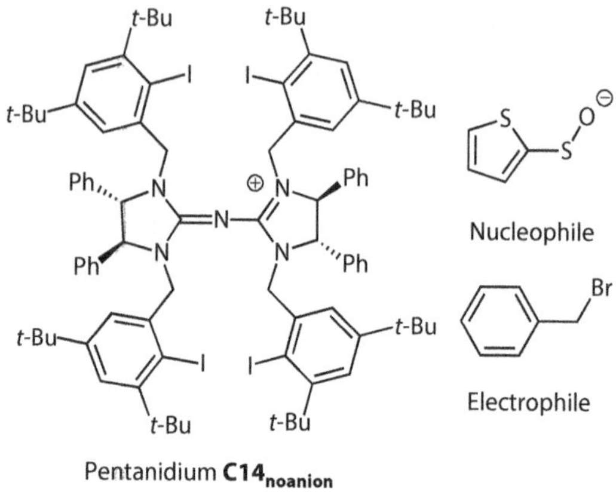

Nucleophile

Electrophile

Pentanidium **C14**$_{\mathbf{noanion}}$

FIGURE 4.9 ONIOM partitioning scheme used in calculations; the *t*-butyl and phenyl groups were included in the low level (UFF); all other atoms were included in the high level (M06/BS1); BS 1 refers to the following basis set combination: High level is 6-31G(d, p) for all atom except Br and I, the LANL2DZ(d, p). (From Check, C.E. et al., *J. Phys. Chem. A,* 105, 34, 8111–8116, 2001. With permission.)

The M06 DFT functional was employed for the high level and molecular mechanics with the universal force field (UFF) was employed for the low level. Subsequent single point calculations using SMD(Et_2O)-M06/BS2 (6-31+G(d, p) on all atoms except Br and I. LANL2DZ(d, p) for Br and I.) were performed. Solvent effects were modeled using the SMD solvation model of Truhlar and Cramer.

The substrates are noncovalently bonded to the catalyst *via* a combination of NCHBs and halogen bonds (Figure 4.10). Calculated results are supported and validated by good agreement between experimental % ee (enantiomeric excess; 91% that

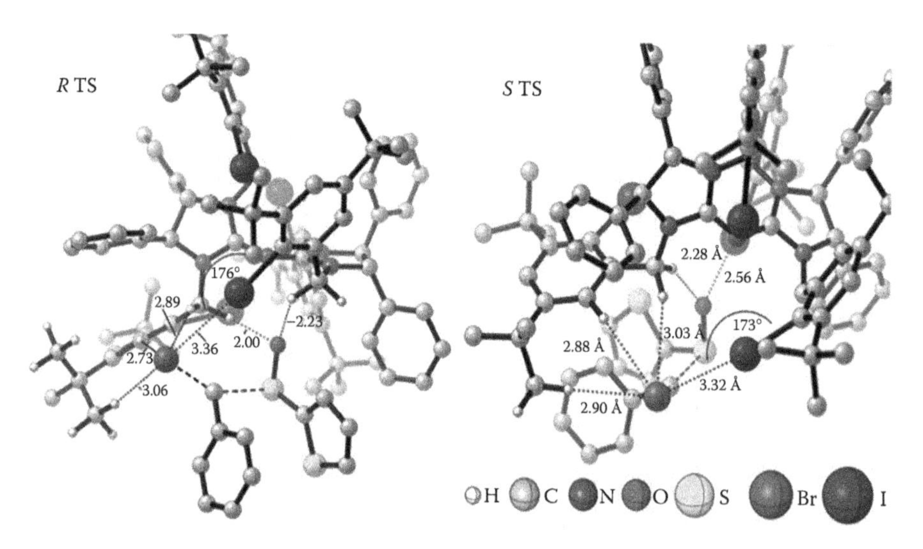

FIGURE 4.10 **(See color insert.)** *R*- and *S*-TS with the lowest Gibbs free energy.

corresponds to experimental ΔΔG‡ of +1.2kcal/mol at −70°C.) and calculated ΔΔG‡ (+1.2kcal/mol at −70°C). Single point calculations with M06-2X that are known to perform better than M06 for kinetics[47] gave similar results (calculated ΔΔG‡ is +0.9kcal/mol at −70°C). The C-H···bromide NCHBs are smaller than the sum of *van der Waals* radii of Br and H (3.10 Å). The leaving bromide is stabilized by three of such hydrogen bonds and a halogen bond in both TS. These halogen bonds are shorter than the sum of *van der Waals radii* of Br and I (3.88 Å). The main difference between these TS is that sulfenate anion is stabilized by two NCHB bonds in *R*-TS, whereas it is stabilized by a NCHB and a halogen bond in the *S*-TS.

(3,-1)-bond critical points could be identified by atoms in molecules (AIM) analysis with multiwfn,[48] thus providing theoretical support for the formation of noncovalent interactions. The potential energy density derived from AIM analysis could be used to estimate the strength of noncovalent interaction. Espinosa proposed that the strength of a hydrogen bond is half of the potential energy density.[49] Tan, Kee, and coworkers used a quartic polynomial function (degree 4) to fit the potential energy density of halogen bonds in model complexes to the interaction energy (self-consistent field, SCF, energy) at various halogen bond distances. Halogen bond strength is then calculated from the potential energy density of the halogen bonds' bond critical points of the TS. The results are tabulated in Table 4.7.

The halogen bonds are generally stronger than the NCHBs. The halogen bond between catalyst's iodine and leaving Br is about 6.5 kcal/mol, whereas the corresponding NCHBs are generally from 1–2 kcal/mol. The highly anionic sulfenate in the TS forms stronger NCHB and halogen bond with the catalyst. The halogen bond between sulfenate and catalyst in the *S* TS is stronger than the corresponding NCHB in the *R* TS (9.5 vs. 3.6 kcal/mol). The sum of NCHBs and halogen bond is larger for *S*-TS than *R*-TS, thus based on the strength of these interactions, *S*-TS, which leads to the minor enantiomer, is favored. The same conclusion is reached when enthalpy is considered (*S*-TS is more stable by 0.83 kcal/mol). Thus, *R*-TS that leads to the major

TABLE 4.7
Bond Length of Hydrogen and Halogen Bonds in *R*-TS and *S*-TS

		Bond Length	Bond Energy[a]	Bond Length	Bond Energy[a]
Donor	**Acceptor**	***R*-TS**		***S*-TS**	
Br	H-C	3.06	−0.9	3.03	−1.1
Br	H-C	2.73	−2.0	2.88	−1.5
Br	H-C	2.89	−1.5	2.90	−1.3
Br	I	3.36	−6.6	3.32	−6.5
O	H-C	2.00	−6.3	2.28	−3.2
O	H-C[b](I[c])	2.23	−3.6	2.56	−9.5
Total		−20.9 kcal/mol		−23.0 kcal/mol	

[a] Bond energies are determined from potential energy density obtained with Multiwfn.[48]
[b] Acceptor in *R*-TS.
[c] Acceptor in *S*-TS.

enantiomer is favored by entropy. Given the difficulty in obtaining accurate entropy especially in solution, the previous calculation should be interpreted with caution. Nevertheless, the feasibility of halogen bond as key interaction in organocatalysis is demonstrated in this report.

4.4 SUMMARY AND OUTLOOK

The efforts of various research groups have culminated in excellent progresses in the utilization of halogen bond in organic syntheses. In our opinion, organocatalytic enantioselective reactions that exploit halogen bondings as controlling element remain a frontier of halogen bond-based organocatalysis. To this end, we anticipate the need for more elaborated studies on the mechanisms of various halogen bond-catalyzed reactions to lay a stronger foundation of its involvement. This would, in turn, provide chemists with inspiration in designing novel halogen bond-based catalysts.

REFERENCES

1. Desiraju, G. R., Ho, P. S., Kloo, L., Legon, A. C., Marquardt, R., Metrangolo, P., Politzer, P., Resnati, G., Rissanen, K. *Pure Appl. Chem.* **2013**, *85* (8), 1711–1713.
2. Clark, T., Hennemann, M., Murray, J., Politzer, P. *J. Mol. Model.* **2007**, *13* (2), 291–296.
3. (a) Corradi, E., Meille, S. V., Messina, M. T., Metrangolo, P., Resnati, G. *Angew. Chem. Int. Ed.* **2000**, *39* (10), 1782–1786; (b) Metrangolo, P., Resnati, G. *Chem. – Eur. J.* **2001**, *7* (12), 2511–2519; (c) Aakeröy, C. B., Fasulo, M., Schultheiss, N., Desper, J., Moore, C. *J. Am. Chem. Soc.* **2007**, *129* (45), 13772–13773; (d) Metrangolo, P., Meyer, F., Pilati, T., Resnati, G., Terraneo, G. *Angew. Chem. Int. Ed.* **2008**, *47* (33), 6114–6127; (e) Vargas, J. A., Emery, D., Mareda, J., Metrangolo, P., Resnati, G., Matile, S. *Angew. Chem. Int. Ed.* **2011**, *50* (49), 11675–11678; (f) Priimagi, A., Cavallo, G., Metrangolo, P., Resnati, G. *Acc. Chem. Res.* **2013**, *46* (11), 2686–2695.
4. (a) Auffinger, P., Hays, F. A., Westhof, E., Ho, P. S. *Proc. Natl. Acad. Sci. U. S. A.* **2004**, *101* (48), 16789–16794; (b) Voth, A. R., Hays, F. A., Ho, P. S. *Proc. Natl. Acad. Sci. U. S. A.* **2007**, *104* (15), 6188–6193; (c) Scholfield, M. R., Zanden, C. M. V., Carter, M., Ho, P. S. *Protein Sci.* **2013**, *22* (2), 139–152.
5. Legon, A. C. *Angew. Chem. Int. Ed.* **1999**, *38* (18), 2686–2714.
6. Erdelyi, M. *Chem. Soc. Rev.* **2012**, *41* (9), 3547–3557.
7. Beale, T. M., Chudzinski, M. G., Sarwar, M. G., Taylor, M. S. *Chem. Soc. Rev.* **2013**, *42* (4), 1667–1680.
8. Bloemink, H. I., Hinds, K., Legon, A. C., Thorn, J. C. *Angew. Chem. Int. Ed.* **1994**, *33* (14), 1512–1513.
9. Ruasse, M.-F. *Adv. Phys. Org. Chem.* **1993**, *28*, 207–291.
10. Bianchini, R., Lenoir, D., Herges, R., Grunenberg, J., Chiappe, C., Lemmen, P. *Angew. Chem. Int. Ed.* **1997**, *36* (12), 1284–1287.
11. Esrafili, M. D., Mahdavinia, G., Javaheri, M., Sobhi, H. R. *Mol. Phys.* **2014**, *112* (8), 1160–1166.
12. Walter, S. M., Kniep, F., Herdtweck, E., Huber, S. M. *Angew. Chem. Int. Ed.* **2011**, *50* (31), 7187–7191.
13. (a) Okumura, S., Lin, C.-H., Takeda, Y., Minakata, S. *J. Org. Chem.* **2013**, *78* (23), 12090–12105; (b) Takeda, Y., Okumura, S., Minakata, S. *Angew. Chem. Int. Ed.* **2012**, *51* (31), 7804–7808.

14. Jungbauer, S. H., Schindler, S., Kniep, F., Walter, S. M., Rout, L., Huber, S. M. *Synlett* **2013**, *24* (20), 2624–2628.
15. Metrangolo, P., Neukirch, H., Pilati, T., Resnati, G. *Acc. Chem. Res.* **2005**, *38* (5), 386–395.
16. (a) Takemoto, Y. *Org. Biomol. Chem.* **2005**, *3* (24), 4299–4306; (b) Taylor, M. S., Jacobsen, E. N. *Angew. Chem. Int. Ed.* **2006**, *45* (10), 1520–1543; (c) Doyle, A. G., Jacobsen, E. N. *Chem. Rev.* **2007**, *107* (12), 5713–5743.
17. Sammakia, T., Latham, H. A. *Tetrahedron Lett.* **1995**, *36* (38), 6867–6870.
18. Schmidt, R. K., Müther, K., Mück-Lichtenfeld, C., Grimme, S., Oestreich, M. *J. Am. Chem. Soc.* **2012**, *134* (9), 4421–4428.
19. Bruckmann, A., Pena, M. A., Bolm, C. *Synlett* **2008**, *2008* (06), 900–902.
20. He, W., Ge, Y.-C., Tan, C.-H. *Org. Lett.* **2014**, *16* (12), 3244–3247.
21. Gao, K., Goroff, N. S. *J. Am. Chem. Soc.* **2000**, *122* (38), 9320–9321.
22. Jungbauer, S. H., Walter, S. M., Schindler, S., Rout, L., Kniep, F., Huber, S. M. *Chem. Commun.* **2014**, *50* (47), 6281–6284.
23. Takeda, Y., Hisakuni, D., Lin, C.-H., Minakata, S. *Org. Lett.* **2015**, *17* (2), 318–321.
24. Rege, P. D., Malkina, O. L., Goroff, N. S. *J. Am. Chem. Soc.* **2002**, *124* (3), 370–371.
25. Kniep, F., Jungbauer, S. H., Zhang, Q., Walter, S. M., Schindler, S., Schnapperelle, I., Herdtweck, E., Huber, S. M. *Angew. Chem. Int. Ed.* **2013**, *52* (27), 7028–7032.
26. Reisman, S. E., Doyle, A. G., Jacobsen, E. N. *J. Am. Chem. Soc.* **2008**, *130* (23), 7198–7199.
27. (a) Palomo, C., Oiarbide, M., Garcia, J. M. *Chem. Soc. Rev.* **2004**, *33* (2), 65–75; (b) Matsuo, J.-I., Murakami, M. *Angew. Chem. Int. Ed.* **2013**, *52* (35), 9109–9118; (c) Kan, S. B. J., Ng, K. K. H., Paterson, I. *Angew. Chem. Int. Ed.* **2013**, *52* (35), 9097–9108; (d) Kitanosono, T., Kobayashi, S. *Adv. Synth. Catal.* **2013**, *355* (16), 3095–3118.
28. Beutner, G. L., Denmark, S. E. *Angew. Chem. Int. Ed.* **2013**, *52* (35), 9086–9096.
29. Denmark, S. E., Griedel, B. D., Coe, D. M., Schnute, M. E. *J. Am. Chem. Soc.* **1994**, *116* (16), 7026–7043.
30. Wong, C. T., Wong, M. W. *J. Org. Chem.* **2004**, *70* (1), 124–131.
31. Curtiss, L. A., Redfern, P. C., Raghavachari, K., Rassolov, V., Pople, J. A. *J. Chem. Phys.* **1999**, *110* (10), 4703–4709.
32. Wang, L., Wah Wong, M. *Tetrahedron Lett.* **2008**, *49* (24), 3916–3920.
33. Phukan, P. *Synth. Commun.* **2004**, *34* (6), 1065–1070.
34. Deuri, S., Phukan, P. *J. Phys. Org. Chem.* **2012**, *25* (12), 1228–1235.
35. Proft, F. D., Ayers, P. W., Geerlings, P. *The Chemical Bonded.* Wiley-VCH Verlag GmbH & Co. KgaA: Weinheim, Germany, **2014**, pp. 233–270.
36. (a) Parr, R. G., Szentpály, L. V., Liu, S. *J. Am. Chem. Soc.* **1999**, *121* (9), 1922–1924; (b) Chattaraj, P. K., Sarkar, U., Roy, D. R. *Chem. Rev.* **2006**, *106* (6), 2065–2091.
37. Denmark, S. E., Kuester, W. E., Burk, M. T. *Angew. Chem. Int. Ed.* **2012**, *51* (44), 10938–10953.
38. Huang, D., Wang, H., Xue, F., Guan, H., Li, L., Peng, X., Shi, Y. *Org. Lett.* **2011**, *13* (24), 6350–6353.
39. Denmark, S. E., Burk, M. T. *Org. Lett.* **2011**, *14* (1), 256–259.
40. Zhou, L., Tan, C. K., Jiang, X., Chen, F., Yeung, Y.-Y. *J. Am. Chem. Soc.* **2010**, *132* (44), 15474–15476.
41. Nakatsuji, H., Sawamura, Y., Sakakura, A., Ishihara, K. *Angew. Chem. Int. Ed.* **2014**, *53* (27), 6974–6977.
42. Johnston, R. C., Cheong, P. H.-Y. *Org. Biomol. Chem.* **2013**, *11* (31), 5057–5064.
43. Zong, L., Ban, X., Kee, C. W., Tan, C.-H. *Angew. Chem. Int. Ed.* **2014**, *53* (44), 11849–11853.
44. Schwan, A. L. *ChemCatChem* **2015**, *7* (2), 226–227.

45. Vreven, T., Morokuma, K., Farkas, Ö., Schlegel, H. B., Frisch, M. J. *J. Comput. Chem.* **2003**, *24* (6), 760–769.
46. Check, C. E., Faust, T. O., Bailey, J. M., Wright, B. J., Gilbert, T. M., Sunderlin, L. S. *J. Phys. Chem. A* **2001**, *105* (34), 8111–8116.
47. Zhao, Y., Truhlar, D. G. *Acc. Chem. Res.* **2008**, *41* (2), 157–167.
48. Lu, T., Chen, F. *J. Comput. Chem.* **2012**, *33* (5), 580–592.
49. Espinosa, E., Molins, E., Lecomte, C. *Chem. Phys. Lett.* **1998**, *285* (3–4), 170–173.

5 Catalysis with Selenium and Sulfur

Yi An Cheng and Ying-Ying Yeung

CONTENTS

5.1 INTRODUCTION

The chemistry of sulfur and selenium is immensely rich. In the past few decades, ongoing research has unearthed new insights into the area of organosulfur and organoselenium chemistry.[1–3] A more recent development, however, is the interest in the utilization of these organic compounds as organocatalysts. The sulfur and selenium atoms in these organic compounds possess properties that predispose them as potentially useful nucleophilic or Lewis base catalysts, including (1) readily available lone pairs for donation to electrophilic reagents and (2) sufficient size to reversibly accommodate additional atoms or groups to form active intermediates during the catalytic process. Other studies have also shed light on other intriguing activation modes in addition to nucleophilic catalysis. Although a number of reviews have documented some earlier studies on organosulfur and organoselenium catalysis,[4–10] this chapter will highlight some of the latest developments in this field. In addition to the reaction strategies and scope, the interesting roles played by the sulfur and selenium in the catalytic mechanism will also be the focal points of the discussion.

5.2 CATALYSIS WITH SELENIUM

5.2.1 Selenides

The use of diphenyl selenide (Ph_2Se) as a catalyst in the chloroamidation reaction of olefins has been explored by Yeung and coworkers.[11] This reaction is illustrated in Scheme 5.1, with *N*-chlorosuccinimide (NCS) serving as the chlorine source and MeCN and water serving as the nucleophilic partners. The protocol was generally tolerant of a variety of olefinic substrates **1**, including those containing acid-sensitive functionalities such as *O*-trityl group. Mechanistically, it has been proposed that the reaction can be triggered first by the activation of the electrophilic Cl on NCS by the Lewis basic selenium on Ph_2Se, represented by intermediate **3**. Transfer of Cl onto the olefin gives the haliranium intermediate **4**. Two successive nucleophilic attacks by MeCN and H_2O then furnish the chloroamide product. Preceding this work, a similar protocol detailing a Lewis acid (BF_3-Et_2O or $SnCl_4$)-catalyzed olefin bromoamidation using the brominating agent *N*-bromosuccinimide (NBS) has also been reported by Corey and coworkers in 2006.[12]

Zhao and coworkers reported a selenide-catalyzed vicinal trifluoromethylthioamination of alkenes as shown in Scheme 5.2.[13] Using selenide **5** as catalyst and *N*-trifluoromethylthiosaccharin (**6**) as the source of SCF_3 electrophile, products such as **7** and **8** could be formed. The scope of the reaction appeared to be broad with respect to both the alkene and nitrile partners. Based on ^{19}F nuclear magnetic resonance (NMR) studies, **6** was found to degrade most rapidly in the presence of both

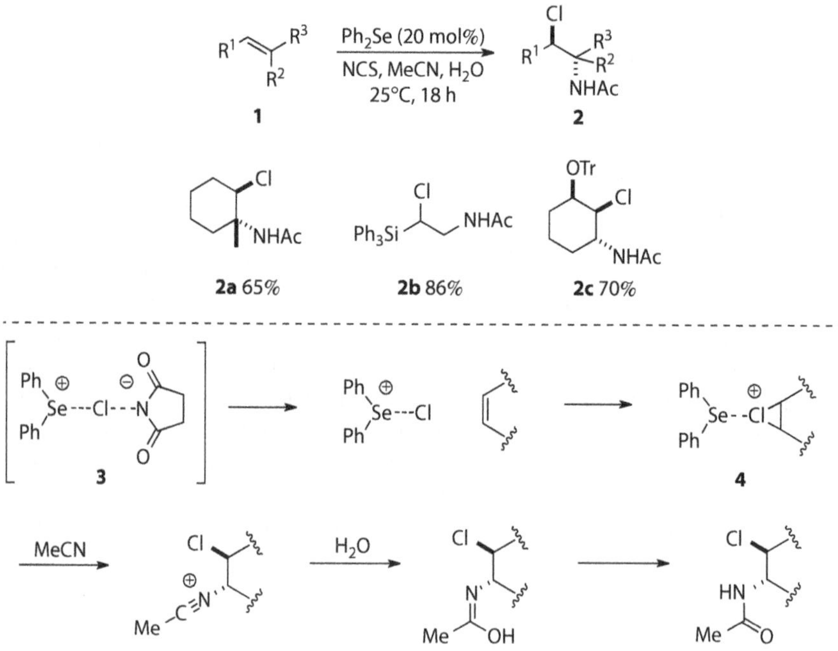

SCHEME 5.1 Ph_2Se-catalyzed chloroamidation of olefin **1** to give **2**.

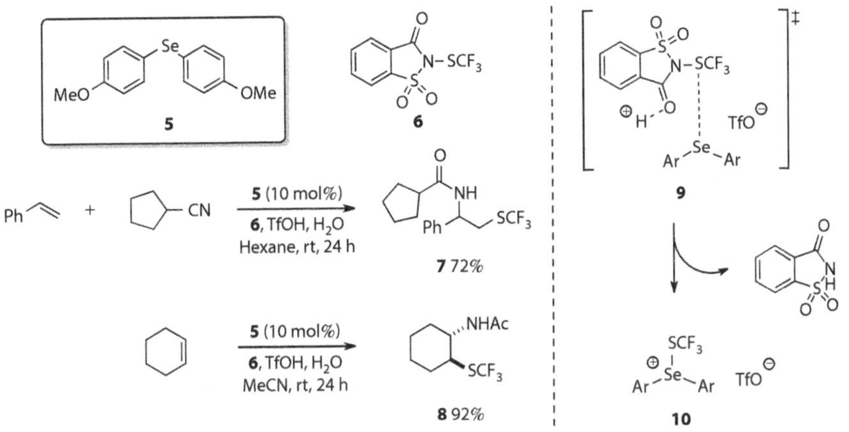

SCHEME 5.2 Trifluoromethylthioamination of alkenes catalyzed by selenide **5**.

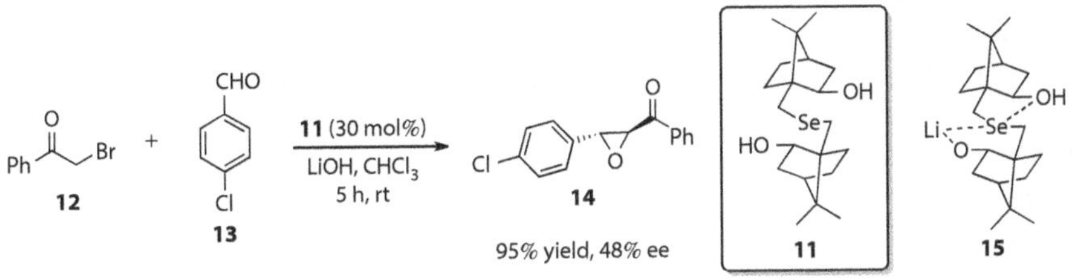

SCHEME 5.3 Asymmetric Darzens reaction catalyzed by C_2-symmetric selenide **11**.

the catalyst **5** and trifluoromethanesulfonic acid, TfOH. The authors proposed that **5** and **6** reacted in the presence of TfOH to give the active species **10**, possibly via the transition state **9**. This is then followed by the transfer of SCF_3 onto the olefin and a consecutive attack by MeCN and H_2O to give the product, tracing a pathway that is generally similar to Yeung's proposals (Scheme 5.1).

Scheme 5.3 depicts an asymmetric Darzens reaction between phenylacyl bromide (**12**) and aldehyde **13** catalyzed by a C_2-symmetric chiral selenide **11**, as reported by Watanabe and coworkers.[14] *Trans*-epoxides such as **14** could be obtained in good to excellent yields, although the enantioselectivities were moderate and typically high catalyst loadings (30 mol%) were required. It is believed that this catalysis involves the intermediate **15** with both Brønsted base (O) and Lewis acid (Li) characteristics.

5.2.2 Diselenides

Breder and coworkers developed an amination of unactivated olefins using diphenyl diselenide $(PhSe)_2$ as the catalyst, shown in Scheme 5.4.[15] In this reaction, the reagent *N*-fluorobenzenesulfonimide (NFSI) played a dual role as the terminal oxidant and as a nitrogen source. Straight-chain and cyclic olefins **16** and **18** subjected to the reaction furnished the respective allylic and vinylic imides **17** and **19** smoothly.

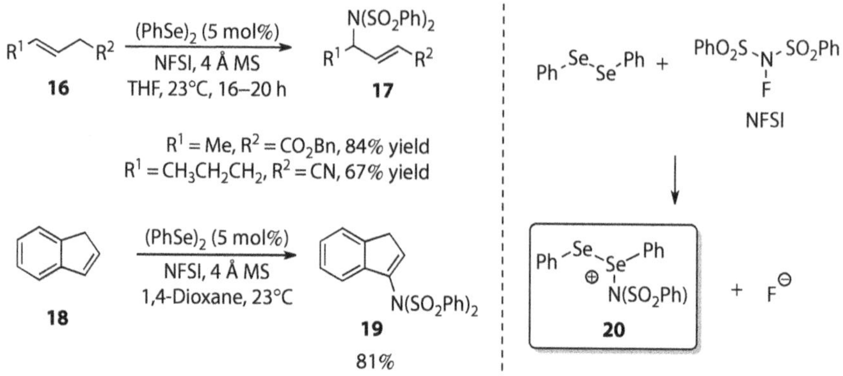

$(PhSe)_2$ (5 mol%)
NFSI, 4 Å MS
THF, 23°C, 16–20 h

16 **17**

R^1 = Me, R^2 = CO_2Bn, 84% yield
R^1 = $CH_3CH_2CH_2$, R^2 = CN, 67% yield

$(PhSe)_2$ (5 mol%)
NFSI, 4 Å MS
1,4-Dioxane, 23°C

18 **19** 81%

NFSI **20**

SCHEME 5.4 $(PhSe)_2$-catalyzed amination of olefins **16** and **18** with NFSI.

The authors ruled out the involvement of an oxidized selenium derivative in the process as the reaction did not proceed when the catalyst $(PhSe)_2$ was substituted with PhSeBr. Further NMR studies demonstrated that NFSI was degraded in the presence of $(PhSe)_2$. Taken together, an intact Se–Se bond in the catalyst was inferred to be crucial for the catalytic process. The authors surmised that $(PhSe)_2$ could react with NFSI to form the intermediate **20** as the active species, which then reacts with the olefin.

Following the report on olefin amination, Breder's group further developed an intramolecular oxidative C(sp^3)-H acyloxylation of *ortho*-allyl benzoic acids **21** to afford isobenzfuranone products **22** with the $(PhSe)_2$/NFSI combination.[16] The reaction is summarized in Scheme 5.5. The isolation of trace amount of side product **23** led the authors to propose that the cyclization process could involve allylic selenation to form a cationic seleniranium intermediate, followed by intramolecular S_N2 attack by the carboxylic group to achieve ring closure.

Another $(PhSe)_2$/NFSI-mediated reaction was reported by Zhao and coworkers,[17] as set out in Scheme 5.6. In the presence of a mixture of bases NaF and pyridine, allylic alcohols **24** underwent amination at the olefin moiety to afford 3-amino allylic alcohols **25**. Removal of the bases led to the generation of α,β-unsaturated aldehydes that arose from decomposition of the 3-amino allylic alcohols **25** under the acidic conditions. Similar to Breder's proposals, $(PhSe)_2$–NFSI intermediate **20** (Scheme 5.4)

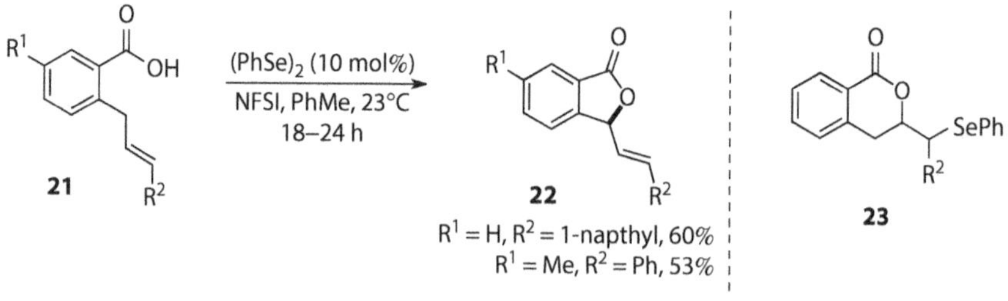

$(PhSe)_2$ (10 mol%)
NFSI, PhMe, 23°C
18–24 h

21 **22** **23**

R^1 = H, R^2 = 1-napthyl, 60%
R^1 = Me, R^2 = Ph, 53%

SCHEME 5.5 $(PhSe)_2$-catalyzed intramolecular oxidative C-H acyloxylation of **21**.

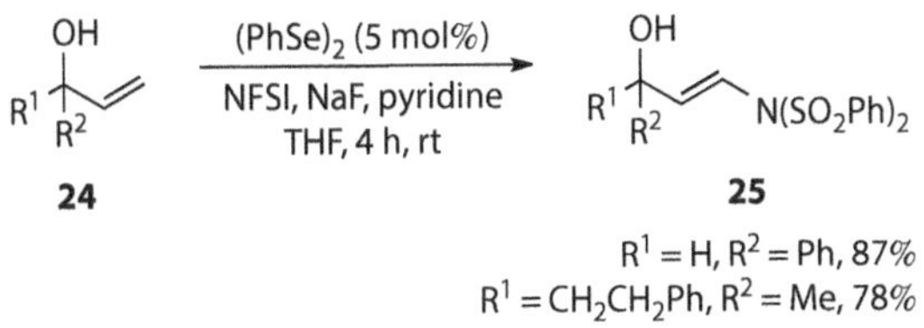

OH

$(PhSe)_2$ (5 mol%)

NFSI, NaF, pyridine

THF, 4 h, rt

24 → **25** ($N(SO_2Ph)_2$)

R^1 = H, R^2 = Ph, 87%

R^1 = CH_2CH_2Ph, R^2 = Me, 78%

SCHEME 5.6 Amination of olefins **24** using $(PhSe)_2$ and NFSI.

could be the key species that interacts with the olefin. The $(PhSe)_2$–NFSI combination was later found to be capable of promoting intramolecular C–H aminations by the same group.[18]

A combination of a substoichiometric amount of diselenide and hydrogen peroxide, H_2O_2, has been utilized by Xu, Yu, and coworkers to promote Baeyer–Villiger oxidations. In separate reports, this combination could effect the conversions of (1) (*E*)-2-methylenecyclobutanones **26** to (*E*)-4-methylenebutanolides **27**[19] and (2) α,β-unsaturated ketones **28** to (*E*)-vinyl esters **29**[20]; both of which are shown in Scheme 5.7. In the latter reaction, the active selenium catalyst could be recovered and recycled, although there was some loss of reaction efficiency with each cycle. More recently, the same team expanded the protocol to cover oxidation of isatins.[21] In contrast to other organoselenium catalysts described previously, the diselenides $(PhSe)_2$ and $(PhCH_2Se)_2$ used in these reactions functioned as precatalysts. The proposed catalytic cycle is believed to involve several organoselenium intermediates: (1) the diselenide precatalyst **30** is first oxidized by H_2O_2 to organoseleninoperoxic

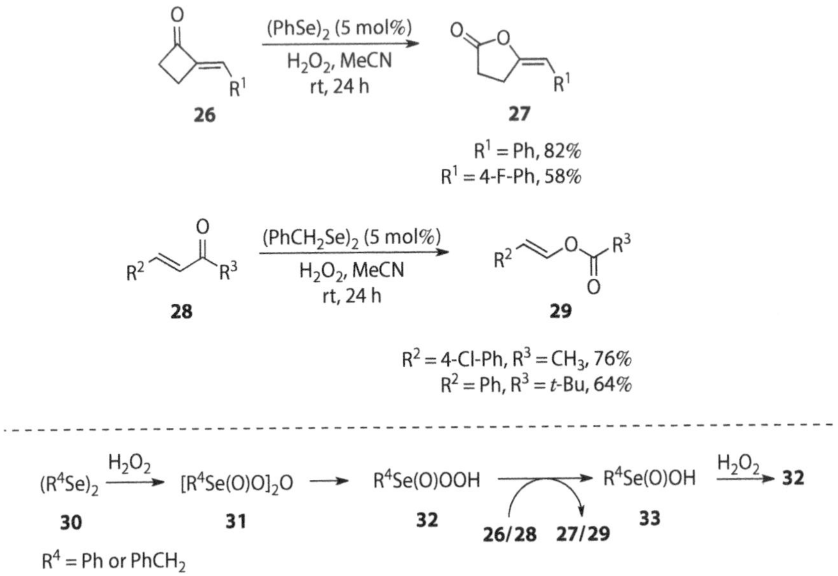

$(PhSe)_2$ (5 mol%)

H_2O_2, MeCN

rt, 24 h

26 → **27**

R^1 = Ph, 82%

R^1 = 4-F-Ph, 58%

$(PhCH_2Se)_2$ (5 mol%)

H_2O_2, MeCN

rt, 24 h

28 → **29**

R^2 = 4-Cl-Ph, R^3 = CH_3, 76%

R^2 = Ph, R^3 = *t*-Bu, 64%

$(R^4Se)_2$ **30** —H_2O_2→ $[R^4Se(O)O]_2O$ **31** → $R^4Se(O)OOH$ **32** → (**26/28** → **27/29**) → $R^4Se(O)OH$ **33** —H_2O_2→ **32**

R^4 = Ph or $PhCH_2$

SCHEME 5.7 Baeyer–Villiger oxidations with diselenide and H_2O_2 and organoselenium intermediates in the oxidation process.

anhydride **31**, which could be detected by ^{77}Se NMR; (2) following hydrolysis, the strongly oxidizing organoseleninoperoxic acid **32** is generated as the active oxidizing species; (3) as one of the oxygen atoms is transferred onto the substrate, **32** is reduced to seleninic acid **33**; and (4) **33** is then oxidized by H_2O_2 to regenerate **32**.

In a related development, the same group described a mild dehydration of aldoximes **35**, leading to the formation of organonitriles **36** using catalytic diselenide **34**/H_2O_2 as depicted in Scheme 5.8.[22] One striking difference between this current protocol and the aforementioned Baeyer–Villiger oxidations (Scheme 5.7) is the use of H_2O_2 in catalytic quantities, whereas the Baeyer–Villiger oxidations employed large excesses of H_2O_2.[19,20] In this case, the authors proposed that the H_2O_2 first oxidized diselenide **34** to selenenic acid **37**, which then condensed to form selenic anhydride **38**. Anhydride **38** could react with the aldoxime **35** to form a mixed anhydride. This mixed anhydride then underwent further rearrangement to form the nitrile product, while regenerating **37** in the process.

Other oxidation reactions using a combination of catalytic diselenide and oxidant have also been carried out. By pairing $(PhSe)_2$ and ammonium persulfate, Santi and coworkers could effect the oxidation of alkynes **39** to 1,2-dicarbonyl derivatives **40** (Scheme 5.9).[23] Subsequently, the group developed a protocol for aldehyde oxidation using a combination of catalytic $(PhSe)_2$ and H_2O_2.[24] In another report, Detty and coworkers presented a method to generate molecular bromine from NaBr in the presence of catalyst **41** and H_2O_2, which was applied to the bromination of substrates

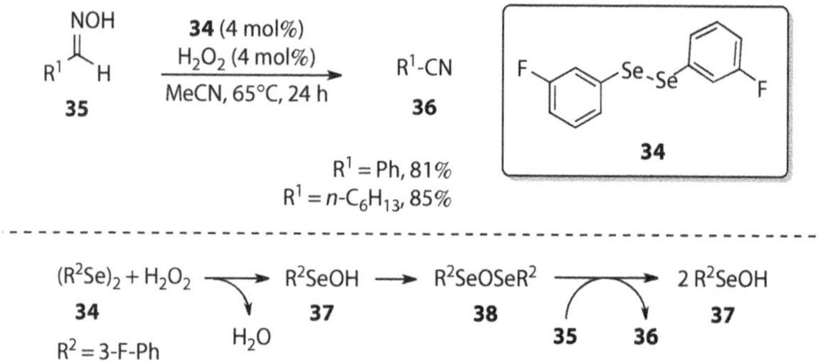

NOH
R1 H
35
34 (4 mol%)
H_2O_2 (4 mol%)
MeCN, 65°C, 24 h
R^1-CN
36
R^1 = Ph, 81%
R^1 = n-C_6H_{13}, 85%
F Se Se F
34

$(R^2Se)_2 + H_2O_2$ → R^2SeOH → $R^2SeOSeR^2$ → 2 R^2SeOH
34 **37** **38** **37**
R^2 = 3-F-Ph H_2O **35** **36**

SCHEME 5.8 Dehydration of aldoximes **35** using a catalytic combination of diselenide **34**/H_2O_2.

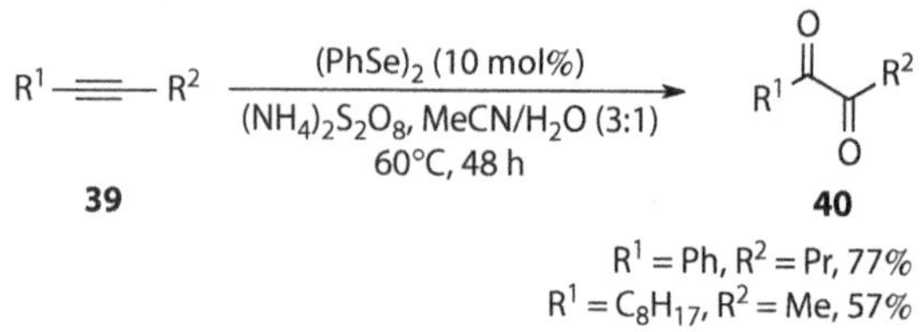

R^1—≡—R^2
39
$(PhSe)_2$ (10 mol%)
$(NH_4)_2S_2O_8$, MeCN/H_2O (3:1)
60°C, 48 h
40
R^1 = Ph, R^2 = Pr, 77%
R^1 = C_8H_{17}, R^2 = Me, 57%

SCHEME 5.9 Oxidation of alkynes **39** using $(PhSe)_2$ catalyst and $(NH_4)_2S_2O_8$.

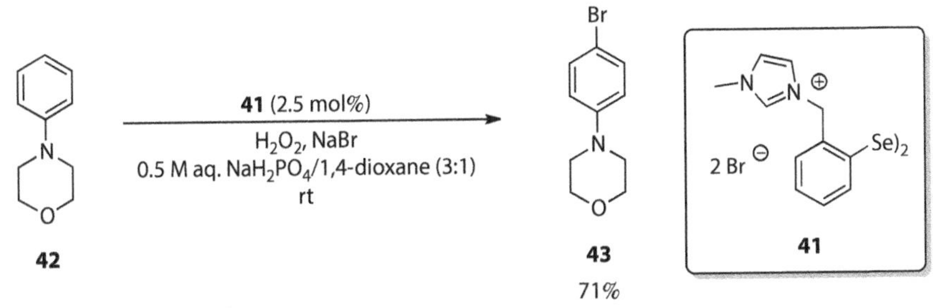

41 (2.5 mol%)
H_2O_2, NaBr
0.5 M aq. NaH_2PO_4/1,4-dioxane (3:1)
rt

42 **43** 71% **41**

2 $Br^{\ominus}$ Se)$_2$

SCHEME 5.10 Bromination under oxidative conditions using ionic diselenide **41** as catalyst.

such as **42** (Scheme 5.10).[25] Notably, all of these reactions could be performed in aqueous medium, which is important in the context of green chemistry development.

In 2010, Wirth and coworkers reported a $(PhSe)_2$-catalyzed cyclization of stilbenecarboxylic acids to isocoumarins using bis(trifluoroacetoxy)iodobenzene, $PhI(OCOCF_3)_2$ as the oxidant.[26] This strategy was later extended to include regioselective cyclization of γ,δ-unsaturated acids **44** to afford 3,6-dihydro-2*H*-pyran-2-ones **45** as shown in Scheme 5.11.[27] In the latter case, $PhI(OCOCF_3)_2$ was found to be the most effective oxidant among various oxidants examined. However, unlike previously described oxidation protocols using diselenide/H_2O_2, no selenium–oxygen intermediates are generated. Instead, it is believed that the reaction between $(PhSe)_2$ and $PhI(OCOCF_3)_2$ can produce a trivalent intermediate **46**, which can then be converted to **47** (detectable by NMR). Interaction of $PhSe^+$ with the olefinic functionality in **44** can induce an electrophilic ring-closing reaction, followed by elimination to give **45**.

In addition, Wirth's group demonstrated that diselenide **48** could catalyze dihydroxylations of olefins such as **49** to give diol **50** with promising enantioselectivities as shown in Scheme 5.12.[28] The catalyst is essentially a dimeric L-selenocysteine, an amino acid that occurs in the enzyme glutathione peroxidase found in living cells.

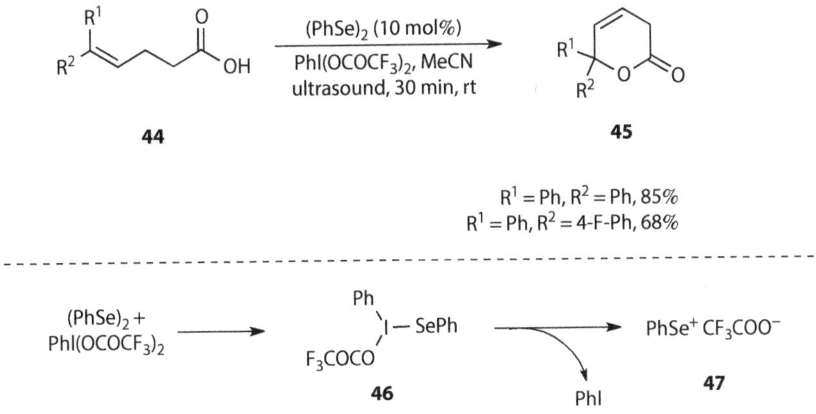

$(PhSe)_2$ (10 mol%)
$PhI(OCOCF_3)_2$, MeCN
ultrasound, 30 min, rt

44 **45**

R^1 = Ph, R^2 = Ph, 85%
R^1 = Ph, R^2 = 4-F-Ph, 68%

$(PhSe)_2$ + $PhI(OCOCF_3)_2$ → Ph(F$_3$COCO)I–SePh **46** → $PhSe^+$ CF_3COO^- **47** + PhI

SCHEME 5.11 Cyclization of γ,δ-unsaturated acids **44** catalyzed by $(PhSe)_2$.

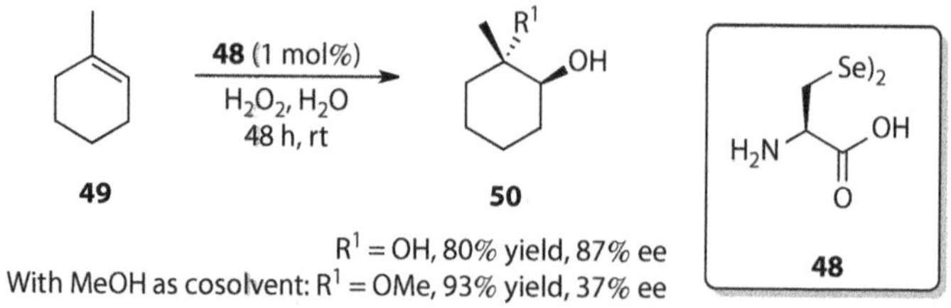

SCHEME 5.12 Dihydroxylation and hydroxymethoxylation with diselenide **48**.

They attempted to extend the reaction scope to hydroxymethoxylations, although generally lower enantioselectivities were attained as compared to the corresponding dihydroxylations. This lackluster performance was ascribed to be a result of a competitive noncatalyzed side reaction. Based on NMR studies, catalyst **48** was found to be oxidized to the peracid in excess H_2O_2, indicating that an oxidative pathway similar to that in Scheme 5.7 could be involved. A similar oxidative protocol to achieve achiral dihydroxylation of cyclohexene has also been reported following this study.[29]

Conventional addition of molecular halogens to alkenes typically gives the vicinal antidihalogenated product. However, *syn*-dichlorinated products such as **52** could be synthesized by Denmark and coworkers in a newly discovered reaction shown in Scheme 5.13.[30] Key ingredients to this reaction include a combination of $(PhSe)_2$ as the precatalyst; benzyltriethylammonium chloride, $BnEt_3NCl$, as the chloride source; and *N*-fluoropyridinium tetrafluoroborate, $[PyF]^+[BF_4^-]$, as the oxidant. Interestingly, omission of the $(PhSe)_2$ precatalyst resulted in a slow formation of the diastereomeric *anti*dichlorinated product, thus illustrating the key role played by $(PhSe)_2$ in not just delivering reaction rate enhancement, but also active involvement in the stereocontrolling steps. The authors believed that the oxidant first converts $(PhSe)_2$ to $PhSeCl_3$, which then loses one chloride to form the electrophile $PhSeCl_2^+$. Interaction between alkene and $PhSeCl_2^+$ forms seleniranium intermediate **53**. This is followed by two successive stereospecific nucleophilic attacks by chloride on the alkenes such as **51** to afford the *syn*-dichloride product.

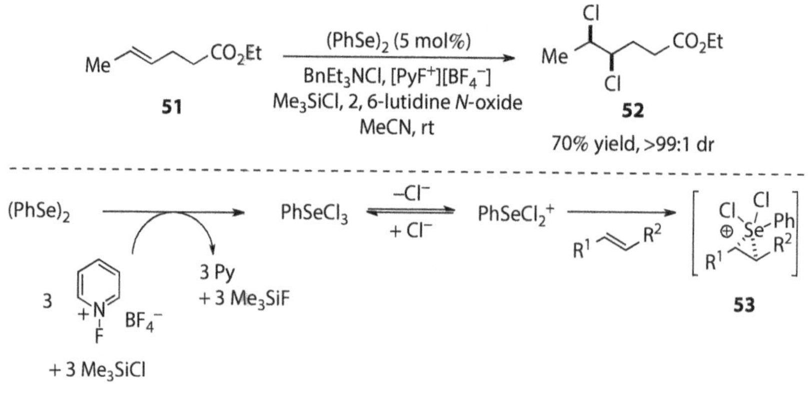

SCHEME 5.13 *Syn*-dichlorination of alkene using precatalyst $(PhSe)_2$.

5.2.3 Other Selenium Catalysts

A collection of selenium-catalyzed enantioselective reactions was featured in the research of Denmark and coworkers. Early efforts by the group have established the viability of achiral selenolactonizations[31] and halolactonizations and etherifications[32] driven by simple Lewis basic catalysts containing selenium, sulfur, or phosphorous as donor atoms such as $(Me_2N)_3PSe$ and Ph_3PS. Further work then demonstrated the stability of enantiomerically enriched thiiranium ions (a three-membered ring species formed by the interaction between olefins and sulfur electrophiles)[33] and the susceptibility of these species to nucleophilic ring opening with preservation of chiral information.[34] On the basis of these studies, the group developed a series of enantioselective alkene thiofunctionalization reactions in very recent years. These reactions were broadly based on 1,1′-binaphthyl-2,2′-diamine-derived selenophosphoramide catalysis.

The first of these thiofunctionalization reactions relates to the formation of sulfenylated tetrahydrofurans **57**-*endo* or tetrahydropyrans **57**-*exo* from alkenols **56** as shown in Scheme 5.14, using *N*-phenylthiophthalimide (**55**) to supply electrophilic sulfur.[35] The intermolecular version using MeOH as the nucleophile was also demonstrated. The selenophosphoramide catalyst **54** is proposed to undergo sulfenylation

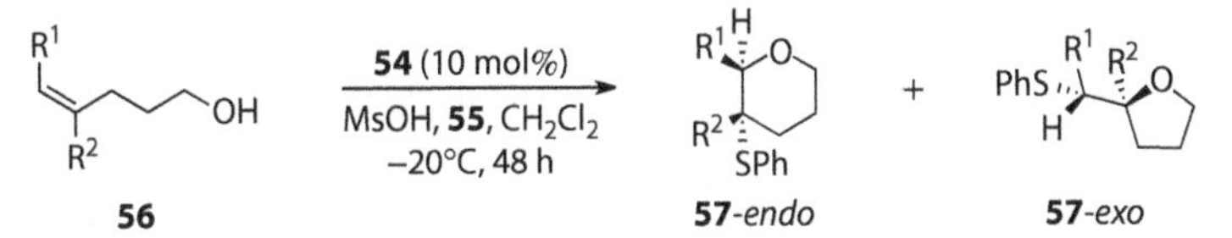

R^1	R^2	**57**-*endo*: yield, ee (%)	**57**-*exo*: yield, ee (%)
Ph	H	78, 82	2, --
H	Ph	<1, --	>84, 24

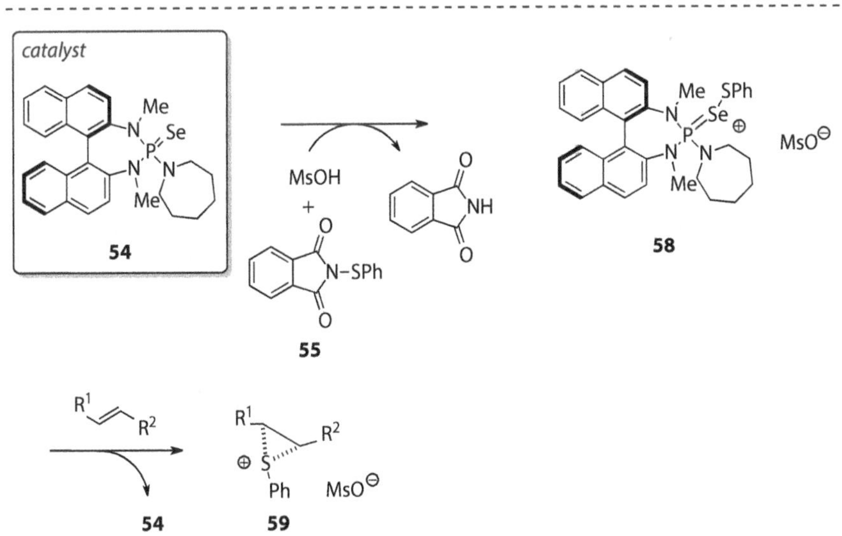

SCHEME 5.14 Cycloetherification of alkenols **56** catalyzed by selenophosphoramide **54**.

first to generate the active catalytic species **58**, which displays a distinctive signature peak in the ^{31}P NMR spectrum. The proposed enantiodetermining step involves the subsequent transfer of the sulfenium from **58** to the alkene to form the thiiranium ion **59**, which is then captured by the nucleophile. Similar mechanistic proposals would be a recurring theme in forthcoming reactions developed by this group. Notably, the catalytically active species could be isolated and studied by X-ray crystallography, and improvements in the *ee*s were achieved by using more sterically hindered sulfenylating agents in a follow-up study on this reaction.[36]

In addition to the C–O bond-forming cycloetherification described earlier, Denmark's group continued to develop similar sulfenylating C–C and C–N bond-forming reactions. With a slightly modified catalyst **60**, sulfenocarbocyclization of alkene **61** to give enantioenriched tetrahydronapthalenes **62** was accomplished as shown in Scheme 5.15.[37,38] In particular, the authors found that this reaction proceeded at similar rates regardless of whether the catalyst was present. It was later discovered that the racemic background reaction was suppressed by the sulfonate ions and phthalimide, thus allowing the enantioselective catalytic pathway to dominate.[39] Other related enantioselective protocols that were successfully worked out by the group include the aminocyclization of olefinic *N*-tosylamines[40] and the sulfenylation of ketone-derived enoxysilanes using a newly developed sulfenylating agent, *N*-phenylthiosaccharin.[41]

Another class of selenium catalyst was described by Kumar and coworkers. Using Br_2 or NBS as the bromine sources, the isoselenazolone **63** catalyzed the bromolactonization of alkenoic acids such as **64** to give lactone **65** (Scheme 5.16).[42] Other reaction types were also demonstrated, including bromoesterification (the intermolecular version of bromolactonization) and the oxidation of secondary alcohols to ketones exemplified by the conversion of **66** to **67**. Mechanistic studies revealed that the selenium (II) catalyst **63** could undergo oxidative addition with Br_2 to give the isoselenazolone (IV) dibromide **68**. The existence of **68** was confirmed by ^{77}Se NMR and mass spectrometry. A *N*-quininamine-substituted isoselenazolone capable of catalyzing the oxidation of thiophenol in the presence of H_2O_2 was also developed by the same group.[43]

Yeung and coworkers developed an enantioselective bromoaminocyclization of trisubstituted olefinic amides **70** that are catalyzed by a monofunctional C_2-symmetric selenium **69** as illustrated in Scheme 5.17.[44] Using *N*-bromophthalimide (NBP) as the bromine source, enantioenriched pyrrolidine products **71** containing two stereogenic centers could be readily furnished. The versatility of the products

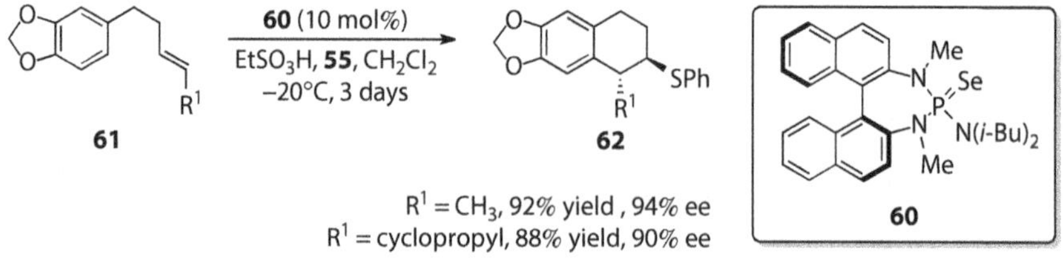

61 → **62**: **60** (10 mol%), $EtSO_3H$, **55**, CH_2Cl_2, −20°C, 3 days

$R^1 = CH_3$, 92% yield, 94% ee
R^1 = cyclopropyl, 88% yield, 90% ee

SCHEME 5.15 Sulfenylcarbocyclization of alkenes **61** catalyzed by selenophosphoramide **60**.

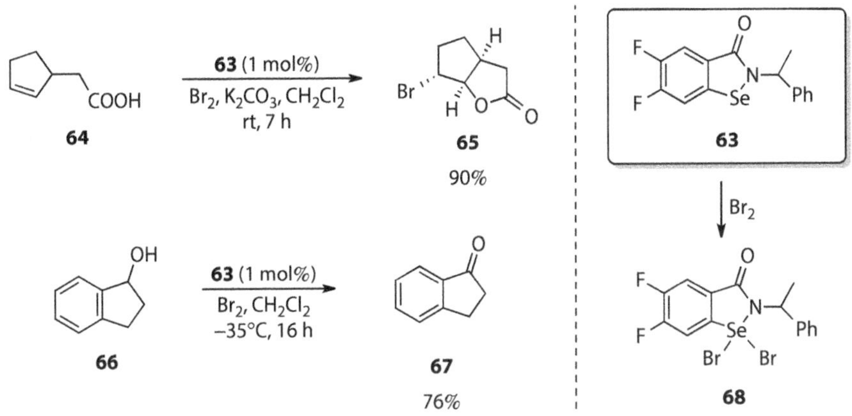

SCHEME 5.16 Bromolactonization and oxidation reactions catalyzed by isoselenazolone **63**.

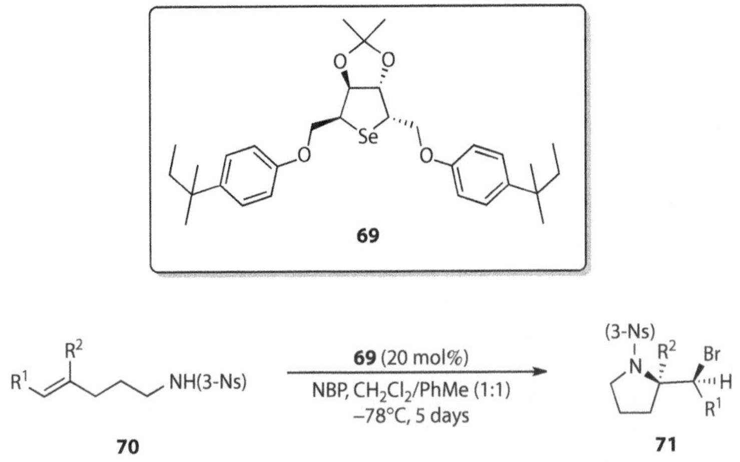

SCHEME 5.17 Enantioselective bromoaminocyclization with catalytic C_2-symmetric selenium **69**.

was further demonstrated by a number of enantiospecific transformations, including a silver salt-mediated rearrangement of the enantioenriched pyrrolidine to give 2,3-disubstituted piperidine.

5.3 CATALYSIS WITH SULFUR

5.3.1 Thiophenes

The nucleophilic character of tetrahydrothiophene (**72**) was harnessed by Clark and coworkers to promote the concise synthesis of substituted furans such as **75**.[45] As shown

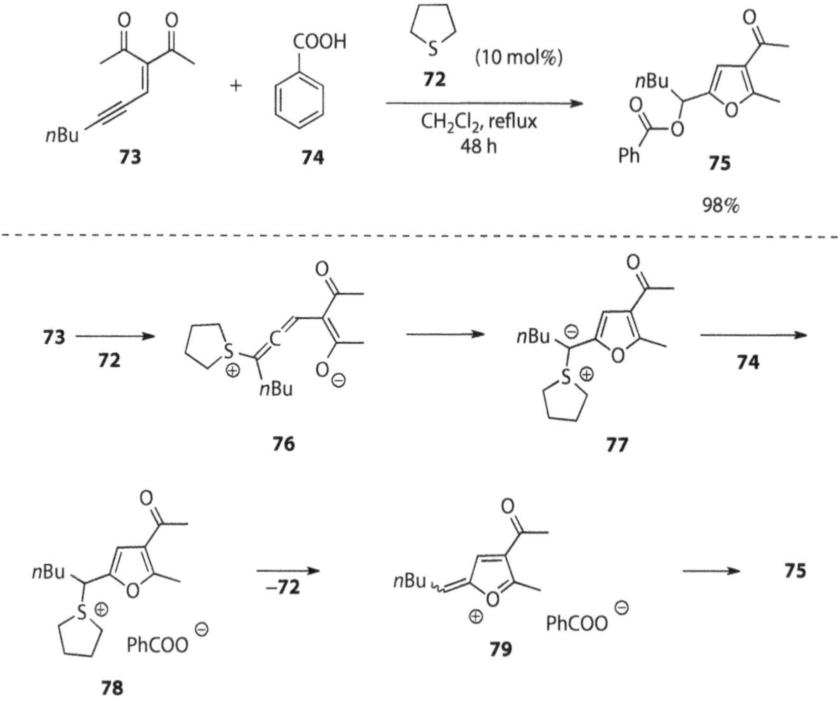

SCHEME 5.18 Tetrahydrothiophene (**72**)-catalyzed formation of substituted furan **75**.

in Scheme 5.18, this was accomplished by the reaction between enyne **73** as the electrophile and benzoic acid (**74**) as the nucleophile. On the other hand, other nonsulfur nucleophilic catalysts, such as 4-dimethylaminopyridine (DMAP) and PBu_3, did not give the desired product. The reaction proceeded equally smoothly when the acid was replaced with other nucleophiles such as aliphatic alcohols, phenol, and 4-nitrobenzenesulfonamide ($NsNH_2$). It is believed that the mechanism follows an orchestrated process with multiple intermediates. The catalyst **72** first attacks the electrophilic alkyne linkage in **73** to form enolate **76**, and this is followed by a cyclization to form sulfur ylide **77**. Protonation of **77** by the acid **74** gives intermediate **78**, which then undergoes a S_N1 pathway that involves (1) the release of tetrahydrothiophene (**72**) to form ion **79** and (2) nucleophilic attack by benzoate on **79** to furnish the product **75**.

Recently, Chein and coworkers developed a synthesis for the chiral tetrahydrothiophene catalyst **80** and applied it to enantioselective epoxidations[46] and aziridinations[47] using a combination of benzyl bromide (**81**) and aryl aldehyde **82** or imine **84** as starting materials. One representative example from each reaction is illustrated in Scheme 5.19. Following these studies, the group reported a broadly similar enantioselective aziridination,[48] with modifications to the reaction including a **80**/*N*-*N*′-diarylurea (2:1) cocatalyst system and the use of cinnamyl bromide in place of **81**. The sulfide catalyst likely reacts with **81** to form a sulfur ylide, which then undergoes a Corey–Chaykovsky reaction with the aldehyde or imine to give the products. A related reaction was reported by Connor and coworkers in which

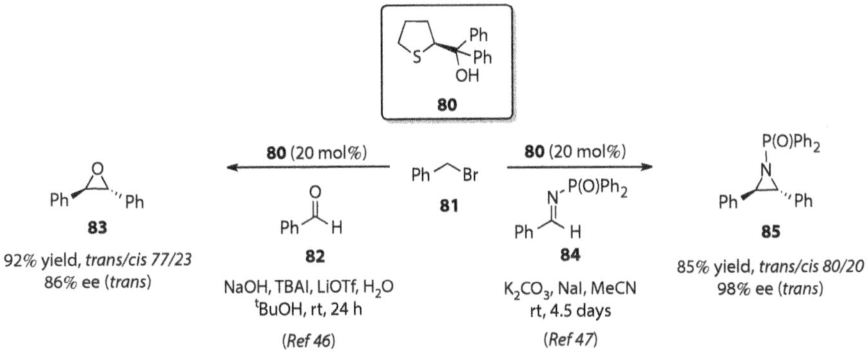

SCHEME 5.19 Enantioselective epoxidation and aziridination catalyzed by chiral tetrahydrothiophene **80**.

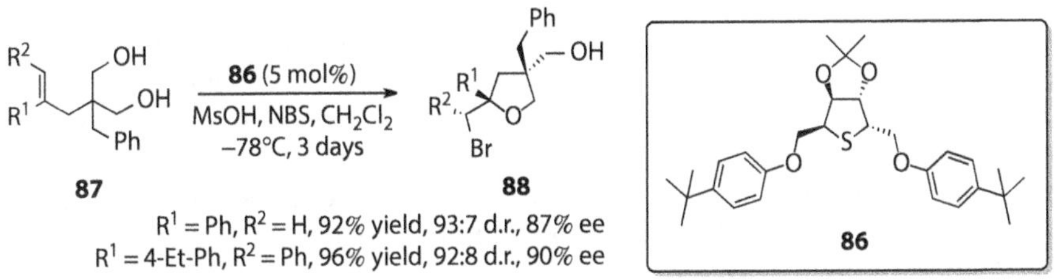

SCHEME 5.20 Enantioselective, desymmetrizing bromoetherification of olefinic diols **87** catalyzed by cyclic sulfide **86**.

catalytic tetrahydrothiophene was used to drive methylene transfer onto ketones to give epoxide products.[49] Preceding these developments, other organosulfide-catalyzed epoxidations have also been documented in earlier years.[50,51]

Another chiral tetrahydrothiophene-based catalysis can be found in the work of Yeung and coworkers as shown in Scheme 5.20.[52] Based on their earlier work C_2-symmetric selenium catalysis (Scheme 5.17),[44] the group designed and synthesized an analogous cyclic sulfide **86**. In the presence of a catalytic quantity of **86** and NBS, olefinic diols **87** underwent a bromoetherification process with a concomitant desymmetrization to give enantioenriched substituted tetrahydrofuran products **88**. The utility of this protocol was further demonstrated in the synthesis of an intermediate of the antifungal drug posaconazole (Noxafil®). A more detailed mechanistic study of this reaction has also been performed.[53]

5.3.2 Thiocarbamates

Successive reports on amino-thiocarbamate-catalyzed enantioselective bromocyclization were published by Yeung and coworkers in recent years. In 2010, the group developed an enantioselective bromolactonization of 1,1-disubstituted olefinic acids **90** to form γ-lactones **91** as shown in Scheme 5.21.[54] The amino-thiocarbamate catalyst **89** incorporated the cinchona alkaloid (+)-cinchonine as the chiral scaffold. Notably, the enantioselectivity deteriorated significantly when the catalyst's

S was replaced with O, or when its N–H was replaced with N–Me, thus indicating the crucial role played by the thiocarbamate moiety. In subsequent studies, the group found that the enantioselectivity could be enhanced by modifying the *N*-aryl or the 6-alkoxy group of the quinoline in the cinchona alkaloid scaffold. A modified amino-thiocarbamate catalyst **92** was thus later applied successfully to the bromoaminocyclization of olefinic sulfonamides **93**, affording enantioenriched pyrrolidines **94** (Scheme 5.22).[55] Further, fine-tuning of the catalyst structure and reaction conditions allowed the protocol to flexibly accommodate other substrates of varying structures and alkene geometries. These included bromolactonizations of 1,2-disubstituted[56,57] and trisubstituted[58] olefinic acids and styrene-type carboxylic acids[59] and bromoaminocyclizations of 1,2-disubstituted olefinic sulfonamides.[60,61]

The authors proposed that the amino-thiocarbamate catalyst might be involved in a dual activation mode in the bromocyclization reaction mechanism as illustrated in Scheme 5.23. The key features of this mechanism include (1) an interaction between the catalyst's Lewis basic S and the electrophilic Br and (2) an acid–base interaction between the quinuclidine N and the nucleophile. Other reaction classes utilizing cinchona alkaloid amino-thiocarbamate catalysis were also developed by Yeung's group. These included the enantioselective bromocyclization of 1,3-dicarbonyl compounds **96** to give highly functionalized furans **97** (Scheme 5.24),[62] and the desymmetrizing bromoetherification of diolefinic diols **99** to give bromoether products **100** (Scheme 5.25).[63] In particular, the products **100** in the latter reaction could undergo a second cyclization catalyzed by Ph_3PS to give spirocycles **101**. Most recently, the group reported a carbamate-catalyzed enantioselective bromolactamization.[64]

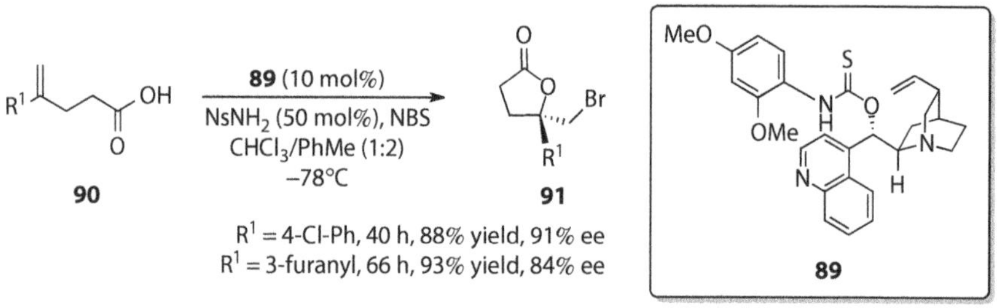

89 (10 mol%)
$NsNH_2$ (50 mol%), NBS
$CHCl_3$/PhMe (1:2)
−78°C

90 → **91**

R^1 = 4-Cl-Ph, 40 h, 88% yield, 91% ee
R^1 = 3-furanyl, 66 h, 93% yield, 84% ee

89

SCHEME 5.21 Enantioselective bromolactonization of olefinic acids **90** catalyzed by amino-thiocarbamate **89**.

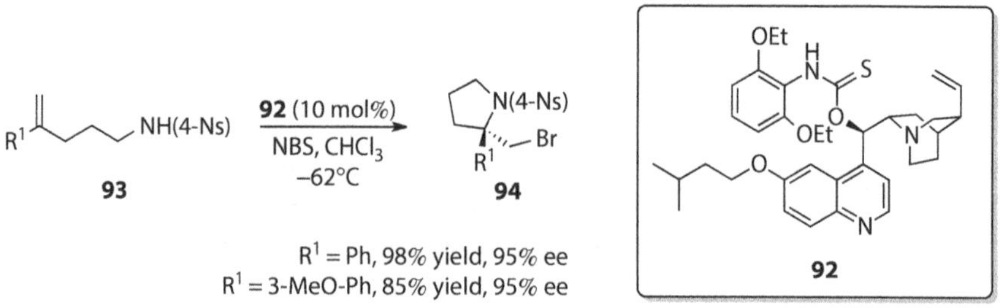

92 (10 mol%)
NBS, $CHCl_3$
−62°C

93 → **94**

R^1 = Ph, 98% yield, 95% ee
R^1 = 3-MeO-Ph, 85% yield, 95% ee

92

SCHEME 5.22 Enantioselective bromoaminocyclization of olefinic sulfonamides **93**.

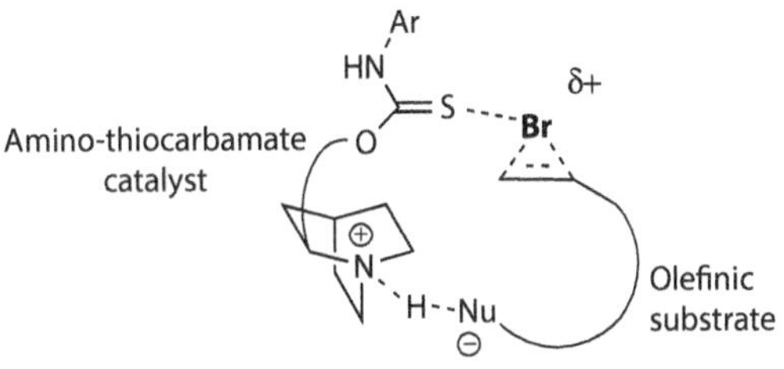

SCHEME 5.23 Proposed mechanism of the amino-thiocarbamate catalysis.

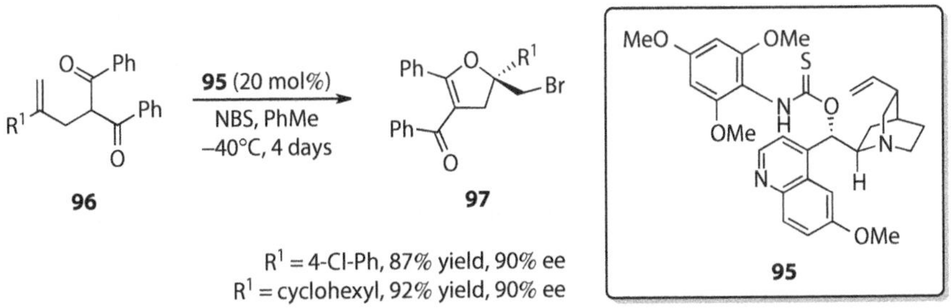

R^1 = 4-Cl-Ph, 87% yield, 90% ee
R^1 = cyclohexyl, 92% yield, 90% ee

SCHEME 5.24 Enantioselective bromocyclization of olefinic 1,3-dicarbonyl compounds **96**.

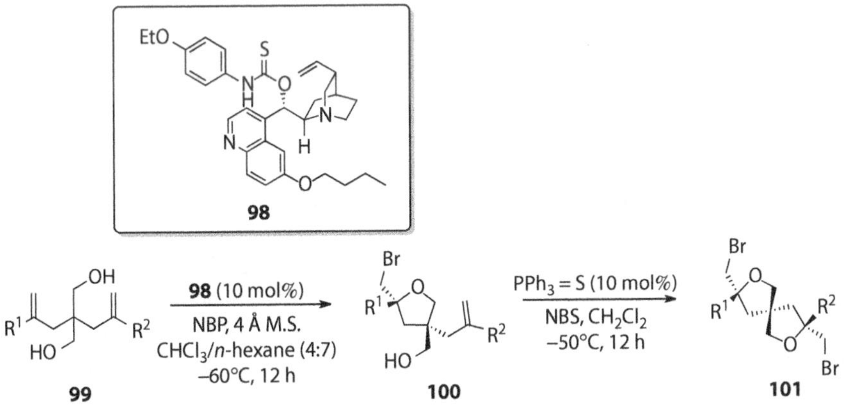

R^1	R^2	**100**: yield, d.r., ee (%)	**101**: yield (%), d.r.
2-Me-Ph	2-Me-Ph	90, 95:5, 92	96, 94:6
Ph	*t*-butyl	91, 94:6, 80	91, 60:40

SCHEME 5.25 Enantioselective, desymmetrizing bromoetherification of diolefinic diols **99** followed by formation of spirocycles **101**.

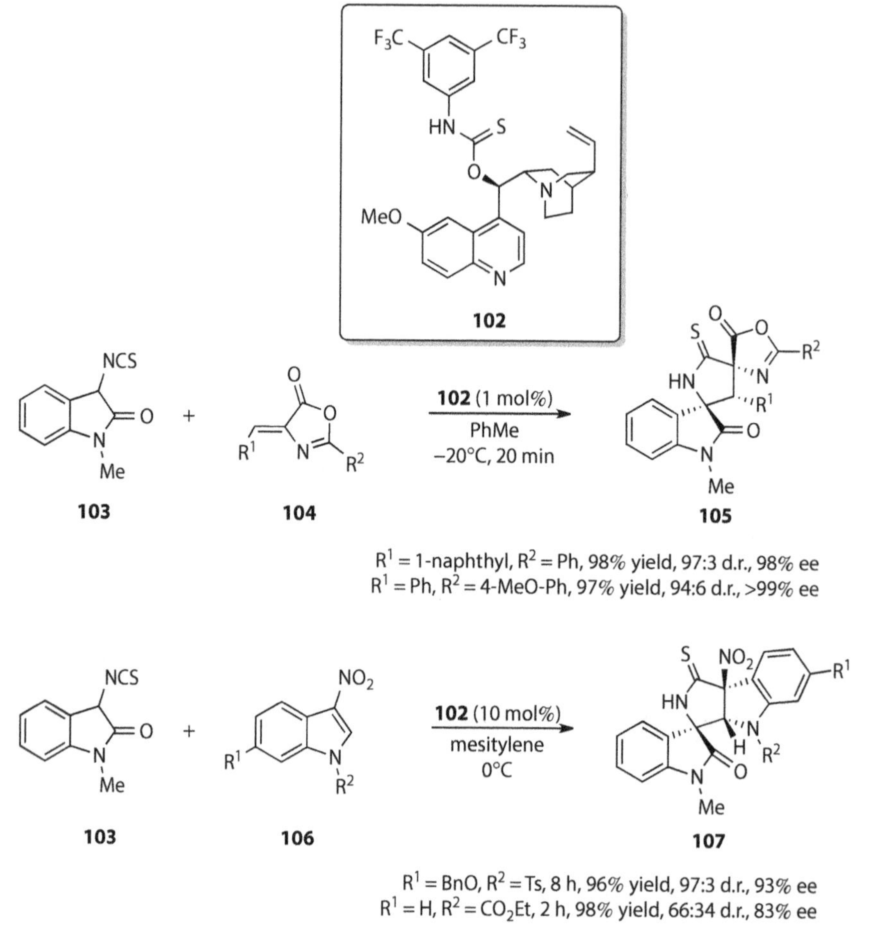

102 (1 mol%)
PhMe
−20°C, 20 min

R^1 = 1-naphthyl, R^2 = Ph, 98% yield, 97:3 d.r., 98% ee
R^1 = Ph, R^2 = 4-MeO-Ph, 97% yield, 94:6 d.r., >99% ee

102 (10 mol%)
mesitylene
0°C

R^1 = BnO, R^2 = Ts, 8 h, 96% yield, 97:3 d.r., 93% ee
R^1 = H, R^2 = CO_2Et, 2 h, 98% yield, 66:34 d.r., 83% ee

SCHEME 5.26 Cascade Michael/cyclization reactions catalyzed by amino-thiocarbamate **102**.

The carbamate catalysis is believed to undergo a different mechanism compared to that of thiocarbamates.

The work of Yuan and coworkers described examples in which the amino-thiocarbamate could be used for nonhalogenation reactions. Scheme 5.26 shows these organic transformations. Using 3-isothiocyanato oxindole **103** as the nucleophile and alkylidene azlactones **104** as the Michael acceptor, amino-thiocarbamate **102** was shown to catalyze a cascade Michael/cyclization reaction to give dispirocyclic thiopyrrolidineoxindole products **105**.[65] The reaction also proceeded smoothly when 3-nitroindoles **106** were used as the Michael acceptors.[66] Both of these reactions were found to be readily scalable without loss in enantioselectivity. It is noteworthy that the catalyst **102** used in these cases contains the highly electron-withdrawing 3,5-bis(trifluoromethyl) *N*-aryl substituent. In contrast, amino-thiocarbamate catalysts with electron-donating substituents generally performed better in Yeung's reactions.

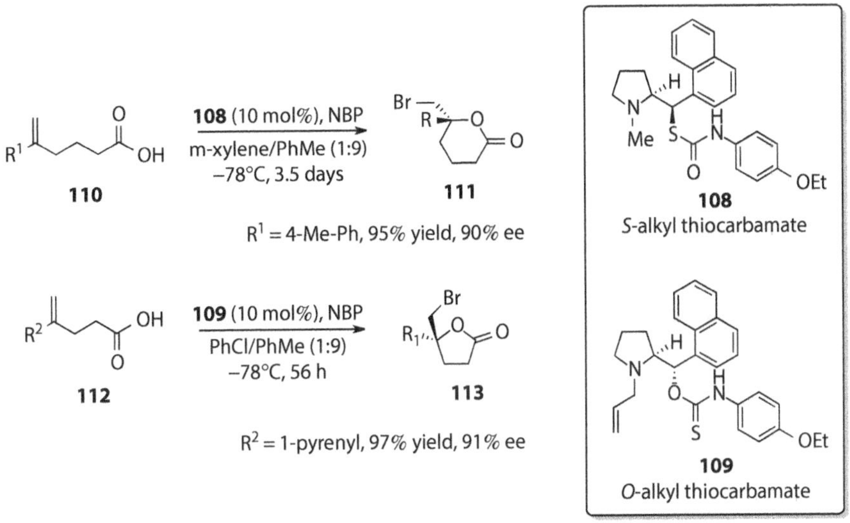

SCHEME 5.27 Enantioselective bromolactonizations with *S*-alkyl and *O*-alkyl thiocarbamates **108** and **109**.

In addition to cinchona alkaloid-based amino-thiocarbamates, Yeung and coworkers presented another solution to achieve enantioselective bromolactonization using the prolinol-derived *S*-alkyl and *O*-alkyl amino-thiocarbamates **108** and **109** as catalysts (Scheme 5.27).[67] In the presence of NBP as the brominating agent, **108** could promote cyclization of hexenoic acid **110** to give γ-lactone **111**. Effective asymmetric cyclization of pentenoic acid **112** to give the γ-lactone **113**, on the other hand, required the use of catalyst **109**.

5.3.3 Sulfinamides

The ability to introduce chirality on the sulfur atom is a defining feature of sulfinamide catalysts. In contrast, other sulfur-based organocatalysts relied on one or more stereocenters located on other atoms to induce enantioselectivity. Following their earlier work on *N*-sulfinyl urea catalysis,[68] Ellman and coworkers recently expanded the reaction scope to cover the addition of cyclohexyl Meldrum's acid **116** to nitroalkenes **115** using low catalyst loadings (0.2–3 mol%).[69] The reaction catalyzed by *N*-sulfinyl urea **114** is shown in Scheme 5.28. In these reactions, the sulfinyl group functioned as a chiral directing group and as an electron-withdrawing substituent. More detailed investigation, however, revealed that the chirality of the diamine moiety of the catalyst did not have a major impact on the enantioselectivity—the *N*-sulfinyl group exerted the dominant stereocontrolling effect. This paved the way to the development of a *N*-sulfinyl catalyst **118** that is chiral solely at the sulfur atom and capable of promoting the addition of α-substituted Meldrum's acids **120** to nitroalkenes **119** (Scheme 5.29).[70] Other *N*-sulfinyl amide-catalyzed enantioselective reactions developed recently by the group include an intermolecular aldol reaction[71] and addition of thioacetic acid to nitroalkenes.[72]

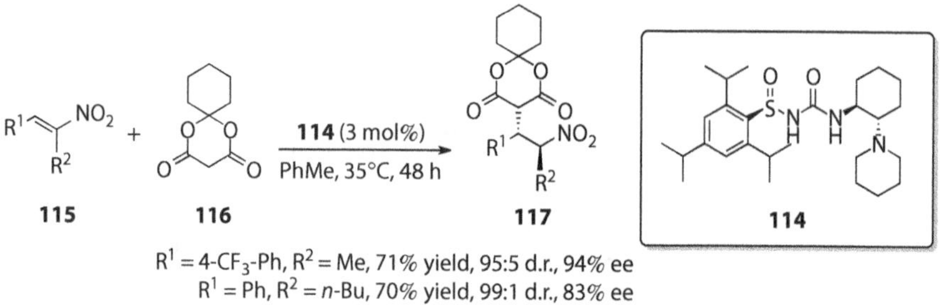
114 (3 mol%)
PhMe, 35°C, 48 h
115
116
117
114
R1 = 4-CF3-Ph, R2 = Me, 71% yield, 95:5 d.r., 94% ee
R1 = Ph, R2 = n-Bu, 70% yield, 99:1 d.r., 83% ee

SCHEME 5.28 Enantioselective addition of cyclohexyl Meldrum's acid **116** to nitroalkenes **115**.

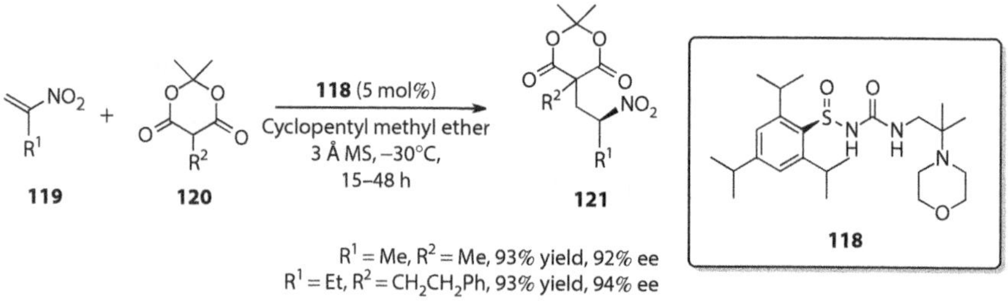
118 (5 mol%)
Cyclopentyl methyl ether
3 Å MS, −30°C,
15–48 h
119
120
121
118
R1 = Me, R2 = Me, 93% yield, 92% ee
R1 = Et, R2 = CH2CH2Ph, 93% yield, 94% ee

SCHEME 5.29 Enantioselective addition of α-substituted Meldrum's acids **120** to nitroalkenes **119** catalyzed by *N*-sulfinyl urea **118**.

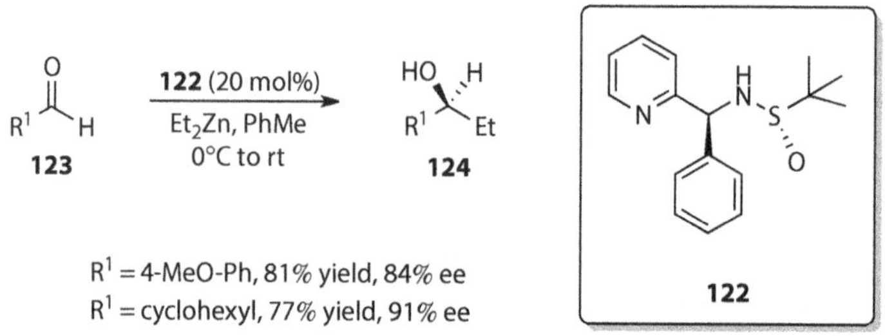
122 (20 mol%)
Et2Zn, PhMe
0°C to rt
123
124
122
R1 = 4-MeO-Ph, 81% yield, 84% ee
R1 = cyclohexyl, 77% yield, 91% ee

SCHEME 5.30 Enantioselective alkylation of aldehydes **123** catalyzed by 2-pyridylsulfinamide **122**.

Prasad and coworkers reported an enantioselective alkylation of aldehydes **123** with diethylzinc catalyzed by 2-pyridylsulfinamide **122**.[73] This reaction, shown in Scheme 5.30, resulted in the formation of alcohols **124** in moderate to good enantioselectvities. As with Ellman's case, it was found that the enantioselectivity is primarily determined by the chirality at the sulfur atom.

A solution to the enantioselective Povarov reaction was introduced by Jacobsen and coworkers, using a combination (2:1) of sulfinamide **125** as catalyst and *ortho*-nitrobenzenesulfonic acid (NBSA) as cocatalyst (Scheme 5.31).[74] This cycloaddition reaction between imine **126** and electron-rich alkenes, such as **127** and **129**, gave enantioenriched

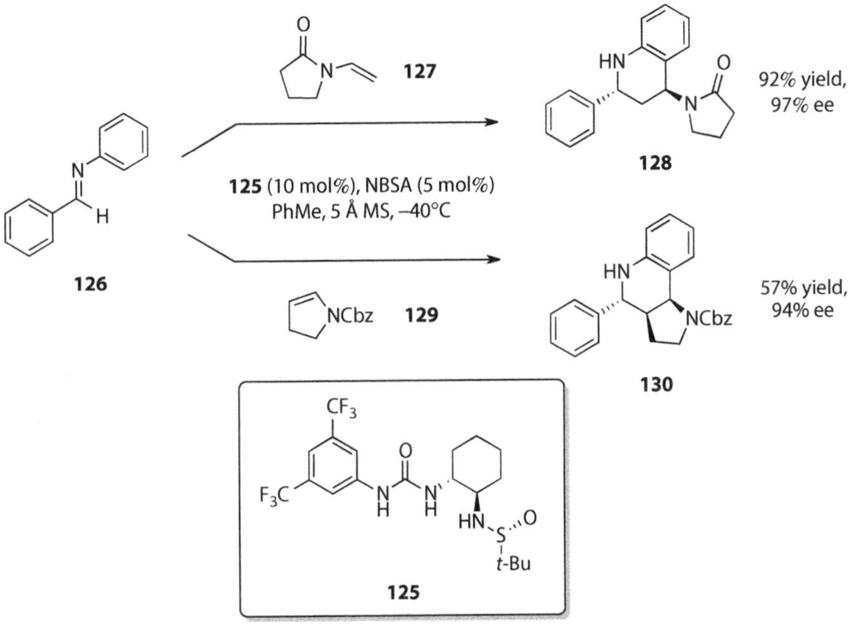

SCHEME 5.31 Enantioselective Povarov reaction between imine **126** and electron-rich alkenes **127** and **129**.

fused-ring heterocyclic products **128** and **130**, respectively. Detailed ^1H NMR and computational analysis revealed that the enantioselectivity was originated from multiple, specific hydrogen bonding interactions that occur between the urea and sulfinamide moieties of the catalyst, the protonated imine substrate and the sulfonate counterion. Following this report, the group developed another protocol for enantioselective protonation of enol silane **132**, using a catalyst system consisting of structurally simpler sulfinamide **131** and 2,4-dinitrobenzenesulfonic acid (2,4-diNBSA) cocatalyst (Scheme 5.32).[75]

Various sulfonamides were also utilized by Sun and coworkers as catalysts for the reduction of unsaturated bonds in the presence of trichlorosilane, $HSiCl_3$. Although the group have had some success in the enantioselective reduction of aromatic ketimines in their earlier work,[76] recent efforts have focused on the reduction of C=C bonds in electron-deficient olefins such as **135** as illustrated in Scheme 5.33.[77,78] Separately, the group reported another sulfonamide-catalyzed enantioselective

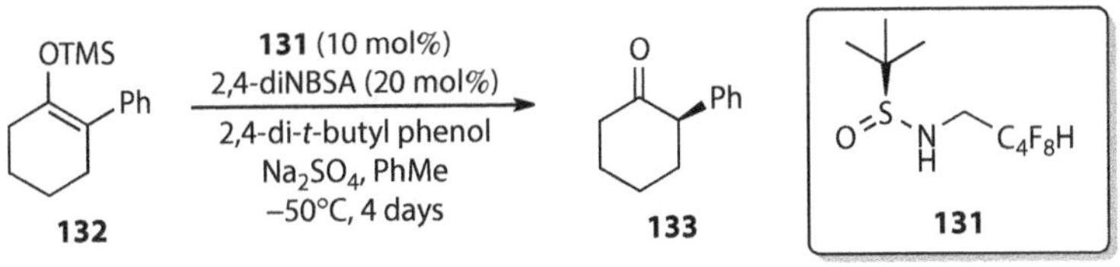

SCHEME 5.32 Enantioselective protonation of enol silanes **132**.

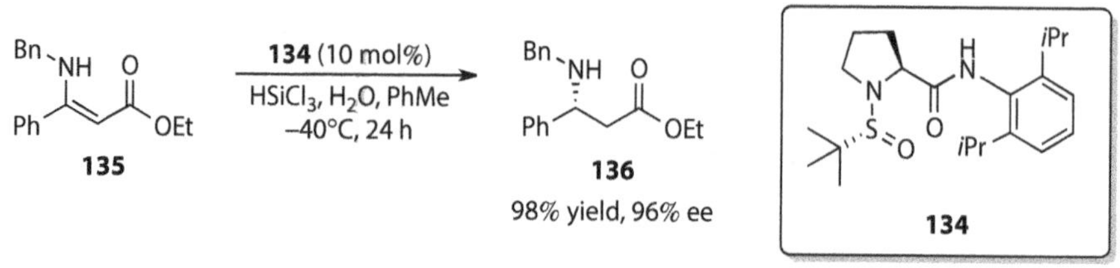

SCHEME 5.33 Enantioselective reduction of electron-deficient olefin **135**.

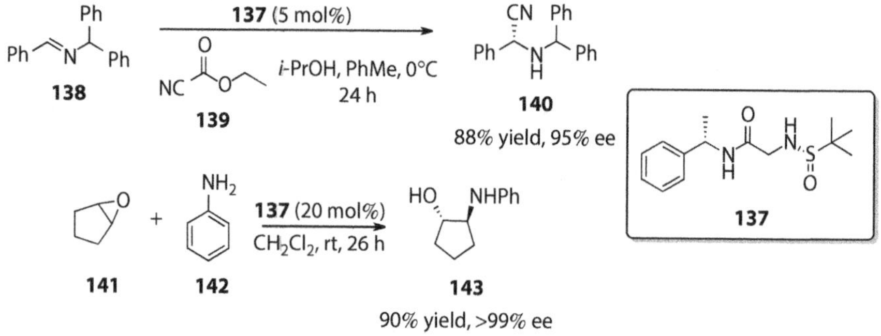

SCHEME 5.34 Enantioselective Strecker reaction and epoxide ring opening catalyzed by **137**.

reduction of 3-aryl-1,4-benzooxazine.[79] It is noteworthy that the sulfinamide catalyst **134** can be derived from the amino acid L-proline.

A more recent development in sulfinamide catalysis came in the work of Khan, Kureshy, and coworkers. Using **137** as the catalyst and ethylcyanoformate (**139**) as the cyanide source, substituted imine **138** could undergo an enantioselective Strecker reaction to give α-amino nitrile **140**.[80] The sulfinyl group appeared to be crucial in the enantioselectivity as omission of this group in the catalyst structure resulted in dramatically lower enantiomeric excess (ee) values. The same catalyst was later employed by the group to accomplish an enantioselective ring opening of epoxide **141** with aniline (**142**) to give chiral β-amino alcohol **143**.[81] Both of these reactions are presented in Scheme 5.34.

5.3.4 Other Sulfur Catalysts

Some recent efforts exploited the Lewis basic nature of triphenylphosphine sulfide (**144**) and used it as a catalyst in halogenation reactions. Gustafson and coworkers reported an electrophilic halogenation of heterocycles and simple arenes such as **145**, using *N*-halosuccinimides as the halogen source (Scheme 5.35).[82] The study was centered on the more challenging chlorination reaction, owing to the relative inertness of NCS as compared to the bromine and iodine counterparts. It was also found that the yield was significantly eroded when the S on the catalyst was replaced with other chalcogens such

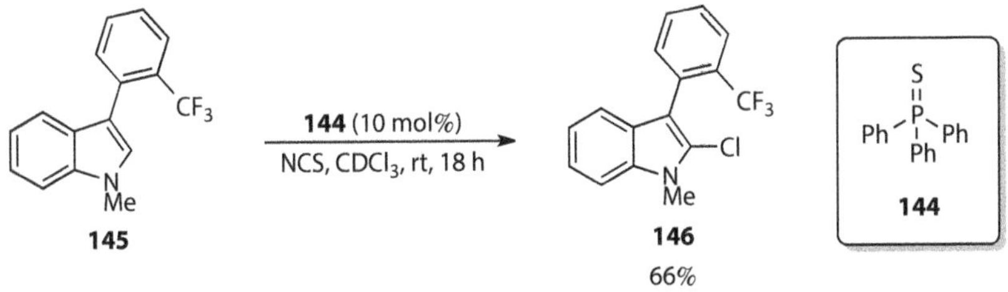

SCHEME 5.35 Triphenylphosphine sulfide-catalyzed chlorination of heterocycle **145**.

as O and Se. Preceding this report, triphenylphosphine sulfide (**144**) has been used by Denmark and coworker as a Lewis base catalyst of halocyclization reactions.[32]

In Yeung's work, the same catalyst promoted the bromocyclization of 1,1- and 1,2-substituted cyclopropylmethyl amides in the presence of 1,3-dibromo-5,5-dimethylhydantoin (DBH) as the bromine source (Scheme 5.36).[83] The cyclization gave the five-membered oxazoline and six-membered oxazine products, respectively.

In addition to the thiocarbamate-catalyzed enantioselective bromoaminocyclizations[55,60–61] (Scheme 5.22), a related haloaminocyclization was also described in a report by Wirth and coworkers.[84] Using thiohydantoin catalyst **149** and *N*-iodosuccinimide (NIS) as the iodinating agent, olefinic amines **150** could undergo cyclization to give *N*-heterocyclic products **151** with moderate to good *ee*s (Scheme 5.37). The choice of additive appeared to control the *exo/endo* cyclization selectivity for some substrates.

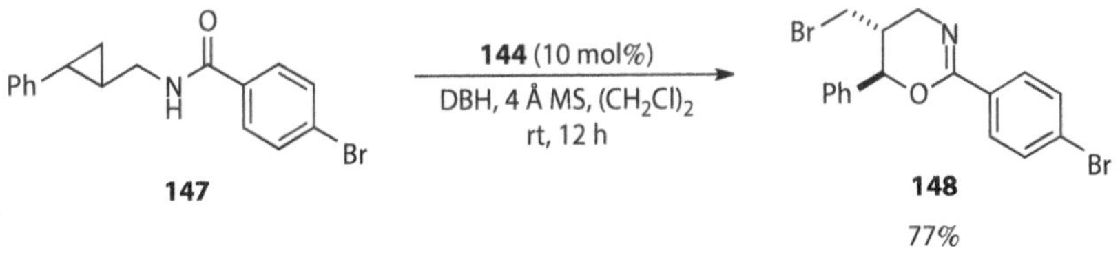

SCHEME 5.36 Bromocyclization of 1,2-substituted cyclopropylmethyl amide **147**.

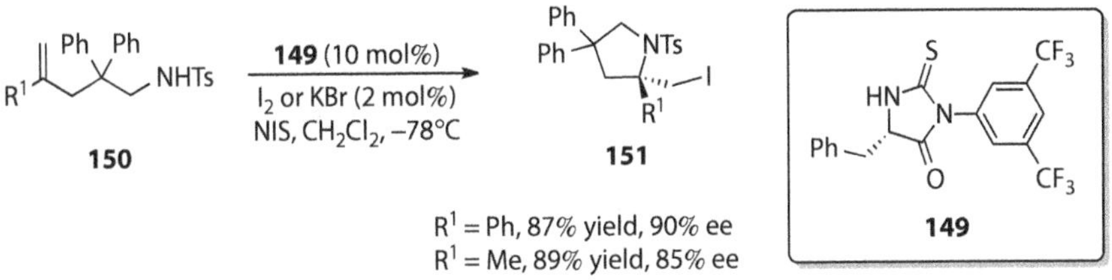

SCHEME 5.37 Enantioselective iodoamination of olefinic amines **150** using thiohydantoin catalyst **149**.

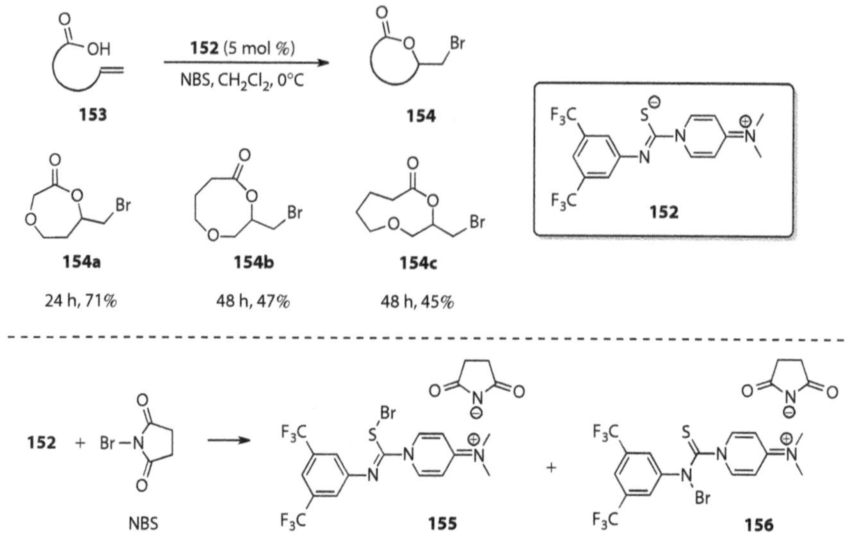

SCHEME 5.38 Medium ring size bromolactonization catalyzed by sulfur-based zwitterion **152**.

An intriguing sulfur-based zwitterionic organocatalyst **152** was employed by Yeung and coworkers in the bromolactonization of long-chain olefinic acids **153** to form medium ring size (seven- to nine-membered) lactones **154** (Scheme 5.38).[85] The catalyst could be formed by a one-step coupling reaction between DMAP and 3,5-bis(trifluoromethyl)phenyl isothiocyanate. Bond length analysis of the X-ray crystal structure of **152** revealed a strong C–S single bond character, indicating that the negative charge primarily resides on the sulfur. Through low-temperature ^{1}H NMR studies, the authors found that **152** existed in equilibrium with its individual components in solution phase and that the equilibrium shifted in response to a temperature decrease to favor the formation of more zwitterion. A 1:1 mixture of **152** and the halogen source NBS gave rise to the S–Br and N–Br complexes **155** and **156**, which are believed to be the active brominating species in the reaction. It is noteworthy that a similar zwitterion could also catalyze the transesterification reaction as reported by Ishihara and coworkers in 2008.[86]

Small molecule thiolates have been used as organocatalysts in the Tishchenko reaction reported by Connon and coworkers.[87] By pairing substoichiometric amounts of thiolate **157** and phenylmagnesium bromide, a Tishchenko homocoupling process could be effected between molecules of aldehyde **159** to afford ester **160** (Scheme 5.39). A cross-Tishchenko coupling, such as that between benzaldehyde (**161**) and ketone **162**, was also possible, although this reaction demanded the use of the more electron-deficient catalyst **158**. In the proposed reaction mechanism, hemithioacetal anion **164** is first generated from the reacting components. Hydride transfer to another carbonyl produces magnesium alkoxide **165** and thioester **166**, which then undergoes the rate-determining coupling process to form the ester product. Notably,

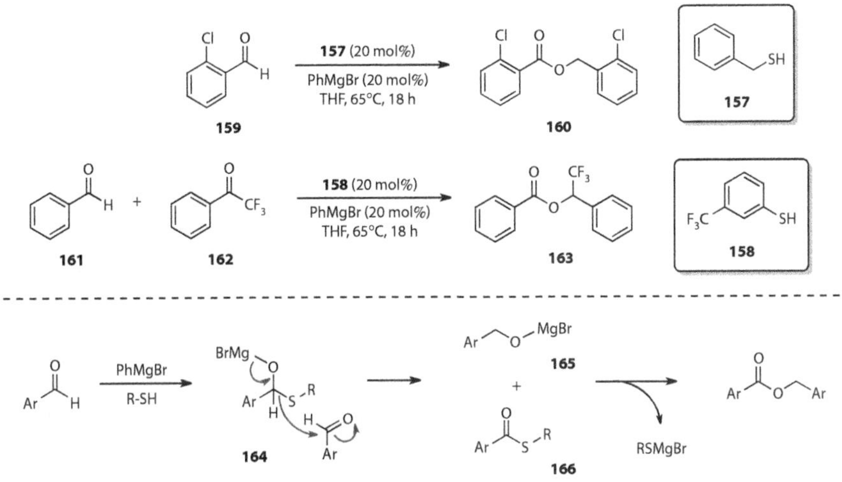

SCHEME 5.39 Thiolate-catalyzed Tishchenko coupling reactions.

later studies by the group included the use of microwave irradiation to shorten the reaction time[88] and also the development of a diselenide/Bu_2Mg-catalyzed variant that can be conducted at room temperature.[89]

Although the vast majority of sulfur catalysis described herein relies on its Lewis basic character, a unique thiyl-based catalyst system was reported by Maruoka and coworkers.[90] Under photolytic conditions, precursor thiol **168** could generate the sulfur radicals. Although binaphthyl-based catalysts such as **167** initially gave modest enantioselectivities, the authors eventually favored catalyst **168** as more bulky substituents could be introduced. This steric bulk around the sulfur appeared to be crucial to induce enantioselectivity. In this thiyl catalytic process shown in Scheme 5.40, vinyl cyclopropanes **169** could react with vinyl ether **170** to furnish substituted cyclopentanes **171** with generally moderate to excellent diastereo- and enantioselectivities. Other studies pertaining to the use of thiol as a catalyst[91] or cocatalyst[92] in radical-type reactions have also been documented.

5.4 CONCLUSION

Building on pioneering works, a great diversity of reactions catalyzed by sulfur and selenium organocatalysts has been developed in these few years; each comes with its own unique story to tell. It is reflective of a growing interest that is undertaken by chemists not just in organosulfur and organoselenium catalysis, but also of the broader concept of organocatalysis in general. With the sustained level of interest and research effort, one can certainly foresee the emergence of many more novel concepts of catalysis in this field in the near future and beyond.

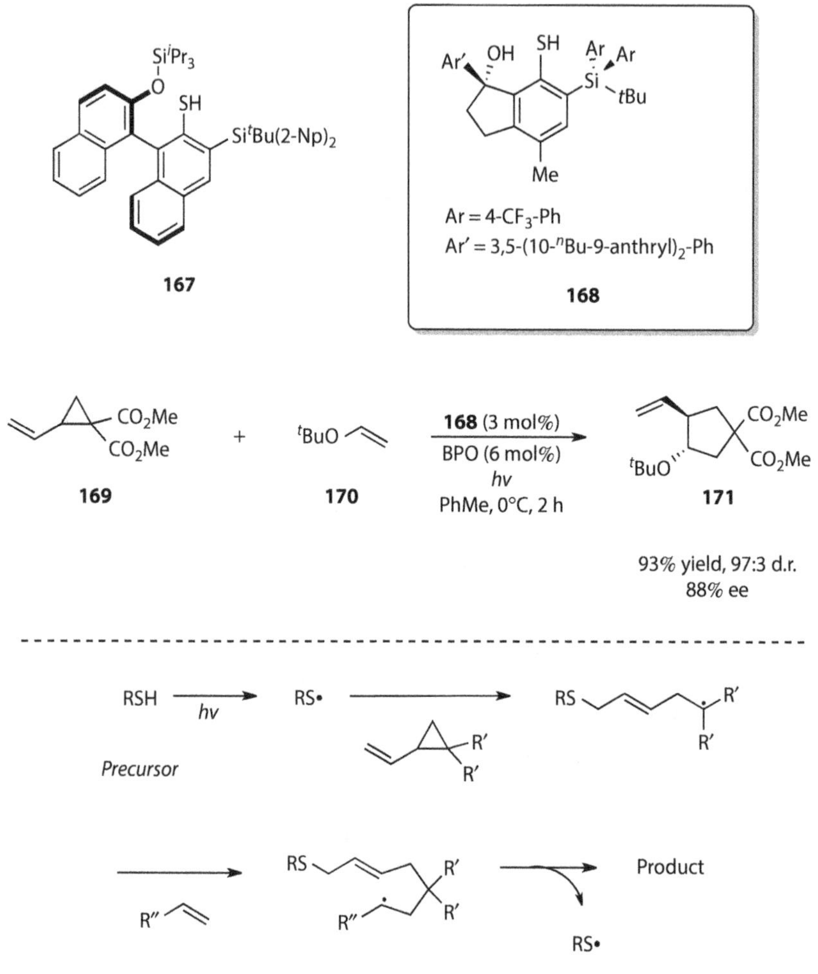

SCHEME 5.40 Enantioselective cyclization catalyzed by a thiyl radical.

REFERENCES

1. Devillanova, F., Du Mont, W.-W. (Eds.). *Handbook of Chalcogen Chemistry: New Perspectives in Sulfur, Selenium and Tellurium*, 2nd ed. RSC Publishing: Cambridge, UK, 2013.
2. Cremlyn, R. J. *An Introduction to Organosulfur Chemistry*. Wiley: Chichester, UK, 1996.
3. Mukherjee, A. J., Zade, S. S., Singh, H. B., Sunoj, R. B. *Chem. Rev.* **2010**, *110*, 4357–4416.
4. McGarrigle, E. M., Myers, E. L., Illa, O., Shaw, M. A., Riches, S. L., Aggarwal, V. K. *Chem. Rev.* **2007**, *107*, 5841–5883.
5. Breder, A., Ortgies, S. *Tetrahedron Lett.* **2015**, *56*, 2843–2852.
6. Arrayás, R. G., Carretero, J. C. *Chem. Commun.* **2011**, *47*, 2207–2211.
7. Santi, C., Santoro, S., Battistelli, B. *Curr. Org. Chem.* **2010**, *14*, 2442–2462.

8. Freudendahl, D. M., Shahzad, S. A., Wirth, T. *Eur. J. Org. Chem.* **2009**, *11*, 1649–1664.
9. Singh, F. V., Wirth, T. *Organoselenium Chemistry: Synthesis and Reactions*. Wirth, T. (Ed.). Wiley-VCH: Weinheim, Germany, 2012, pp. 321–360.
10. Freudendahl, D. M., Santoro, S., Shahzad, S. A., Santi, C., Wirth, T. *Angew. Chem. Int. Ed.* **2009**, *48*, 8409–8411.
11. Tay, D. W., Tsoi, I. T., Er, J. C., Leung, G. Y. C., Yeung, Y.-Y. *Org. Lett.* **2013**, *15*, 1310–1313.
12. Yeung, Y.-Y., Gao, X., Corey, E. J. *J. Am. Chem. Soc.* **2006**, *128*, 9644–9645.
13. Luo, J., Zhu, Z., Liu, Y., Zhao, X. *Org. Lett.* **2015**, *17*, 3620–3623.
14. Watanabe, S., Hasebe, R., Ouchi, J., Nagasawa, H., Kataoka, T. *Tetrahedron Lett.* **2010**, *51*, 5778–5780.
15. Trenner, J., Depken, C., Weber, T., Breder, A. *Angew. Chem. Int. Ed.* **2013**, *52*, 8952–8956.
16. Krätzschmar, F., Kaβel, M., Delony, D., Breder, A. *Chem. Eur. J.* **2015**, *21*, 7030–7034.
17. Deng, Z., Wei, J., Liao, L., Huang, H., Zhao, X. *Org. Lett.* **2015**, *17*, 1834–1837.
18. Zhang, X., Guo, R., Zhao, X. *Org. Chem. Front.* **2015**, 2, 1334–1337.
19. Yu, L., Wu, Y., Cao, H, Zhang, X., Shi, X., Luan, J., Chen, T., Pan, Y., Xu, Q. *Green Chem.* **2014**, *16*, 287–293.
20. Zhang, Y., Ye, J., Yu, L., Shi, X., Zhang, M., Xu, Q., Lautens, M. *Adv. Synth. Catal.* **2015**, *357*, 955–960.
21. Yu, L., Ye, J., Zhang, X., Ding, Y., Xu, Q. *Catal. Sci. Tech.* **2015**, *5*, 4830–4838.
22. Yu, L., Li, H., Zhang, X., Ye, J., Liu, J., Xu, Q., Lautens, M. *Org. Lett.* **2014**, *16*, 1346–1349.
23. Santoro, S., Battistelli, B., Gjoka, B., Si, C.-W. S., Testaferri, L., Tiecco, M., Santi, C. *Synlett* **2010**, *9*, 1402–1406.
24. Sancineto, L., Tidei, C., Bagnoli, L., Marini, F., Lenardão, E. J., Santi, C. *Molecules* **2015**, *20*, 10496–10510.
25. Alberto, E. E., Braga, A. L., Detty, M. R. *Tetrahedron* **2012**, *68*, 10476–10481.
26. Shahzad, S. A., Venin, C., Wirth, T. *Eur. J. Org. Chem.* **2010**, *18*, 3465–3472.
27. Singh, F. V., Wirth, T. *Org. Lett.* **2011**, *13*, 6504–6507.
28. Santi, C., Lorenzo, R. D., Tidei, C., Bagnoli, L., Wirth, T. *Tetrahedron* **2012**, *68*, 10530–10535.
29. Yu, L., Wang, J., Chen, T., Wang, Y., Xu, Q. *Appl. Organomet. Chem.* **2015**, *28*, 652–656.
30. Cresswell, A. J., Eey, S. T.-C., Denmark, S. E. *Nat. Chem.* **2015**, *7*, 146–152.
31. Denmark, S. E., Collins, W. R. *Org. Lett.* **2007**, *9*, 3801–3804.
32. Denmark, S. E., Burk, M. T. *Proc. Natl. Acad. Sci. U.S.A.* **2010**, *107*, 20655–20660.
33. Denmark, S. E., Collins, W. R., Cullen, M. D. *J. Am. Chem. Soc.* **2009**, *131*, 3490–3492.
34. Denmark, S. E., Vogler, T. *Chem. Eur. J.* **2009**, *15*, 11737–11745.
35. Denmark, S. E., Kornfilt, D. J. P., Vogler, T. *J. Am. Chem. Soc.* **2011**, *133*, 15308–15311.
36. Denmark, S. E., Hartmann, E., Komfilt, D. J. P., Wang, H. *Nat. Chem.* **2014**, *6*, 1056–1064.
37. Denmark, S. E., Jaunet, A. *J. Am. Chem. Soc.* **2013**, *135*, 6419–6422.
38. Denmark, S. E., Jaunet, A. *J. Org. Chem.* **2014**, *79*, 140–171.
39. Denmark, S. E., Chi, H. M. *J. Am. Chem. Soc.* **2014**, *136*, 3655–3663.
40. Denmark, S. E., Chi, H. M. *J. Am. Chem. Soc.* **2014**, *136*, 8915–8918.
41. Denmark, S. E., Rossi, S., Webster, M. P., Wang, H. *J. Am. Chem. Soc.* **2014**, *136*, 13016–13028.
42. Balkrishna, S. J., Prasad, C. D., Panini, P., Detty, M. R., Chopra, D., Kumar, S. *J. Org. Chem.* **2012**, *77*, 9541–9552.
43. Balkrishna, S. J., Kumar, S., Azad, G. K., Bhakuni, B. S., Panini, P., Ahalawat, N., Tomar, R. S., Detty, M. R., Kumar, S. *Org. Biomol. Chem.* **2014**, *12*, 1215–1219.

44. Chen, F., Tan, C. K., Yeung, Y.-Y. *J. Am. Chem. Soc.* **2013**, *135*, 1232–1235.
45. Clark, J. S., Boyer, A., Aimon, A., García, P. E., Lindsay, D. M., Symington, A. D. F., Danoy, Y. *Angew. Chem. Int. Ed.* **2012**, *51*, 12128–12131.
46. Wu, H.-Y., Chang, C.-W., Chien, R.-J. *J. Org. Chem.* **2013**, *78*, 5788–5793.
47. Huang, M.-T., Wu, H.-Y., Chien, R.-J. *Chem. Commun.* **2014**, *50*, 1101–1103.
48. Wang, S.-H., Chien, R.-J. *Tetrahedron.* **2016**, *72*, 2607–2615. doi:10.1016/j.tet.2014.12.063.
49. Kavanagh, S. A., Piccinini, A., Connon, S. J. *Adv. Synth. Catal.* **2010**, *352*, 2089–2093.
50. Aggarwal V. K., Winn, C. L. *Acc. Chem. Res.* **2004**, *37*, 611–620.
51. Davis, R. L., Stiller, J., Naicker, T., Jiang, H., Jørgensen, K. A. *Angew. Chem. Int. Ed.* **2014**, *53*, 7406–7426.
52. Ke, Z., Tan, C. K., Chen, F., Yeung, Y.-Y. *J. Am. Chem. Soc.* **2014**, *136*, 5627–5630.
53. Ke, Z., Tan, C. K., Liu, Y., Lee, K. G. Z., Yeung, Y.-Y. *Tetrahedron.* **2016**, *72*, 2683–2689. doi:10.1016/j.tet.2015.09.016.
54. Zhou, L., Tan, C. K., Jiang, X., Chen, F., Yeung, Y.-Y. *J. Am. Chem. Soc.* **2010**, *132*, 15474–15476.
55. Zhou, L., Chen, J., Tan, C. K., Yeung Y.-Y. *J. Am. Chem. Soc.* **2011**, *133*, 9164–9167.
56. Tan, C. K., Zhou, L., Yeung, Y.-Y. *Org. Lett.* **2011**, *13*, 2738–2741.
57. Tan, C. K., Le, C., Yeung, Y.-Y. *Chem. Commun.* **2012**, *48*, 5793–5795.
58. Tan, C. K., Er, J. C., Yeung, Y.-Y. *Tetrahedron Lett.* **2014**, *55*, 1243–1246.
59. Chen, J., Zhou, L., Tan, C. K., Yeung, Y.-Y. *J. Org. Chem.* **2012**, *77*, 999–1009.
60. Chen, J., Zhou, L., Yeung, Y.-Y. *Org. Biomol. Chem.* **2012**, *10*, 3808–3811.
61. Zhou, L., Tay, D. W., Chen, J., Leung, G. Y. C., Yeung, Y.-Y. *Chem. Commun.* **2013**, *49*, 4412–4414.
62. Zhao, Y., Jiang, X., Yeung, Y.-Y. *Angew. Chem. Int. Ed.* **2013**, *52*, 8597–8601.
63. Tay, D. W., Leung, G. Y. C., Yeung, Y.-Y. *Angew. Chem. Int. Ed.* **2014**, *53*, 5161–5164.
64. Cheng, Y. A., Yu, W. Z., Yeung, Y.-Y. *Angew. Chem. Int. Ed.* **2015**, *54*, 12102–12106.
65. Han, W.-Y., Li, S.-W., Wu, Z.-J., Zhang, X.-M., Yuan, W.-C. *Chem. Eur. J.* **2013**, *19*, 5551–5556.
66. Zhao, J.-Q., Zhou, M.-Q., Wu, Z.-J., Wang, Z.-H., Yue, D.-F., Xu, X.-Y., Zhang, X.-M., Yuan, W.-C. *Org. Lett.* **2015**, *17*, 2238–2241.
67. Jiang, X., Tan, C. K., Zhou, L., Yeung, Y.-Y. *Angew. Chem. Int. Ed.* **2012**, *51*, 7771–7775.
68. Robak, M. T., Trincado, M., Ellman, J. A. *J. Am. Chem. Soc.* **2007**, *129*, 15110–15111.
69. Kimmel, K. L., Weaver, J. D., Ellman, J. A. *Chem. Sci.* **2012**, *3*, 121–125.
70. Kimmel, K. L., Weaver, J. D., Lee, M., Ellman, J. A. *J. Am. Chem. Soc.* **2012**, *134*, 9058–9061.
71. Robak, M. T., Herbage, M. A., Ellman, J. A. *Tetrahedron* **2011**, *67*, 4412–4416.
72. Kimmel, K. L., Robak, M. T., Thomas, S., Lee, M., Ellman, J. A. *Tetrahedron* **2012**, *68*, 2704–2712.
73. Prasad, K. R., Revu, O. *Tetrahedron* **2013**, *69*, 8422–8428.
74. Xu, H., Zuend, S. J., Woll, M. G., Tao, Y., Jacobsen, E. N. *Science* **2010**, *327*, 986–990.
75. Beck, E. M., Hyde, A. M., Jacobsen, E. N. *Org. Lett.* **2011**, *13*, 4260–4263.
76. Pei, D., Wang, Z., Wei, S., Zhang, Y., Sun, J. *Org. Lett.* **2006**, *8*, 5913–5915.
77. Wu, X., Li, Y., Wang, C., Zhou, L., Lu, X., Sun, J. *Chem. Eur. J.* **2011**, *17*, 2846–2848.
78. Liu, X.-W., Yan, Y., Wang, Y.-Q., Wang, C., Sun, J. *Chem. Eur. J.* **2012**, *18*, 9204–9207.
79. Liu, X.-W., Wang, C., Yan, Y., Wang, Y.-Q., Sun, J. *J. Org. Chem.* **2013**, *78*, 6276–6280.
80. Saravanan, S., Khan, N. H., Kureshy, R. L., Abdi, S. H. R., Bajaj, H. C. *ACS Catal.* **2013**, *3*, 2873–2880.
81. Kumar, M., Kureshy, R. I., Saravanan, S., Verma, S., Jakhar, A., Khan, N. H., Abdi, S. H. R., Bajaj, H. C. *Org. Lett.* **2014**, *16*, 2798–2801.
82. Maddox, S. M., Nalbandian, C. J., Smith, D. E., Gustafson, J. L. *Org. Lett.* **2015**, *17*, 1042–1045.
83. Wong, Y.-C., Ke, Z., Yeung, Y.-Y. *Org. Lett.* **2015**, *17*, 4944–4947.

84. Mizar, P., Burrelli, A., Günther, E., Söftje, M., Farooq, U., Wirth, T. *Chem. Eur. J.* **2014**, *20*, 13113–13116.
85. Cheng, Y. A., Chen, T., Tan, C. K., Heng, J. J., Yeung, Y.-Y. *J. Am. Chem. Soc.* **2012**, *134*, 16492–16495.
86. Ishihara, K., Niwa, M., Kosugi, Y. *Org. Lett.* **2008**, *11*, 2187–2190.
87. Cronin, L., Manoni, F., O'Connor, C. J., Connon, S. J. *Angew. Chem. Int. Ed.* **2010**, *49*, 3045–3048.
88. O'Connor, C. J., Manoni, F., Curran, S. P., Connon, S. J. *New J. Chem.* **2011**, *35*, 551–553.
89. Curran, S. P., Connon, S. J. *Org. Lett.* **2012**, *14*, 1074–1077.
90. Hashimoto, T., Kawamata, Y., Maruoka, K. *Nat. Chem.* **2014**, *6*, 702–705.
91. Donner, C. D., Casana, M. I. *Tetrahedron Lett.* **2012**, *53*, 1105–1107.
92. Chou, C.-M., Guin, J., Mück-Lichtenfeld, C., Grimme, S., Studer, A. *Chem. Asian J.* **2011**, *6*, 1197–1209.

6 Use of Phosphine Oxides as Catalysts and Precatalysts

Zhiqi Lao and Patrick H. Toy

CONTENTS

6.1 INTRODUCTION

In the field of organic chemistry, phosphine oxides have generally been viewed as undesirable nuisances because they are often formed as reaction by-products or result from undesired oxidation of a phosphine catalyst or ligand. This is especially true in the context of organic synthesis with regard to the Wittig (Equation 6.1) and Mitsunobu (Equation 6.2) reactions, because a stoichiometric amount of a phosphine oxide, typically $Ph_3P{=}O$ (**1**), is produced alongside the desired product in these transformations. Although both of these reactions have found great utility in organic synthesis over the years due to their general reliability and predictable stereoselectivity, their use can sometimes be impractical due to problems associated with product isolation from a complex reaction mixture that contains large amounts of by-products.[1] However, this review aims to cast phosphine oxides in a more favorable light by summarizing their use as catalysts and in cases where they are formed but can be converted *in situ* back into reactive species as precatalysts. The research outlined in the following section is organized around several general applications of phosphine oxides: (a) catalysis of the transformation of carbon–oxygen double bonds into carbon–nitrogen double bonds,

(b) catalysis based on their Lewis basicity, (c) catalysis of metal–halogen exchange reactions, (d) conversion into halogenation reagents, and (e) generation and *in situ* reduction in catalytic Wittig and related reactions.

$$R^1R^2C{=}O + Ph_3P{=}CR^3R^4 \longrightarrow R^1R^2C{=}CR^3R^4 + \underset{\mathbf{1}}{Ph_3P{=}O} \quad (6.1)$$

$$R^1CH(OH)R^2 + Nu{-}H \xrightarrow[Ph_3P \;\rightarrow\; Ph_3P{=}O\ (\mathbf{1})]{EtO_2C{-}N{=}N{-}CO_2Et \;\rightarrow\; EtO_2C{-}NH{-}NH{-}CO_2Et} R^1CH(Nu)R^2 \quad (6.2)$$

6.2 PHOSPHINE OXIDES AS CATALYSTS FOR TRANSFORMATION OF C=O BONDS INTO C=N BONDS

6.2.1 Catalytic Carbodiimide Synthesis

A very early example of the use of a phosphine oxide as a catalyst in organic synthesis was the observation that such compounds could be used to convert isocyanates into carbodiimides (Figure 6.1).[2] It was reported that **2a** was effective in catalytically transforming 4-nitrophenyl isocyanate (**3**) into the corresponding carbodiimide **4** with concomitant formation of carbon dioxide in quantitative yield, and that diisocyanate **5** could be converted into polycarbodiimide **6**. Phosphine oxide **2a** was described as being a more active catalyst in these reactions than the more readily available **2b**. Observed evidence mechanistically supported the notion that these reactions precede via a phosphinimide intermediate that reacts as a nucleophile with a second molecule of the starting electrophilic isocyanate.[3–6] At around this time, it was reported that **2a** could also be used catalytically in an aza-Wittig reaction between **5** and an aldehyde (**7**, R = Ph) to form diimine **8**.[7] In this reaction, the phosphinimide formed from **5** reacted with the aldehyde group of **7**, rather than an isocyanate group. In addition, when a related reaction of benzoyl isocyanate (**9**) was studied, 1,3,5-oxadiazine **10** was the major product formed.[8]

6.2.2 Catalytic Aza-Wittig Reactions

Marsden and coworkers, subsequently, reported intramolecular versions of the catalytic aza-Wittig reaction for heteroaromatic compound synthesis using commercially available **2c**.[9] In this work, biaryl isocyanates **11** could be converted into phenanthridines **12**, and aryl isocyanates **13** could be transformed into benzoxazoles **14** directly in refluxing toluene together with the simultaneous release of carbon dioxide (Figure 6.2). These reactions, presumably, proceed via phosphinimide intermediates that react intramolecularly with the carbonyl groups to form the obtained cyclic products.

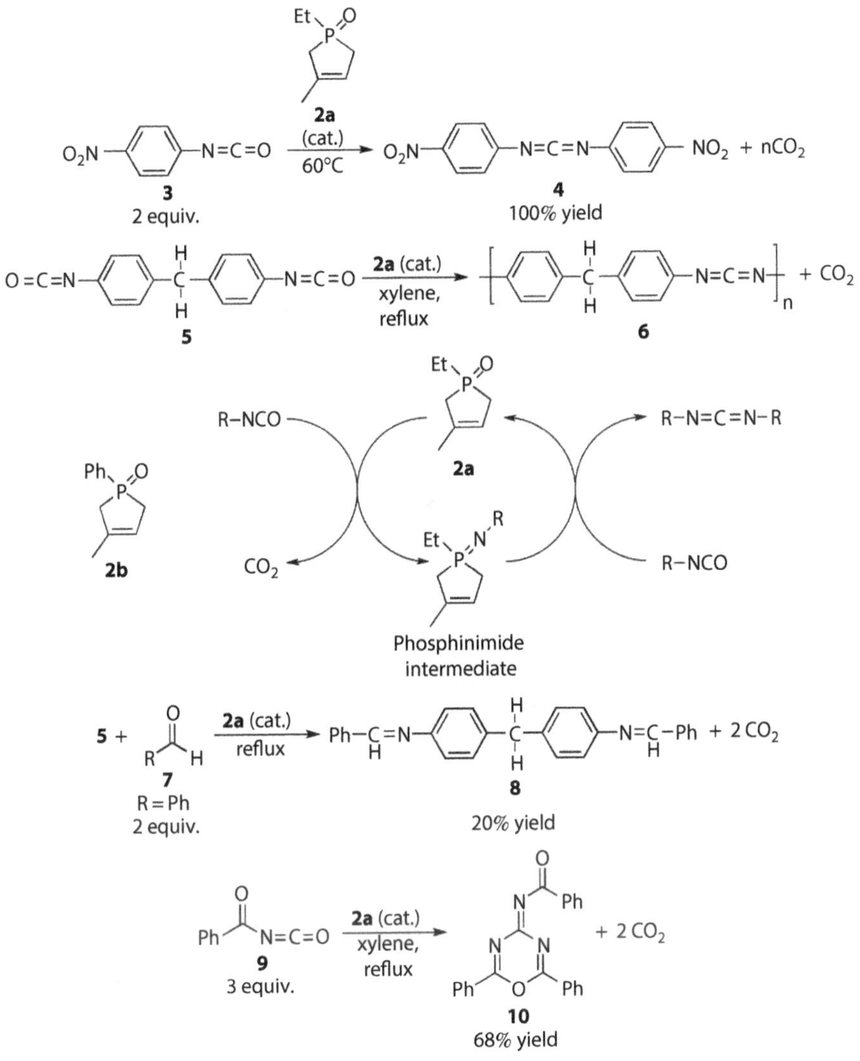

FIGURE 6.1 Phosphine oxide-catalyzed carbodiimide synthesis and related reactions.

Ding et al. most recently reported the application of similar reactions for the conversion of carboxylic acid derivatives **15** into 1,4-benzodiazepin-5-ones **16** that were catalyzed by **2c** (Figure 6.3).[10] The overall transformations proceed via acyl azide intermediates **17** that are not purified but instead used directly in thermal Curtis rearrangement reactions that afford isocyanates **18**. These were, in turn, treated *in situ* with catalytic **2c** to afford the final products **16**. It was proposed that intermediate **19** were formed and cyclized to produce the final product **16**. This work is somewhat similar to their prior research in which traditional intramolecular aza-Wittig reactions were used to convert aryl azides **20** into 4(*3H*)-quinazolinones **21** via phosphorimine intermediates **22** (Figure 6.4).[11] These

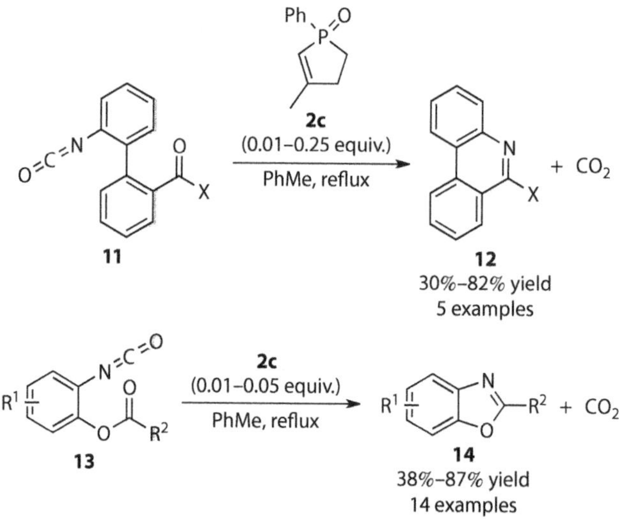

FIGURE 6.2 Phosphine oxide-catalyzed intramolecular aza-Wittig reactions.

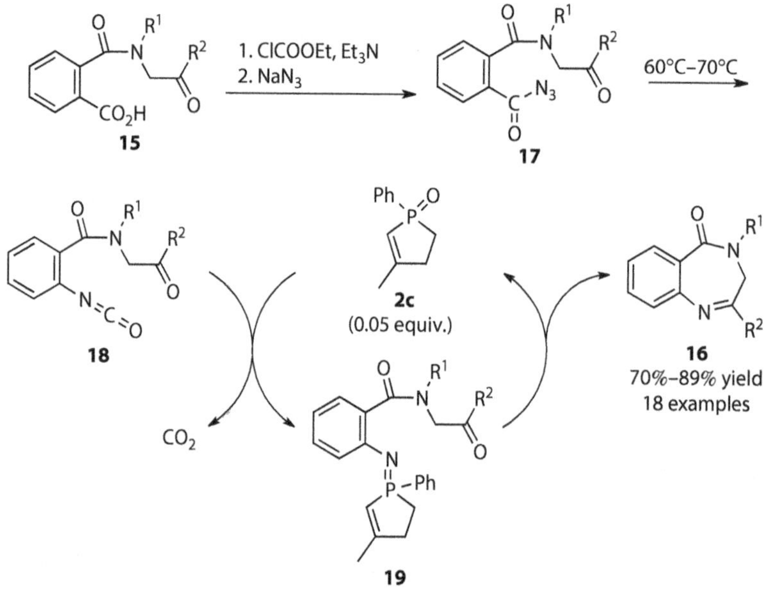

FIGURE 6.3 Synthesis of 1,4-benzodiazepin-5-ones catalyzed by **2c**.

reactions were performed using a substoichiometric amount of Ph_3P, and the by-product **1** was reduced *in situ* by the combination of $Ti(O\text{-}iPr)_4$ and tetramethyldisiloxane at high temperature in order to regenerate Ph_3P for further reaction cycles. Thus, in this initial research, the actual catalyst was Ph_3P and **1** was formally a precatalyst. Additional examples of this strategy will be discussed in Section 6.6.

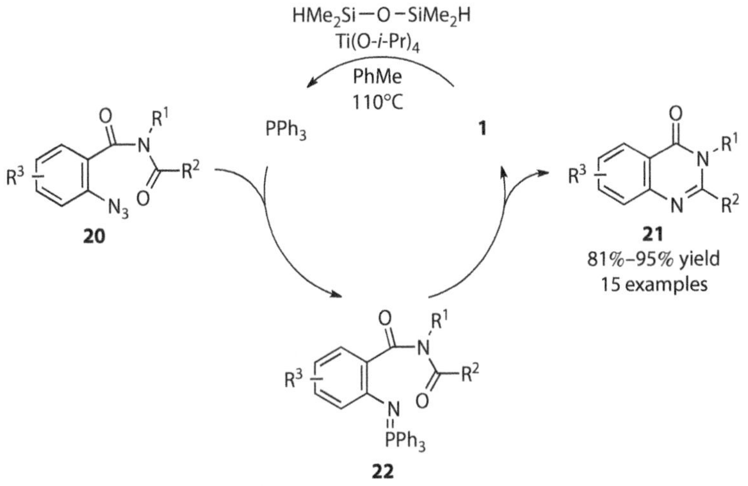

HMe$_2$Si–O–SiMe$_2$H
Ti(O-*i*-Pr)$_4$
PhMe
110°C
PPh$_3$
1
20
21
81%–95% yield
15 examples
22

FIGURE 6.4 Aza-Wittig reactions using catalytic phosphine.

6.3 PHOSPHINE OXIDES AS LEWIS BASE CATALYSTS

6.3.1 Catalytic Nucleophilic Addition Reactions to Aldehydes, Imines, Ketones, and Epoxides

Early research regarding nucleophilic Lewis base catalysis of aldehyde allylation reactions using allyltrichlorosilanes identified phosphoramides as good catalysts,[12] and based on this observation, phosphine oxides have also been examined in this context. An early, if not the first, example of such an application of a phosphine oxide as a nucleophilic Lewis base catalyst was reported by Berrisford and coworkers when they described the use of **1** with Bu_4NI in an aldehyde (**7**, R = Ph) allylation reaction to form homoallylic alcohol **23** (Equation 6.3).[13] In this reaction, **1** was observed to be a more effective catalyst than *N*,*N*-dimethylformamide (DMF), which required a full equivalent to be added for the reactions to go to completion under similar reaction conditions.

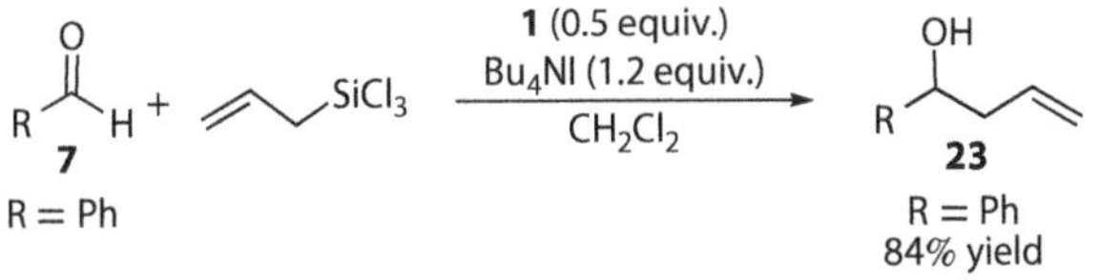

7 (R = Ph) + allyl-SiCl$_3$ $\xrightarrow[\text{CH}_2\text{Cl}_2]{\textbf{1}\text{ (0.5 equiv.)},\ \text{Bu}_4\text{NI (1.2 equiv.)}}$ **23** (R = Ph, 84% yield) (6.3)

Similarly, Kobayashi and coworkers reported allylation reactions of *N*-acylhydrazones (**24**) that are catalyzed by phosphine oxides.[14] In their studies, they examined a wide range of phosphine oxides, including **1** and tethered bisphosphine oxide **25**, and referred to their role in these reactions as *neutral coordinate-organocatalysts* (Figure 6.5). Although **25** was found to be the most efficient catalyst for the synthesis of homoallylic hydrazines **26** in these reactions, a full equivalent of it was used so

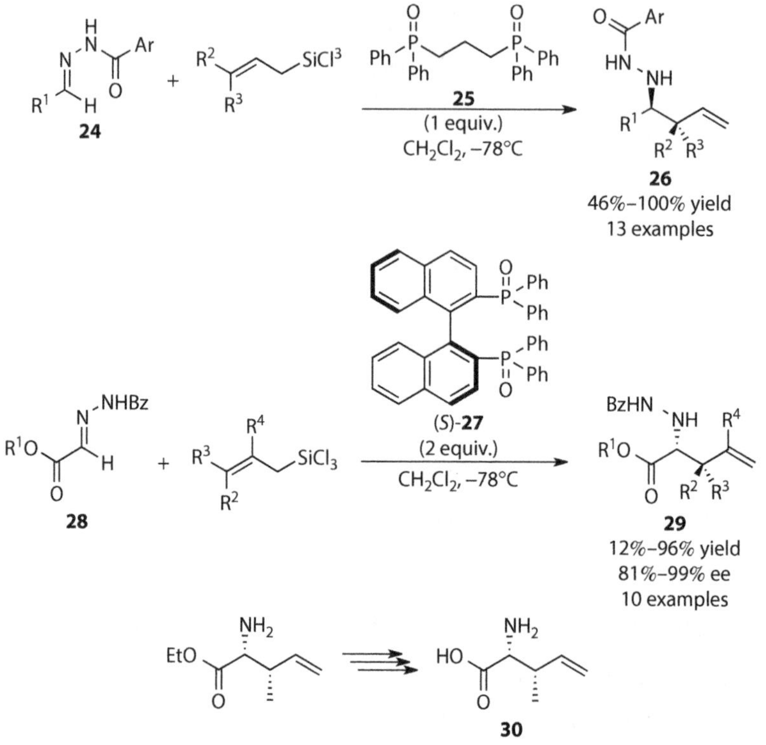

FIGURE 6.5 Allylation reactions catalyzed by **25** and **27**.

that the reactions proceeded efficiently. They subsequently reported the first application of a chiral phosphine oxide as a catalyst for enantioselective reactions when they described the use of chiral BINAP dioxide (*S*)-**27**, readily prepared by H_2O_2 oxidation of the corresponding BINAP, in a related asymmetric transformation.[15] In this asymmetric reaction, α-hydrazino esters **28** were used as the substrates, resulting in the formation of highly enantiomerically enriched homoallylic amines **29**. Although two full equivalents of (*S*)-**27** were necessary for high reaction efficiency, the utility of this enantioselective reaction was demonstrated by the synthesis of D-alloisoleucine (**30**) from an allylation reaction product. It should be noted that the early research published before 2010 regarding asymmetric catalysis using chiral phosphine oxides has been comprehensively reviewed by Benaglia and Rossi.[16]

Shortly after this latter report by Kobayashi's group appeared, Nakajima and coworkers published their findings regarding the use of (*S*)-**27** as a catalyst in asymmetric reactions. In a pair of early reports, they described asymmetric ring-opening reactions of *meso*-epoxides **31** with tetrachlorosilane to form enantiomerically enriched 1,2-chlorohydrins **32** (Figure 6.6).[17,18] Significantly, these reactions required only 0.1 equivalents of catalyst (*S*)-**27**, and presumably this reacted first with $SiCl_4$ to generate an active Lewis acid species. In addition, they reported that the *meso*-epoxides **31** could be generated *in situ*, and thus 1,2-chlorohydrins **32** could be synthesized in a one-pot procedure using the corresponding alkenes as

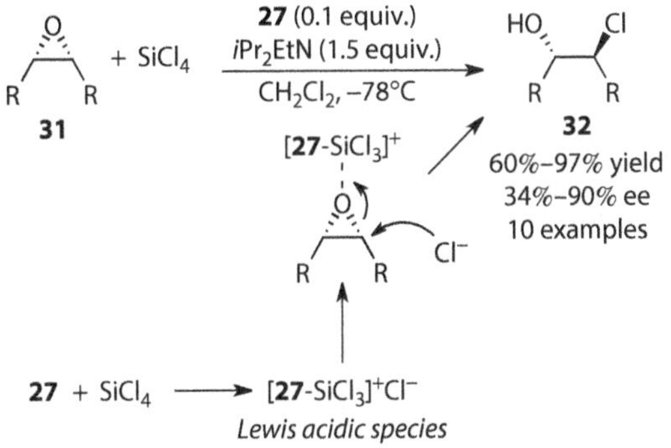

FIGURE 6.6 Enantioselective ring opening of *meso*-epoxides catalyzed by (*S*)-**27**.

starting materials, with similar enantioselectivities for the ring-opening process. For a comprehensive review of organic reactions mediated by tetrachlorosilane, see the recent report by Benaglia and coworkers.[19]

They also studied allylation reactions of aldehydes **7** with methallyltrichlorosilane to form homoallylic alcohols **33** (Figure 6.7).[20,21] These reactions also required only 0.1 equivalents of catalyst (*S*)-**27**. Good yields and enantioselectivities were obtained in these reactions when aryl aldehydes were used as the reaction substrate (53%–79% ee), but alkyl aldehydes and an enal did not react with much stereoselectivity (5%–32% ee). Enantioselective Abramov-type reactions were also reported to be catalyzed by (*S*)-**27**.[22] In these reactions, aldehydes **7** were reacted with $P(OEt)_3$ to afford α-hydroxyphosphonates **34** with modest stereoselectivities. ^{31}P-NMR analysis was used to study the mechanism of these reactions, and it supported the formation

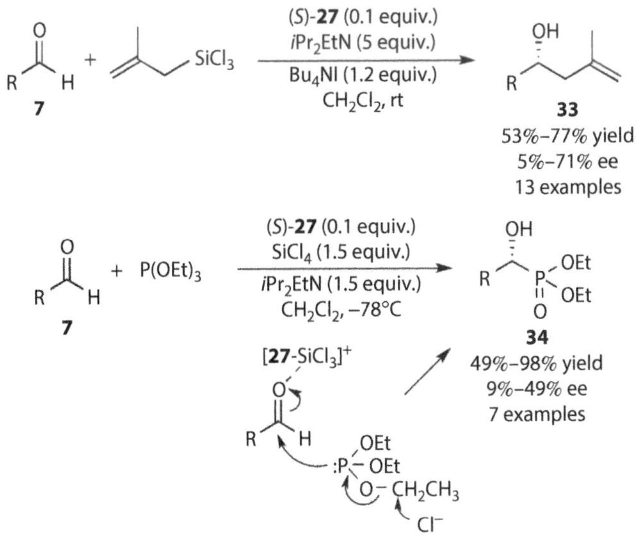

FIGURE 6.7 Enantioselective allylation and Abramov-type reactions catalyzed by (*S*)-**27**.

of an adduct between (*S*)-**27** and $SiCl_4$, as in the epoxide-opening reactions described earlier, that activates the aldehyde **7** substrate toward nucleophilic attack by the phosphorous atom.

Nakajima's group then focused its attention toward the use of (*S*)-**27** to catalyze various enantioselective aldol reactions, and they have recently summarized their research in this area in a review[23] and have written an account that describe their research on enantioselective double aldol reactions.[24] Initially, they examined the reactions of trichlorosilyl enol ethers with aldehydes (Figure 6.8).[21,25] A range of aldehydes **7** were reacted with cyclic trichlorosilyl enol ether **35** to afford mixtures of *anti*-**36** and *syn*-**36**. Product yields were generally very high, except for when an alkyl aldehyde was used. However, in all cases, high *anti:syn* selectivity was observed (6–25:1), and *anti*-**36** was obtained with a high level of enantioselectivity (55%–96% ee). Related reactions using a silyl ketene acetal were also studied.[26] In these reactions, trimethylsilyl ketene acetal **37** was used instead of **35**, and it was proposed that the role of (*S*)-**27** in these reactions was to activate tetrachlorosilane in order to generate the actual hyperactive Lewis acid catalyst, as in the epoxide-opening and Abramov-type reactions discussed earlier. High yields of products **38** were obtained with moderate levels of enantioselectivity.

Nakajima's group then moved on to study the use of (*S*)-**27** to catalyze enantioselective direct aldol reactions in which trichlorosilyl enol ethers were generated *in situ*. For example, reactions of aldehydes **7** with cyclohexanone derivative **39** in the presence of catalytic (*S*)-**27**, $SiCl_4$, and *i*Pr_2EtN led to selective formation of products *anti*-**40** with moderate enantioselectivity (Figure 6.9).[27] Aryl aldehydes afforded high yields and stereoselectivities in these reactions, but the lowest yield and stereoselectivity were observed when cinnamaldehyde was the substrate. In addition, reactions between two aldehydes, with one being branched **41**, were also reported. Diols **42** were isolated in high yields with moderate enantiomeric excesses after direct reduction of the initial aldol reaction product with $NaBH_4$. They subsequently observed

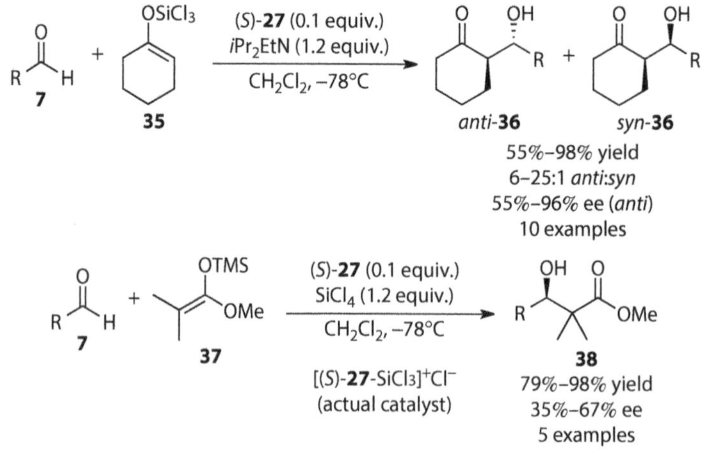

FIGURE 6.8 Enantioselective aldol reactions of aldehydes catalyzed by (*S*)-**27**.

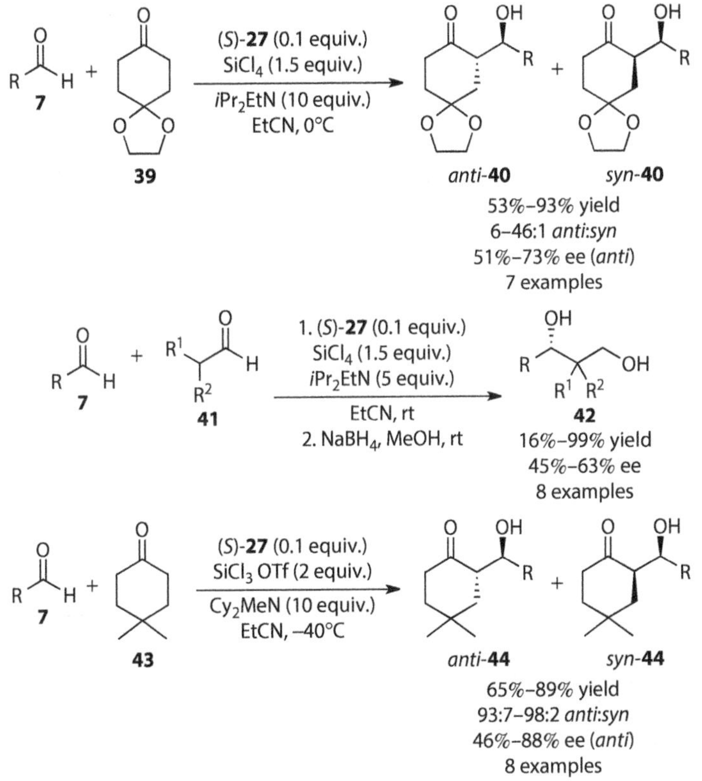

FIGURE 6.9 Enantioselective direct aldol reactions with aldehyde acceptors catalyzed by (*S*)-**27**.

that better results in terms of stereoselectivity could be obtained when $SiCl_4$ was replaced by $SiCl_3OTf$, and when the more bulky base Cy_2MeN was used instead of $i Pr_2EtN$.[28] In these reactions, dimethyl-substituted cyclohexanone **43** was used as the aldol donor to afford mixtures of products *anti*-**44** and *syn*-**44**.

Examples of enantioselective double direct aldol reactions between a ketone and an aldehyde were also reported (Figure 6.10).[29] For example, two equivalents of aldehydes **7** and one equivalent of methyl ketone **45** afforded high yields of mixtures of diastereomers *chiro*-**46** and *meso*-**46** with generally good enantioselectivity. In these reactions, $SiCl_4$ was used instead of $SiCl_3OTf$. Finally, unsymmetrical ketones **47** underwent double aldol reactions with aldehydes **7** to generate linear products **48** with two (when R^1 = H) or three stereocenters.[30] Interestingly, only two of the possible diastereomers were formed, with the major product **48** having the relative configuration determined by X-ray crystallographic analysis to be as indicated.

Although the aforementioned direct aldol reactions involved the use of aldehyde acceptors/electrophiles that react with ketone or aldehyde donors/nucleophiles, Nakajima's group has also studied enantioselective direct aldol reactions in which

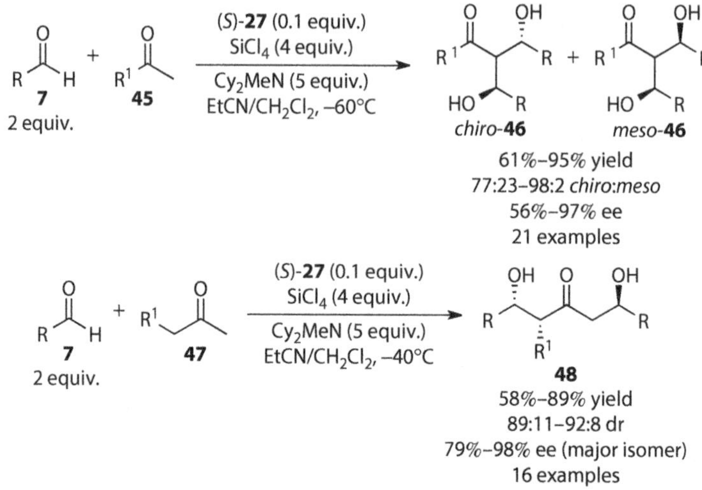

7 (2 equiv.) + 45 → chiro-46 + meso-46

(S)-**27** (0.1 equiv.)
$SiCl_4$ (4 equiv.)
Cy_2MeN (5 equiv.)
$EtCN/CH_2Cl_2$, −60°C

61%–95% yield
77:23–98:2 *chiro:meso*
56%–97% ee
21 examples

7 (2 equiv.) + 47 → 48

(S)-**27** (0.1 equiv.)
$SiCl_4$ (4 equiv.)
Cy_2MeN (5 equiv.)
$EtCN/CH_2Cl_2$, −40°C

58%–89% yield
89:11–92:8 dr
79%–98% ee (major isomer)
16 examples

FIGURE 6.10 Enantioselective direct double aldol reactions catalyzed by (*S*)-**27**.

a ketone group serves as the acceptor/electrophile catalyzed by (*S*)-**27** (Figure 6.11). Their first reported examples of such reactions involved the use of methyl ketones **45** and cyclohexanone (**49**) as the starting materials to afford products **50** in high yields with moderate diastereoselectivities and high enantioselectivities.[31] In these reactions, **49** was first treated with (*S*)-**27**, $SiCl_3OTf$, and iPr_2EtN to generate the corresponding trichlorosilyl enol ether prior to addition of **45**. Chiral phosphine oxides (*S*)-tol-BINAPO ((*S*)-**51**), (*S*)-SEGPHOSO ((*S*)-**52**), (*S*)-H_8-BINAPO ((*S*)-**53**), and (*R*,*R*)-DIOPO ((*R*,*R*)-**54**) were also examined as catalysts in such reactions, and it was observed that (*S*)-**51** and (*S*)-**52** gave similar results as (*S*)-**27** had given. Catalyst (*S*)-**53** afforded lower yield, and (*R*,*R*)-**54** resulted in both lower yield and enantioselectivity. Interestingly, monodentate **1** was not an effective catalyst. This seems to indicate that bidentate bisphosphine oxides are required for catalysis of such aldol reactions between ketones. Intramolecular versions of these reactions have also been reported, and the best catalyst for these transformations was found to be (*S*)-4,4′-TMS_2-BINAPO ((*S*)-**55**).[32] Using this catalyst,initial diketones **56** could be converted directly into cyclohexanone derivatives **57** in high yields and enantioselectivities through initial enolization of the required methyl ketone group.

More recently, the Nakajima group has synthesized numerous analogs of (*R*,*R*)-**54** and has examined their utility as enantioselective catalysts.[33] For example, (*R*,*R*)-**58** was prepared from a common intermediate that was used to synthesize (*R*,*R*)-**54**, and it was found to be the most enantioselective catalyst prepared and generally afforded similar yields and marginally higher enantioselectivities of products **34** than (*S*)-**27** in the same Abramov-type reactions that were discussed previously (Figure 6.12). Only when cinnamaldehyde was used as the substrate was both the yield and stereoselectivity were higher with (*S*)-**27**. In addition, they have collaborated with Takahashi's group and synthesized atropisomeric chiral diene-based bisphosphine

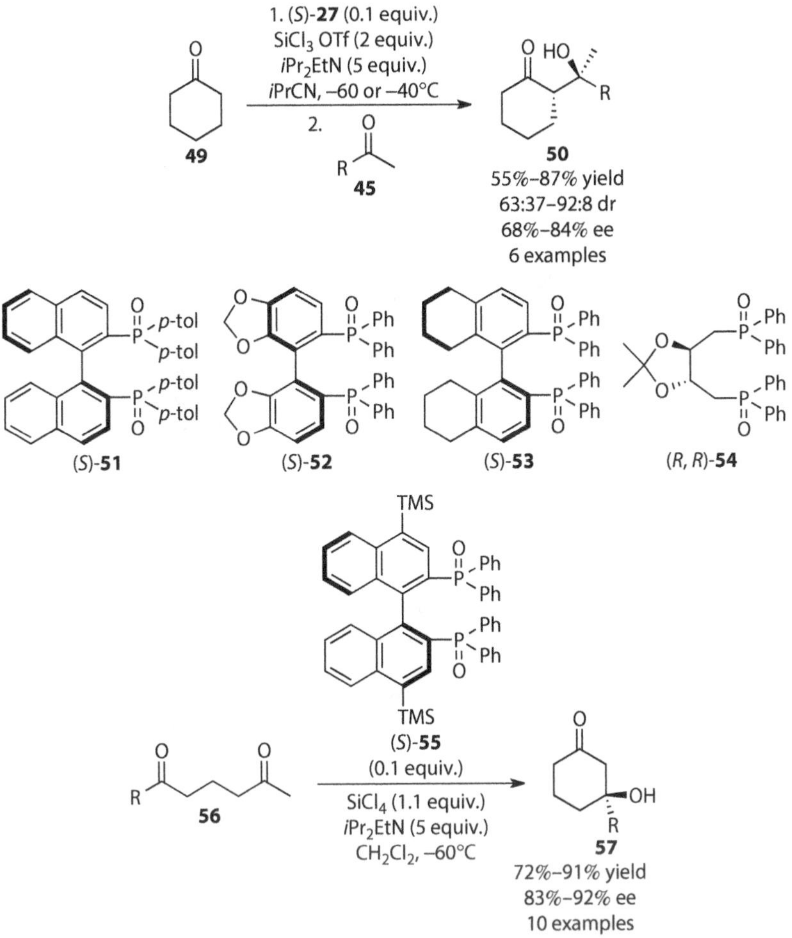

FIGURE 6.11 Enantioselective direct aldol reactions with ketone acceptors.

oxide (*M*)-(+)-**59** and have used this as a catalyst for asymmetric allylation reactions.[34] In the three examples in which the same substrates were used, (*M*)-(+)-**59** afforded higher yield and similar, if not, or better enantioselectivity than (*S*)-**27** of product **33** (R^1 = Me, Ph or *t*-Bu). In one case, (*S*)-**27** produced no appreciable amount of the desired product, but (*M*)-(+)-**59** resulted in 89% yield being obtained. Thus, at least in some cases (*M*)-(+)-**59** appears to be a much more active catalyst than (*S*)-**27** when sterically bulky substrates are used.

The Nakajima group finally studied enantioselective Morita–Baylis–Hillman reactions between aldehydes **7** and aryl vinyl ketones **60** to form chiral adducts **61** (Figure 6.13).[35] They screened six chiral catalysts, (*S*)-**27** and (*S*)-**51**–(*S*)-**55**, and found that (*S*)-**52** was optimal. It was proposed that chlorinated silyl ether intermediate **62** was involved in these reactions, and that elimination of the chloride and desilylation afforded the final obtained products **61**. Cinnamaldehyde was again the

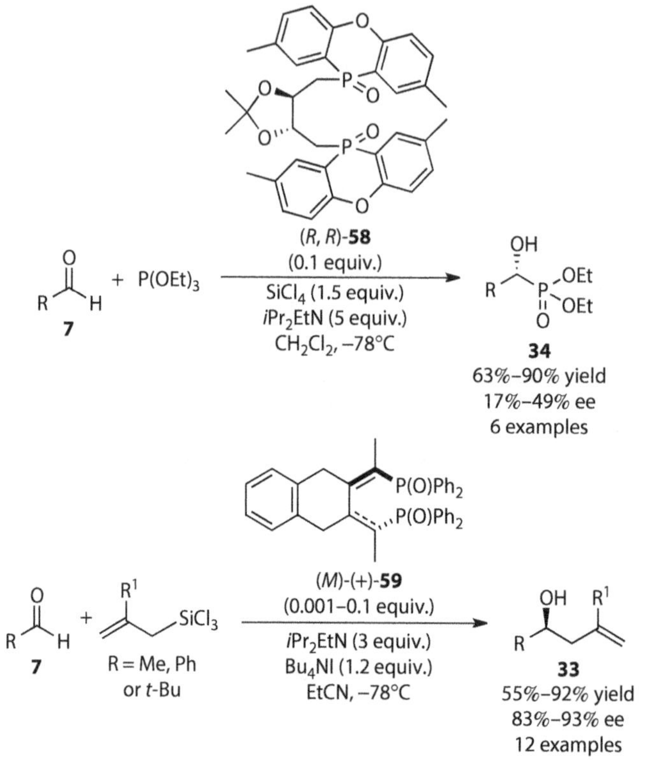

(R, R)-**58** (0.1 equiv.)
$SiCl_4$ (1.5 equiv.)
iPr$_2$EtN (5 equiv.)
CH_2Cl_2, −78°C

7 + P(OEt)$_3$ → **34**
63%–90% yield
17%–49% ee
6 examples

(M)-(+)-**59** (0.001–0.1 equiv.)
iPr$_2$EtN (3 equiv.)
Bu$_4$NI (1.2 equiv.)
EtCN, −78°C

7 + R = Me, Ph or t-Bu → **33**
55%–92% yield
83%–93% ee
12 examples

FIGURE 6.12 Improved enantioselective catalysts (*R*,*R*)-**58** and (*M*)-(+)-**59**.

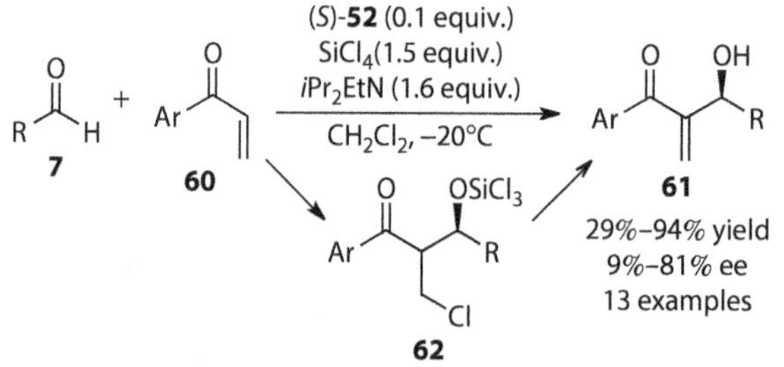

(S)-**52** (0.1 equiv.)
$SiCl_4$(1.5 equiv.)
iPr$_2$EtN (1.6 equiv.)
CH_2Cl_2, −20°C

7 + **60** → **62** → **61**
29%–94% yield
9%–81% ee
13 examples

FIGURE 6.13 Enantioselective Morita–Baylis–Hillman reactions catalyzed by (*S*)-**52**.

worst substrate of the aldehydes **7** examined in these reactions, affording the lowest yield and enantioselectivity.

Benaglia and his collaborators have also been active in this area of research, and most of their work has focused on using chiral bisphosphine oxide (*S*)-**63**, prepared by H_2O_2 oxidation of the corresponding industrially manufactured bisphosphine, as an enantioselective catalyst (Figure 6.14). In their first report, they described the use of (*S*)-**63** as a catalyst in highly enantioselective allylation reactions of aldehydes **7**

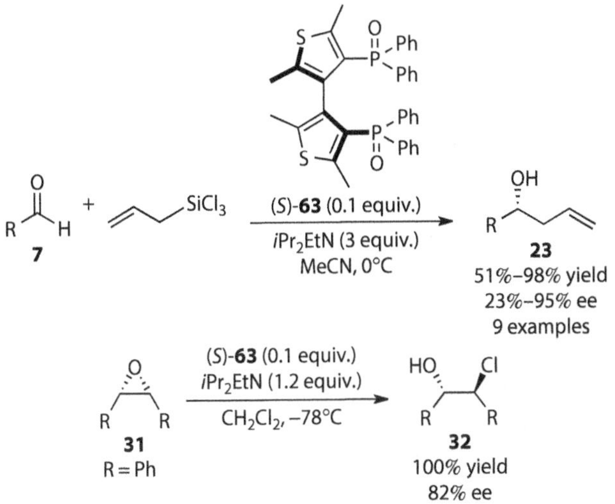

FIGURE 6.14 Enantioselective reactions catalyzed by (*S*)-**63**.

to produce homoallylic alcohols **23** and in the ring opening of *cis*-stilbene oxide (**31**, R = Ph) to form chlorohydrin **32** (R = Ph).[36] As in the previous research by others, the allylation reactions afforded very high yields and enantioselectivities with aromatic aldehydes but worked less well when the substrates were cinnamaldehyde and dihydrocinnamaldehyde. For the ring opening of **31** (R = Ph), catalyst (*S*)-**63** was slightly less enantioselective than (*S*)-**27** (82% ee vs. 90% ee).

After their initial report, Benaglia and coworkers examined the utility of (*S*)-**63** in various asymmetric aldol reactions of thioesters (Figure 6.15).[37] Initially, they studied direct aldol reactions of trifluroethyl thioesters **64** and obtained mixtures of

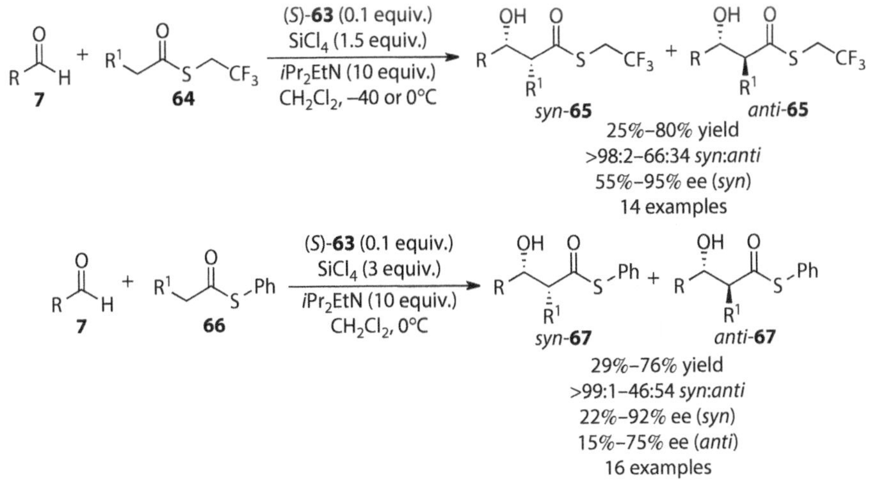

FIGURE 6.15 Enantioselective direct aldol reactions of thioesters catalyzed by (*S*)-**63**.

syn-**65** and *anti*-**65**. In these reactions, **7** was the limiting reagent and two equivalents of **64** were used, and very high *syn*-selectivity was observed in most cases. They subsequently changed the thioester group of their substrates to increase the acidity of the α-protons in order to increase reactivity.[38] Thus, **64** was replaced by **66**. When catalysts (*S*)-**27** and (*S*)-**63** were compared head-to-head in identical reactions of **66**, the results were somewhat mixed. Isolated yields of products *syn*-**67** and *anti*-**67** were similar with the two catalysts, but the *syn*:*anti* and enantiomeric ratios varied with (*S*)-**63**, affording higher ratios in some cases and lower ratios in others. In addition, there was no obvious evidence regarding increased reactivity of thio esters **66** compared to **64**. In fact, the reactions of **66** generally required longer times than those of **64** to achieve similar product yields.

In addition, direct aldol reactions between aldehydes and ketones, similar to those previously studied by Nakajima et al., were also studied by Benaglia and coworkers using (*S*)-**63** as the catalyst (Figure 6.16). For example, the results obtained in reactions of aldehydes **7** with **49** to form mixtures of *anti*-**36** and *syn*-**36** were similar to those that were reported by Nakajima in related reactions.[39] Specifically, aromatic aldehydes afforded better product yields and stereoselectivities than aliphatic aldehydes, and cinnamaldehyde was also a poor substrate. The results regarding the direct double aldol reactions between aldehydes **7** and methyl ketones **45** catalyzed by (*S*)-**63** were generally poorer than when (*S*)-**27** was used.[40] In these reactions, the diol aldol reaction products were directly acetylated, and *chiro*-**68** and *meso*-**68** were the actual compounds analyzed. Thus, as chiral binapthyl-based catalysts have been very widely studied with great success and are widely available, it seems that (*S*)-**27** might be a better starting point than (*S*)-**63** for those who are interested in undertaking this type of asymmetric catalysis research.

Although (*S*)-**27** and (*S*)-**63** have been the most widely studied chiral phosphine oxides in the context of catalysis by far, others that are based on different frameworks

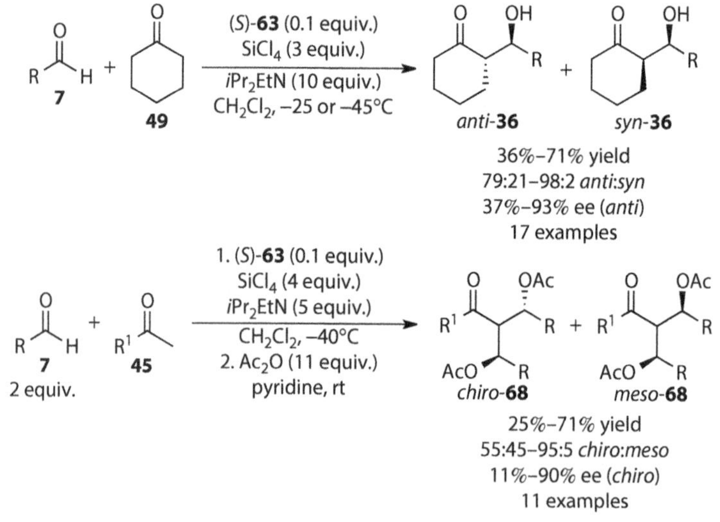

FIGURE 6.16 Enantioselective direct aldol reactions of ketones catalyzed by (*S*)-**63**.

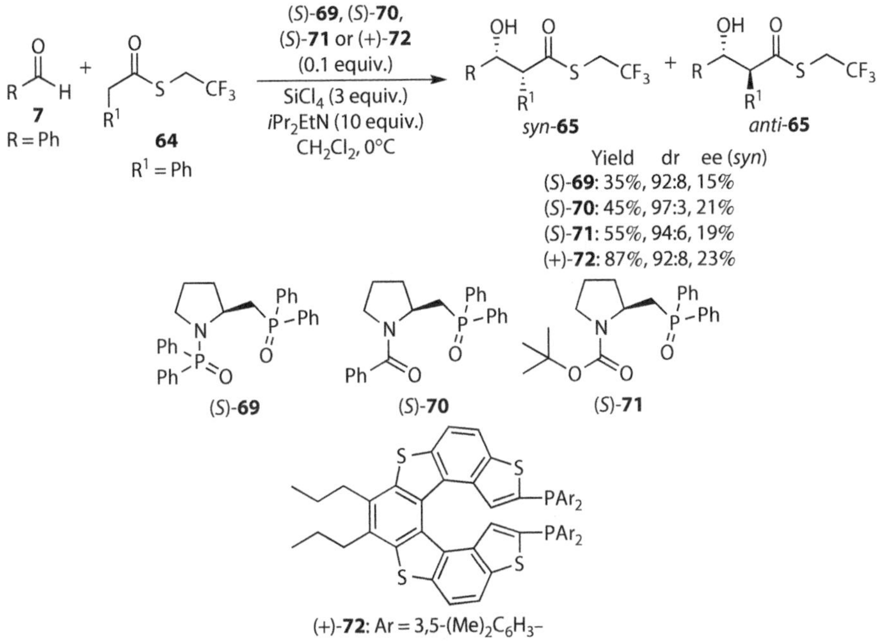

FIGURE 6.17 Enantioselective aldol reactions catalyzed by (*S*)-**69**, (*S*)-**70**, (*S*)-**71**, and (+)-**72**.

have also been recently examined. For example, Benaglia and coworkers have synthesized and studied a series of (*S*)-proline-based phosphine oxides, phosphonates and phosphoramides, and compounds that contain mixtures of functional groups (Figure 6.17).[41] The phosphine oxide containing molecules synthesized and studied were (*S*)-**69**, (*S*)-**70**, and (*S*)-**71** (Figure 6.17). They were examined as catalysts for the aldol reactions between **7** (R = Ph) and **64** (R^1 = Ph) but were found to be less efficient and stereoselective than the other catalysts studied. In fact, although good distereoselectivities were observed, all the new catalysts that were examined performed worse than (*S*)-**63** in terms of product yield and stereoselectivity in this study. Furthermore, the groups of Cauteruccio and Benaglia have reported the use of chiral tetrathiahelicene diphosphine oxide (+)-**72** in this same reaction with similar results.[42] Note that the absolute configuration of (+)-**72** was not shown in the original publication.

Ready and coworkers reported chiral allene-containing bisphosphine oxide **73** as a catalyst for epoxide ring opening reactions using $SiCl_4$ similar to those described previously by Nakajima's group (Figure 6.18).[43] *Meso*-epoxides **31** were converted to 1,2-chlorohydrins **32** in very high yields with high enantioselectivities. Significantly, much less of catalyst **73** could be used compared to when (*S*)-**27** was used (0.001–0.02 vs. 0.1 equivalents), indicating higher activity.

Recently, Ding's research group has described the use of chiral spiro[4,4]-1,6-nonadiene-based bisphosphine oxide (*R*)-**74** as an enantioselective catalyst of

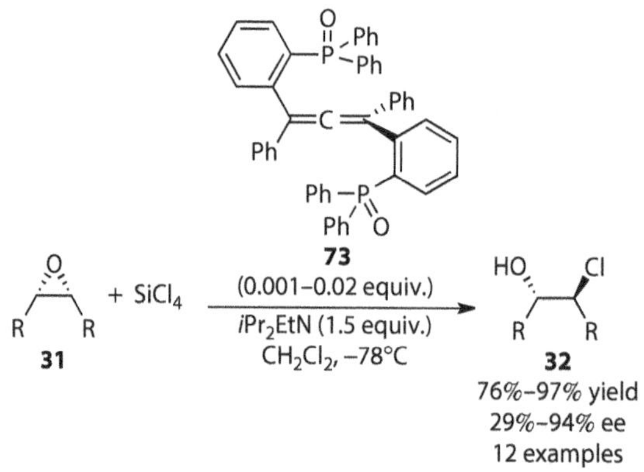

FIGURE 6.18 Enantioselective epoxide-opening reactions catalyzed by **73**.

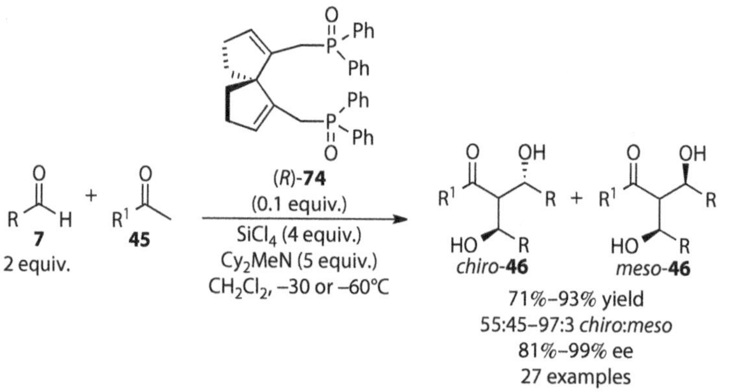

FIGURE 6.19 Enantioselective direct double aldol reactions catalyzed by (*R*)-**74**.

asymmetric direct double aldol reactions similar to those reported by Nakajima and coworkers (Figure 6.19).[44] Specifically, a wide range of aldehydes **7** were reacted with a variety of methyl ketones **45** to afford mixtures of diols *chiro*-**46** and *meso*-**46** in high yields, diastereoselectivities, and excellent enantioselectivities. Importantly, in a direct side-by-side comparison, (*R*)-**74** afforded higher yield (84 vs. 62% yield), diastereoselectivity (91:9 vs. 75:25 dr), and enantioselecitivity (94 vs. 70% ee) than (*S*)-**27** with the same substrates using the same reaction conditions.

Zhang and coworkers have described the synthesis of what they refer to as axially chiral C_{10}-BridgePHOS oxide (*S*)-**75** and its use as a catalyst in asymmetric aldehyde allylation reactions (Figure 6.20).[45] A variety of aldehydes **7** were reacted with allyl trichlorosilane in the presence of (*S*)-**75** to form homoallylic alcohols **23**. Although the enantioselectivities that were reported using (*S*)-**75** might be considered to be similar to those that were reported previously with (*S*)-**27**, it should be noted that high yields were reported using only 0.02 equivalents of (*S*)-**75** compared to 0.1 equivalents of (*S*)-**27** that were used in the previous reactions. Furthermore, these

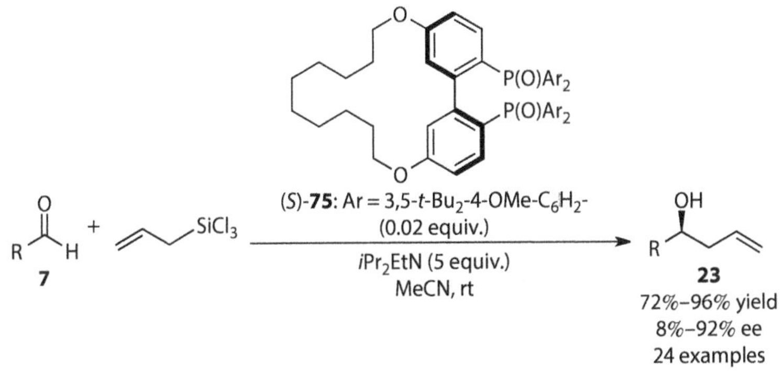

(S)-**75**: Ar = 3,5-t-Bu$_2$-4-OMe-C$_6$H$_2$-
(0.02 equiv.)
iPr$_2$EtN (5 equiv.)
MeCN, rt

23
72%–96% yield
8%–92% ee
24 examples

FIGURE 6.20 Enantioselective allylation reactions catalyzed by (*S*)-**75**.

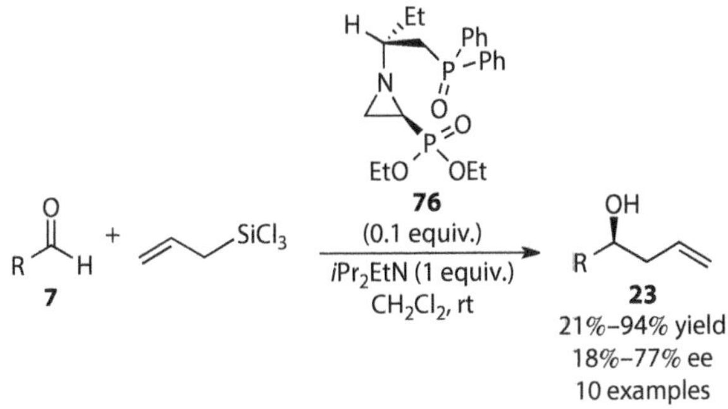

76
(0.1 equiv.)
iPr$_2$EtN (1 equiv.)
CH$_2$Cl$_2$, rt

23
21%–94% yield
18%–77% ee
10 examples

FIGURE 6.21 Enantioselective allylation reactions catalyzed by **76**.

new reactions did not involve the use of the additive Bu_4NI and were performed in acetonitrile rather than dichloromethane. An interesting aspect of these results is that the products **23** obtained in this study were of the opposite stereochemical configuration than those that were produced in the work by Nakajima's group even though the catalysts had similar configuration.

Finally, Dogan and coworkers reported the use of chiral aziridine-based phosphine oxide **76** as a catalyst in the same allylation reactions of aldehydes **7** (Figure 6.21).[46] In these reactions, the products **23** had the same stereochemical configuration as reported by Zhang's group, but 0.1 equivalents of catalyst **76** were used to produce lower product yields with lower levels of enantioselectivity.

6.3.2 Catalytic Reduction Reactions

It is clearly evident from the previous section that the use of phosphine oxides as Lewis bases to activate silyl reagents in nucleophilic addition reactions has been widely studied with notable achievements, especially in the context of asymmetric catalysis. An extension of this theme was reported by Nakajima and coworkers when they described the activation $HSiCl_3$ in conjugate reduction reactions (Figure 6.22).[47]

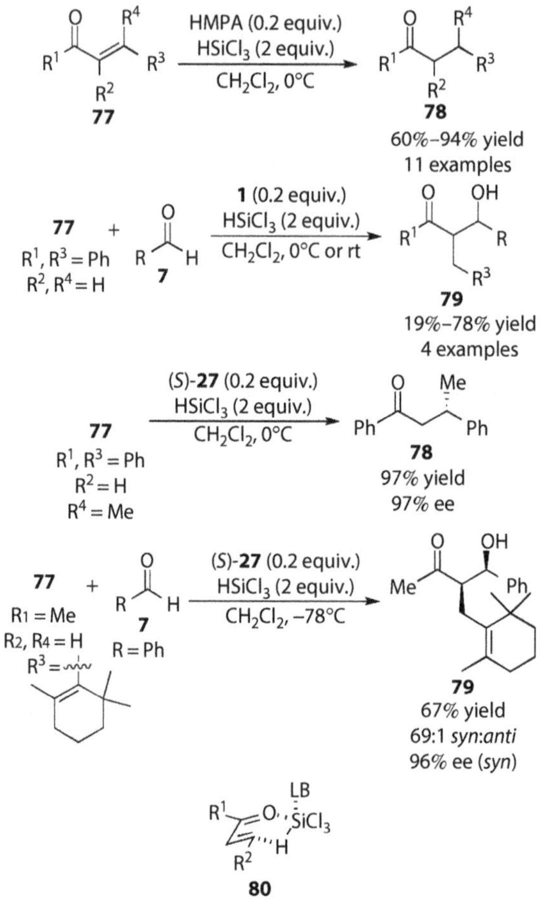

77 → HMPA (0.2 equiv.), $HSiCl_3$ (2 equiv.), CH_2Cl_2, 0°C → 78
60%–94% yield
11 examples

77 + 7 (R^1, R^3 = Ph; R^2, R^4 = H) → **1** (0.2 equiv.), $HSiCl_3$ (2 equiv.), CH_2Cl_2, 0°C or rt → 79
19%–78% yield
4 examples

77 (R^1, R^3 = Ph; R^2 = H; R^4 = Me) → (*S*)-**27** (0.2 equiv.), $HSiCl_3$ (2 equiv.), CH_2Cl_2, 0°C → 78
97% yield
97% ee

77 + 7 (R1 = Me; R2, R4 = H; R = Ph) → (*S*)-**27** (0.2 equiv.), $HSiCl_3$ (2 equiv.), CH_2Cl_2, −78°C → 79
67% yield
69:1 *syn:anti*
96% ee (*syn*)

80

FIGURE 6.22 Conjugate reduction of enones and enantioselective reactions catalyzed by (*S*)-**27**.

Initially, they used the phosphoramide HMPA as a catalyst for the selective reduction of enones **77** to ketones **78**, and then they used **1** as a catalyst in reductive aldol reactions of **77** to form β-hydroxy ketones **79**. Based on the success of these reactions, they subsequently used (*S*)-**27** in the enantioselective reduction of **77** to afford ketone **78**, and in the reductive aldol reaction for the conversion of enone **77** into aldol product **79**. Very high levels of stereoselectivity were observed in both of these reactions, and it was proposed that these transformations proceeded via a cyclic six-membered transition state such as **80**. Importantly, aldehydes **7** were found not to be reduced by the combination of $HSiCl_3$ and a phosphine oxide, and therefore the reductive aldol reactions could be performed simply by mixing both substrates **77** and **7** together at the beginning of the reactions.

Nakajima's group further studied the substrate scope of the stereoselective reductive aldol reactions between enones **77** and aldehydes **7** to produce adducts **79** that

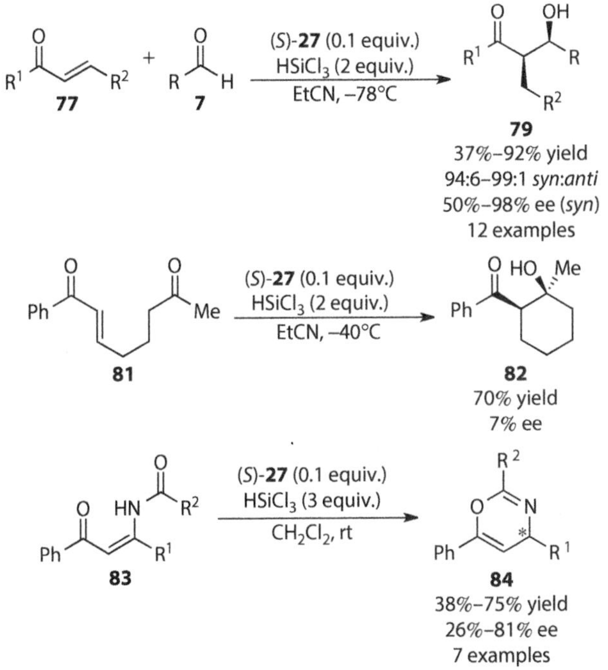

FIGURE 6.23 Enantioselective reactions catalyzed by (*S*)-**27**.

were catalyzed by (*S*)-**27**, and they even reported an intramolecular example in the conversion of **81** into **82**, albeit with low enantioselectivity (Figure 6.23).[48] For the reaction of **81**, they reported that a chiral bisquinoline *N*,*N*′-dioxide afforded better enantioselectivity but lower product yield (41% yield, 55% ee). Finally, they showed that *N*-acylated β-amino enones **83** could participate in similar reactions and could be transformed into 4*H*-1,3-oxazines **84** with modest stereoselectivity.[49] In these later reactions, (*S*)-**27** proved to be a more stereoselective catalyst than others that were examined, including (*S*)-**52** and (*R*,*R*)-**54**.

After the initial reports by Nakajima's group, Zhou and coworkers put an interesting twist on these selective conjugate reduction reactions by forming the enone substrates **77** *in situ* using Wittig reactions that produced **1** as a by-product (Figure 6.24).[50] In this way, the waste formed in one reaction served as a catalyst in a subsequent transformation. Thus, aldehydes **7** could be converted into ketones **78** in a one-pot process. Importantly, they noted that only enones were reduced in these reactions, and enoates were unreactive. They subsequently extended this concept to the conversion of ethyl trifluoropyruvate (**85**) into α-CF_3 γ-keto esters **86**,[51] and most recently to the transformation of 1,2-dicarbonyl compounds **87** into substituted furans **88**.[52] In these last reactions, MeOH was added in the third step to generate HCl, which induced cyclization to form the final products.

Toy and coworkers extended this concept further by going back one step and by generating the phosphorane for the Wittig reactions *in situ* (Figure 6.25).[53] In this way, aldehydes **7** could be converted in a one-pot process into aldol products **79**, incorporating

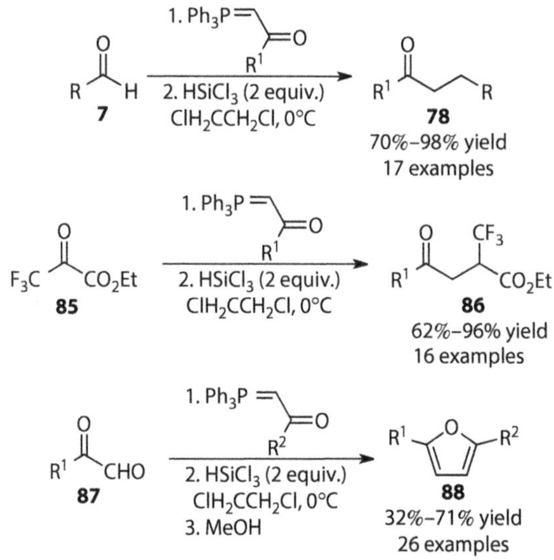

FIGURE 6.24 Conjugate reduction reactions catalyzed by **1** generated *in situ*.

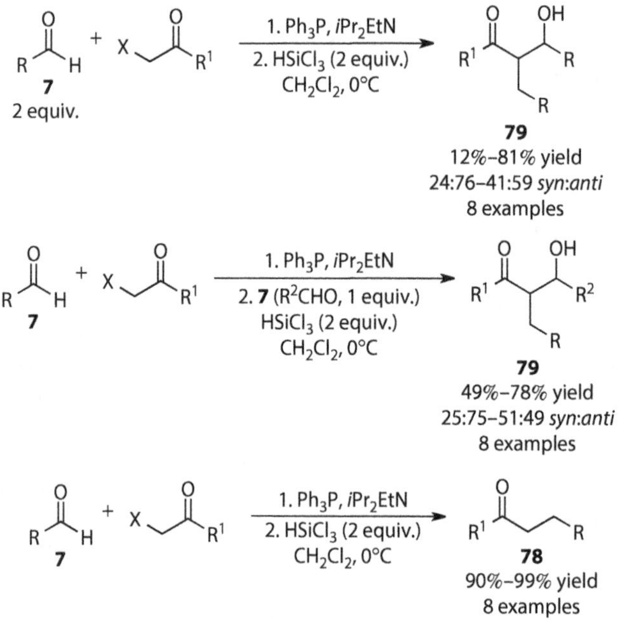

FIGURE 6.25 One-pot Wittig reaction followed by phosphine oxide-catalyzed reactions.

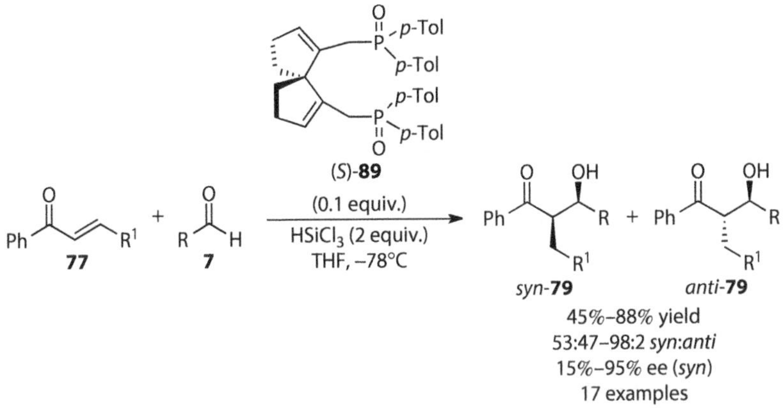

(S)-**89**
(0.1 equiv.)
$HSiCl_3$ (2 equiv.)
THF, −78°C

77 + **7** → *syn*-**79** + *anti*-**79**

45%–88% yield
53:47–98:2 *syn*:*anti*
15%–95% ee (*syn*)
17 examples

FIGURE 6.26 Stereoselective reductive aldol reactions catalyzed by (*S*)-**89**.

one or two different aldehyde building blocks. In addition, they reported that a recyclable bifunctional polymer-supported phosphine could be used in the one-pot transformations of aldehydes **7** to produce either ketones **78** or aldol products **79** that were easy to purify.[54]

Ding's group finally and most recently reported enantioselective reductive aldol reactions of **77** to form mixtures of *syn*-**78** and *anti*-**78** that were catalyzed by (*S*)-**89** (Figure 6.26).[55] In this project, they screened several analogs of their previously reported chiral phosphine oxide (*R*)-**74** and found that when its phenyl groups were replaced with *p*-tolyl groups to afford (*S*)-**89**, higher levels of stereoselectivity could be achieved in the reactions studied.

Carbon–nitrogen double bonds have also been reduced by $HSiCl_3$ using a variety of Lewis base activators, especially chiral ones for asymmetric reactions. For details of such reactions, see the extensive reviews written by Benaglia[56] and Jones.[57] In this regard, Benaglia and coworkers have described what may be the only example of such a transformation using a phosphine oxide.[58] In their reaction, (*S*)-**63** was used to catalyze the reduction of imine **90** enantioselectively to form chiral amine **91** (Equation 6.4). Catalyst (*S*)-**63** was found to be inferior to other catalysts that were examined in this study and was not studied further. The best catalyst for this transformation was a chiral picolinamide that was derived from ephedrine (91% yield, 99% ee), and such chiral amides generally seem to be better catalysts for asymmetric imine reduction reactions than phosphine oxides.

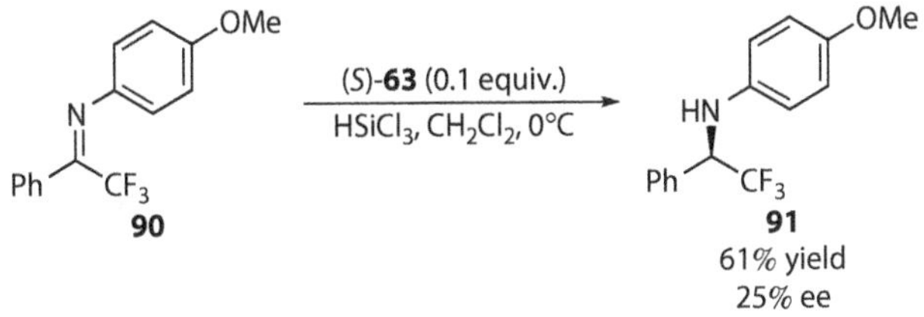

90 → (*S*)-**63** (0.1 equiv.), $HSiCl_3$, CH_2Cl_2, 0°C → **91** (6.4)

61% yield
25% ee

6.4 CATALYTIC METAL–HALOGEN EXCHANGE REACTIONS

Uchiyama and coworkers have reported the first example of the use of a phosphine oxide to catalyze a zinc–iodine exchange reactions (Figure 6.27).[59] Specifically, exchange reactions of polyfluroaryl iodides such as **92** and dimethylzinc could be catalyzed by **1** to produce allylated or propargylated products **93** in good overall yields. However, although the mechanism by which **1** acts as a catalyst is unclear, it was more effective than pyridine, dimethyl sulfoxide, DMPU, and even tertiary phosphines and phosphites in this role.

6.5 PHOSPHINE OXIDES AS PRECATALYSTS IN HALOGENATION REACTIONS

Dihalophosphonium salts **94** are versatile reagents for organic synthesis in which the phosphorous atom is in the same oxidation state as in **1**. This realization led to the observation by Masaki and Fukui that when **1** was treated with oxallyl chloride or oxallyl bromide, phosphonium salts **94** are obtained with simultaneous evolution of carbon dioxide and carbon monoxide (Equation 6.5).[60]

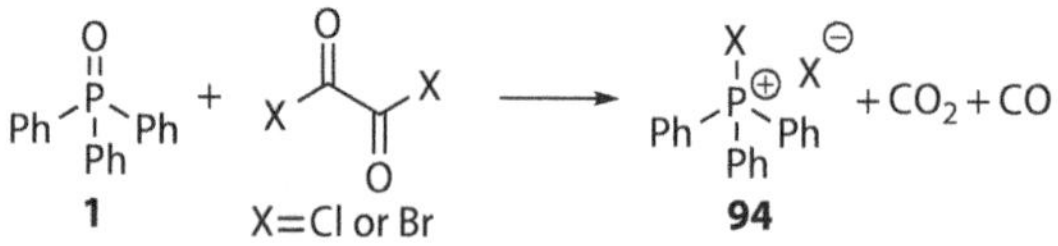

$$Ph_3P{=}O\ (\mathbf{1}) + XC(O)C(O)X\ (X{=}Cl\text{ or }Br) \longrightarrow Ph_3P^{\oplus}X\ X^{\ominus}\ (\mathbf{94}) + CO_2 + CO \tag{6.5}$$

This reaction has subsequently been exploited to allow for reagents **94** to be generated *in situ* from substoichiometric quantities of **1** for further reaction. This is possible because many of the reactions that utilize **94** as a reagent produce **1** as a by-product. Thus, in these processes, the role of **94** can be viewed as that of a catalyst, and thus **1** serves as a precatalyst. Much of this research has been summarized by Xia and Toy in a manuscript in which they utilized this reaction to recycle a phosphine oxide,[61] and Rutjes et al. have reviewed some of this research in the context of bypassing phosphine oxide waste.[62]

The first example of the use of **1** as a precatalyst in this context was reported by Denton and coworkers when they described catalytic Appel reactions that converted alcohols **95** into the corresponding alkyl halides **76** (Figure 6.28).[63,64] As alcohols **95** can react with oxally halides, the reactions were performed by separating these incompatible reactants for as long as possible. This was achieved by dissolving a

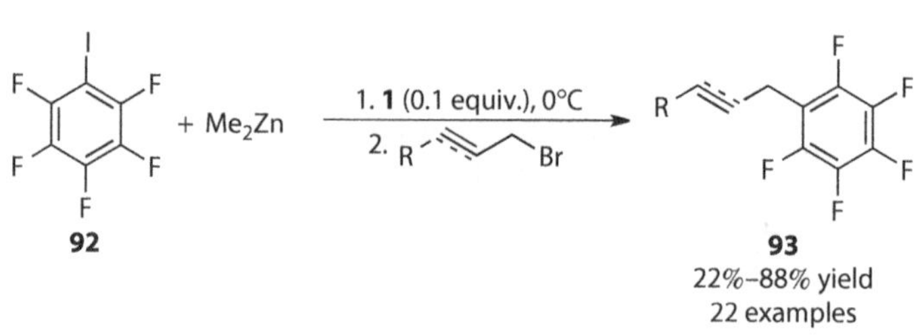

FIGURE 6.27 Zinc-iodide exchange reactions catalyzed by **1**.

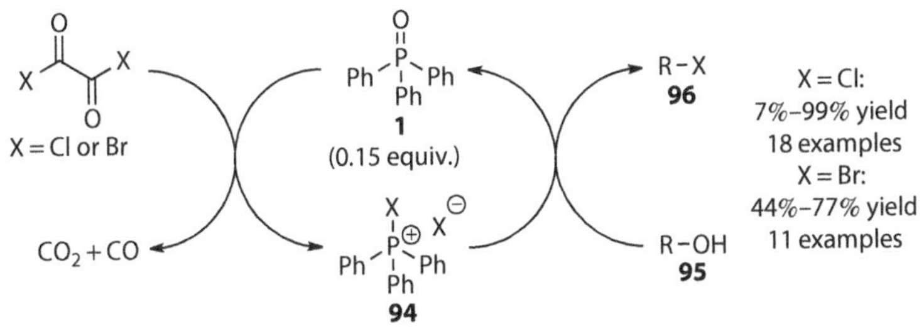

FIGURE 6.28 Catalytic Appel reactions involving the conversion of **1** into **94**.

substoichiometric amount of **1**, by adding the appropriate amount of oxallyl halide to convert this to **94**, and then subsequently by adding solutions of both the oxallyl halide and **95** simultaneously over a period of hours. These reactions resulted in inversion of stereochemistry at the carbinol center, and product yields of these reactions were on the low side of what is typical for Appel reactions. Primary and activated alcohols afforded the highest yields, whereas hindered alcohols performed poorly. For example, cyclohexanol afforded only 7% yield in the chlorination reaction and did not react at all in the bromination reaction.

Denton's group subsequently extended this concept to epoxide dichlorination reactions (conversion of **97** to **98**),[65] the transformation of oximes to nitriles (conversion of **99** to **100**),[66] and deoxydichlorination reactions of aldehydes (conversion of **7** to **101**) (Figure 6.29).[67] Note that in the reactions of epoxides **97**, an equivalent of a sterically hindered pyridine base was required for the products **98** to be isolated in high yield. They even reported that the use of a heterogeneous polystyrene-supported version of **1** greatly facilitated product purification from such reactions.[68]

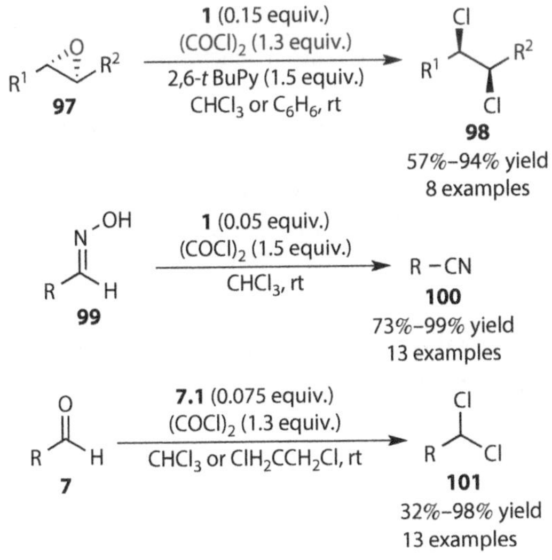

FIGURE 6.29 Catalytic reactions involving the conversion of **1** into **94**.

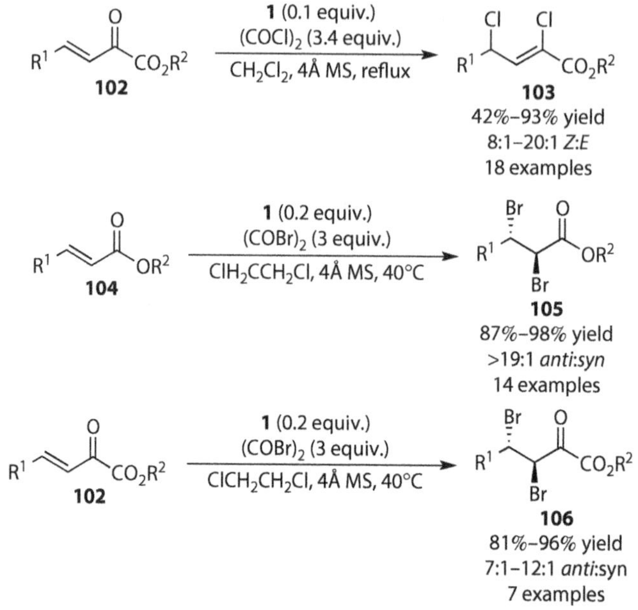

FIGURE 6.30 Stereoselective halogenation reactions involving the conversion of **1** into **94**.

Xu and coworkers have applied this concept to reactions of more complex substrates in a series of stereoselective reactions (Figure 6.30). For example, they found that conjugated unsaturated α-ketoesters **102** underwent reaction with *in situ* generated **94** to form dichlorinated products **103** with high degrees of *Z*-selectivity.[69] It should be noted that these reactions proceeded only when the alkenyl substituent was an aryl group. When a substrate with an alkyl group at this position was subjected to the reaction conditions, no desired product was obtained. They subsequently reported bromination reactions of **104** using similar reaction conditions to form brominated products **105** with generally very high levels of diastereoselectivity.[70] Unlike the previous reactions, these were not limited to cinnamic acid derivatives, and examples of reactions with substrates having alkyl group substituents at the alkene position were reported to produce high yields of **105**. Interestingly, when substrates **102** were subjected to these reaction conditions, dibrominated α-ketoesters **106** were obtained instead of products that were analogous to **103**.

6.6 PHOSPHINE OXIDES AS PRECATALYSTS IN WITTIG AND RELATED REACTIONS

As mentioned earlier, another way in which phosphine oxides can function as a precatalyst is when they are reduced *in situ* back to the phosphine oxidation state in reactions in which they are formed as by-products. In such reactions, the phosphine is the actual catalyst, and the starting materials and products are all stable to the reduction conditions. An example of this strategy applied in aza-Wittig reactions is illustrated in Figure 6.4, but the first example of Wittig reactions to form carbon–carbon double bonds that

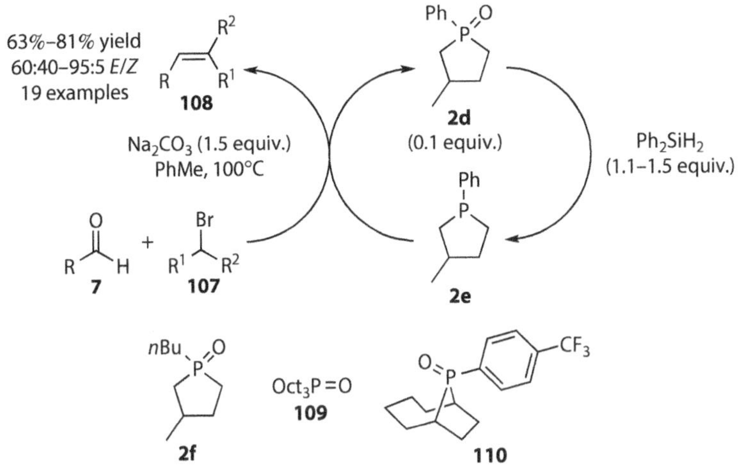

FIGURE 6.31 Catalytic Wittig reactions involving phosphine oxides reported by O'Brien.

involved a phosphine oxide precatalyst was reported by O'Brien and coworkers who reported several versions of such reactions in a series of publications (Figure 6.31). They initially reported the use of **2d** (prepared by the reduction of **2c**) as the precatalyst that was added at the beginning of the reactions.[71] Once this was reduced *in situ* by Ph_2SiH_2, the corresponding phosphine **2e** reacted with electron-withdrawing group-activated alkyl halides **107** and Na_2CO_3 to form the actual reactive phosphorus ylides that participated in the Wittig reactions of aldehydes **7** to form alkene products **108** and **2d** as the by-product. They subsequently reported that soluble organic base iPr$_2$EtN was a good replacement for Na_2CO_3 in such reactions,[72] and that the addition of 4-nitrobenzoic acid facilitated the phosphine oxide reduction step using $PhSiH_3$.[73] Usage of this combination of 4-nitrobenzoic acid and $PhSiH_3$ for phosphine oxide reduction allowed reactions starting with **2f** to be conducted at room temperature, and for acyclic phosphine oxides **1** and **109** to be used as well, albeit at elevated temperature. They most recently found that by using **110** as the precatalyst in conjunction with NaOCO$_2$*t*Bu (a slow release form of NaO*t*Bu) as the precursor for the required base, catalytic Wittig reactions could be performed using semi- or nonstabilized ylides.[74]

Werner's research group has also been active in this area of research and reported the first example of a catalytic enantioselective Wittig reaction (Figure 6.32).[75] This reaction involved the intramolecular cyclization of **111** to form **112**. A variety of phosphines were examined as the catalyst, and (*S*,*S*)-**113** ((*S*,*S*)-Me-DuPhos) was found to provide the best combination of reactivity and stereoselectivity (69% yield, 62% ee). In these reactions, butylene oxide was used as a base precursor, and they were performed using microwave irradiation (MWI). They subsequently reported the scope and limitations of such microwave-assisted catalytic Wittig reactions using Bu_3P, $PhSiH_3$, and butylene oxide,[76,77] and reactions with the combination of **114**, $HSi(OMe)_3$, and Na_2CO_3 using conventional heating.[78] They finally reported the first base-free Wittig reactions using diethyl maleate (**115**) as the starting material to form products **116** that were catalyzed by Bu_3P.[79] In these transformations, initial reaction between **115**

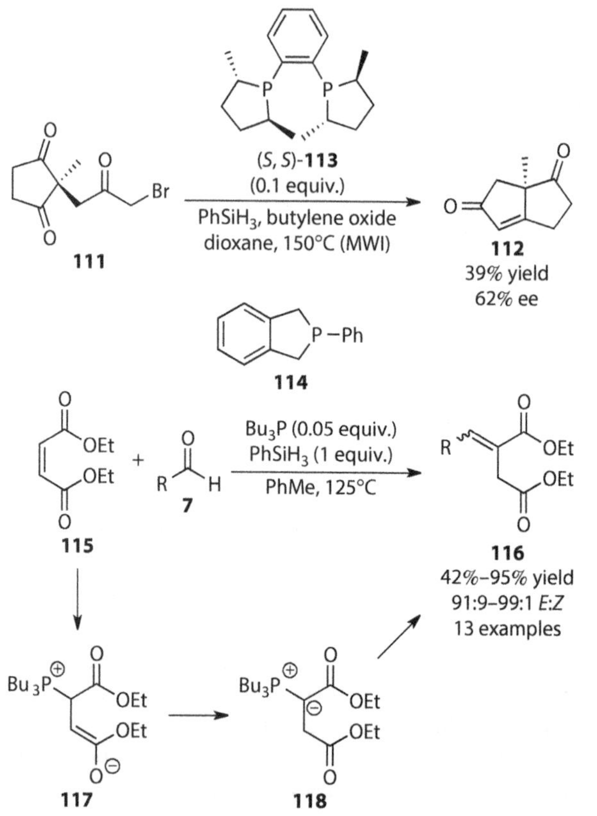

(*S*, *S*)-**113**
(0.1 equiv.)
$PhSiH_3$, butylene oxide
dioxane, 150°C (MWI)

111

112
39% yield
62% ee

P–Ph

114

Bu_3P (0.05 equiv.)
$PhSiH_3$ (1 equiv.)
PhMe, 125°C

115 **7** **116**
42%–95% yield
91:9–99:1 *E*:*Z*
13 examples

117 **118**

FIGURE 6.32 Catalytic Wittig reaction involving phosphine oxides reported by Werner.

and Bu_3P generated zwitterion **117** that underwent internal proton transfer to generate ylide **118**. This in turn reacted with aldehyde **7** to form product **116**.

At about the same time as the last report by Werner and coworkers appeared, Tsai and Lin published similar catalytic Wittig reactions that were based on their previous research regarding related noncatalytic Wittig reactions (Figure 6.33).[80] They started with Michael acceptors **119** to generate products **120** stereoselectively using **2d** as the precatalyst with Et_3N as the base and 4-nitrobenzoic acid as an acidic additive.

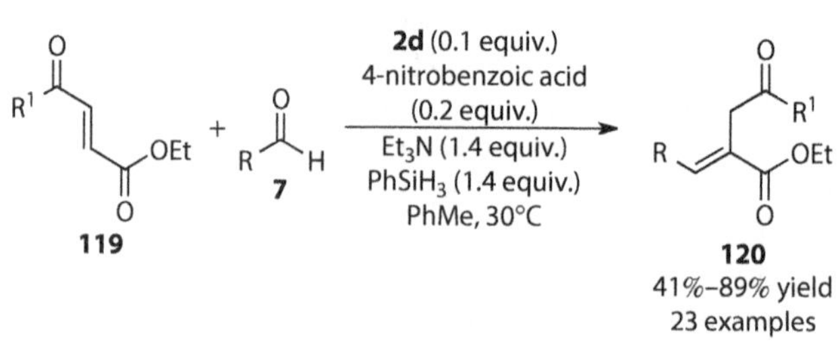

2d (0.1 equiv.)
4-nitrobenzoic acid
(0.2 equiv.)
Et_3N (1.4 equiv.)
$PhSiH_3$ (1.4 equiv.)
PhMe, 30°C

119 **7** **120**
41%–89% yield
23 examples

FIGURE 6.33 Catalytic Wittig reactions involving **2d** as the precatalyst.

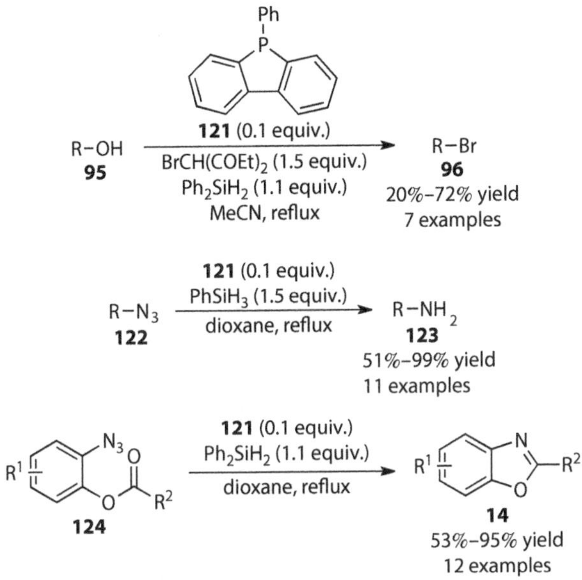

FIGURE 6.34 Reactions catalyzed by **121**.

Similar reactions have been reported by van Delft and coworkers using phosphole **121** as the catalyst (Figure 6.34). In their first report, they used **121** as a catalyst for Appel reactions using $BrCH(CO_2Et)_2$ as the brominating reagent and Ph_2SiH_2 as the reductant.[81] Using this reaction system, relatively unhindered alcohols **95** could be converted into the corresponding alkyl bromides **96** in good yields. They subsequently reported using **121** to catalyze Staudinger reactions of azides **122** to form amines **123** in high yields.[82] Of all the substrates examined, only 2-azidophenol did not reduce cleanly, and it afforded a complex mixture of products. They finally extended these catalytic Staudinger reactions to aza-Wittig reactions, exemplified by the conversion of aryl azides **124** into benzoazoles **14**.[83] Note that the catalytic Appel and aza-Wittig reactions reported by the van Delft research group are conceptually different from those reported by the Denton and Marsden groups, respectively, whose reactions utilized P(V) catalysts rather than a P(III) catalyst.

Ashfield and coworkers later extended the catalytic Staudinger reactions of van Delft's group by coupling such reactions to amide formation in what they referred to as a Staudinger ligation process (Figure 6.35).[84] In these reactions, an azide **122** was mixed with a carboxylic acid **125** in the presence of catalytic Ph_3P and $PhSiH_3$ to afford amides **126** in high yields.

Fourmy and Voituriez very recently used **121** as a catalyst for reactions between α-ketoesters **127** and diisopropyl azodicarboxylate (DIAD, **128**) to form cyclic adducts **129** in reactions that were presumed to involve Huisgen zwitterion **130** (Figure 6.36).[85] These reactions were based on previously reported work by others that were performed using Ph_3P as a reagent and involved the use of an electron-poor phosphoric acid and $i Pr_2EtN$ as additives to facilitate the phosphine oxide reduction step.

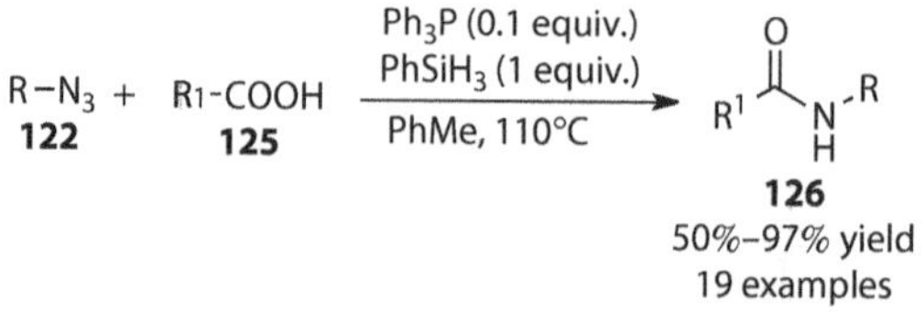

FIGURE 6.35 Catalytic Staudinger ligation reactions.

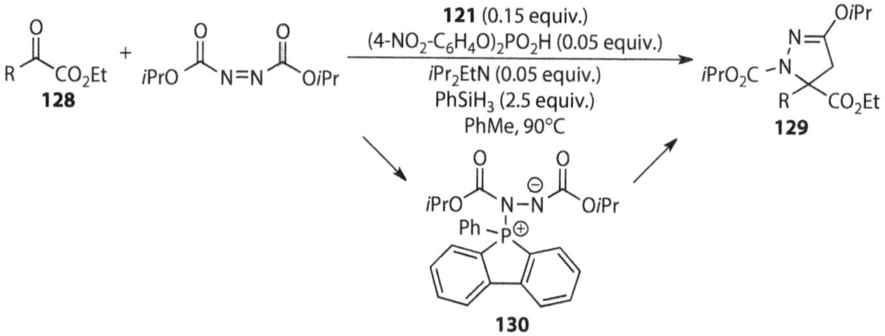

FIGURE 6.36 Cycloaddition reactions catalyzed by **121**.

6.7 SUMMARY AND OUTLOOK

As highlighted in Sections 6.2 and 6.3, phosphine oxides are useful catalysts for the transformation of carbon–oxygen double bonds into carbon–nitrogen double bonds and for the activation of a wide variety of silicon-based reagents by virtue of their Lewis basicity. Furthermore, their role in catalyzing metal–halogen exchange reactions is virtually unexplored. Considering the vast literature regarding chiral phosphines, it is easy to envision that there are many opportunities to develop chiral phosphine oxides as enantioselective catalysts, and research into this topic has really only just scratched the surface. Sections 6.4 and 6.5 highlight the emerging concept of converting phosphine oxide by-products *in situ* back into their original forms, and thereby allowing many reactions involving phosphonium salts and phosphines to be performed catalytically instead of stoichiometrically. As new and improved methods for phosphine oxide reduction are developed, it seems likely that virtually any reactions involving phosphines to be performed catalytically in the future. Thus, it seems that the outlook for phosphine oxide research is indeed bright, and many reactions involving them will be performed in new ways, with broadened applicability.

REFERENCES

1. Constable, D. J. C., Dunn, P. J., Hayler, J. D., Humphrey, G. R., Leazer, J. J. L., Linderman, R. J. Lorenz et al. *Green Chem.* **2007**, *9*, 411–420.
2. Campbell, T. W., Monagle, J. J. *J. Am. Chem. Soc.* **1962**, *84*, 1493.

3. Monagle, J. J., Campbell, T. W., McShane, H. F., Jr. *J. Am. Chem. Soc.* **1962**, *84*, 4288–4295.
4. Monagle, J. J. *J. Org. Chem.* **1962**, *27*, 3851–3855.
5. Monagle, J. J., Mengenhauser, J. V. *J. Org. Chem.* **1966**, *31*, 2321–2324.
6. Appleman, J. O., DeCarlo, V. J. *J. Org. Chem.* **1967**, *32*, 1505–1507.
7. Campbell, T. W., Monagle, J. J., Foldi, V. S. *J. Am. Chem. Soc.* **1962**, *84*, 3673–3677.
8. McGrew, L. A., Sweeny, W., Campbell, T. W., Foldi, V. S. *J. Org. Chem.* **1964**, *29*, 3002–3004.
9. Marsden, S. P., McGonagle, A. E., McKeever-Abbas, B. *Org. Lett.* **2008**, *10*, 2589–2591.
10. Wang, L., Qin, R.-Q., Yan, H.-Y., Ding, M.-W. *Synthesis* **2015**, *47*, 3522–3528.
11. Wang, L., Wang, Y., Chen, M., Ding, M.-W. *Adv. Synth. Catal.* **2014**, *356*, 1098–1104.
12. Kobayashi, S., Nishio, K. *Tetrahedron Lett.* **1993**, *34*, 3453–3456.
13. Short, J. D., Attenoux, S., Berrisford, D. J. *Tetrahedron Lett.* **1997**, *38*, 2351–2354.
14. Ogawa, C., Konishi, H., Sugiura, M., Kobayashi, S. *Org. Biomol. Chem.* **2004**, *2*, 446–448.
15. Ogawa, C., Sugiura, M., Kobayashi, S. *Angew. Chem. Int. Ed.* **2004**, *43*, 6491–6493.
16. Benaglia, M., Rossi, S. *Org. Biomol. Chem.* **2010**, *8*, 3824–3830.
17. Tokuoka, E., Kotani, S., Matsunaga, H., Ishizuka, T., Hashimoto, S., Nakajima, M. *Tetrahedron: Asymmetry* **2005**, *16*, 2391–2392.
18. Kotani, S., Furusho, H., Sugiura, M., Nakajima, M. *Tetrahedron* **2013**, *69*, 3075–3081.
19. Rossi, S., Benaglia, M., Genoni, A. *Tetrahedron* **2014**, *70*, 2065–2080.
20. Nakajima, M., Kotani, S., Ishizuka, T., Hashimoto, S. *Tetrahedron Lett.* **2005**, *46*, 157–159.
21. Kotani, S., Hashimoto, S., Nakajima, M. *Tetrahedron* **2007**, *63*, 3122–3132.
22. Nakanishi, K., Kotani, S., Sugiura, M., Nakajima, M. *Tetrahedron* **2008**, *64*, 6415–6419.
23. Kotani, S., Sugiura, M., Nakajima, M. *Chem. Rec.* **2013**, *13*, 362–370
24. Kotani, S., Sugiura, M., Nakajima, M. *Synlett* **2014**, *25*, 631–640.
25. Kotani, S., Hashimoto, S., Nakajima, M. *Synlett* **2006**, *17*, 1116–1118.
26. Shimoda, Y., Tando, T., Kotani, S., Sugiura, M., Nakajima, M. *Tetrahedron: Asymmetry* **2009**, *20*, 1369–1379.
27. Kotani, S., Shimoda, Y., Sugiura, M., Nakajima, M. *Tetrahedron Lett.* **2009**, *50*, 4602–4605.
28. Kotani, S., Aoki, S., Sugiura, M., Nakajima, M. *Tetrahedron Lett.* **2011**, *52*, 2834–2836.
29. Shimoda, Y., Kotani, S., Sugiura, M., Nakajima, M. *Chem. Eur. J.* **2011**, *17*, 7992–7995.
30. Shinoda, Y., Kubo, T., Sugiura, M., Kotani, S., Nakajima, M. *Angew. Chem. Int. Ed.* **2013**, *52*, 3461–3464.
31. Aoki, S., Kotani, S., Sugiura, M., Nakajima, M. *Chem. Commun.* **2012**, *48*, 5524–5526.
32. Kotani, S., Aoki, S., Sugiura, M., Ogasawara, M., Nakajima, M. *Org. Lett.* **2014**, *16*, 4802–4805.
33. Ohmaru, Y., Sato, N., Mizutani, M., Kotani, S., Sugiura, M., Nakajima, M. *Org. Biomol. Chem.* **2012**, *10*, 4562–4570.
34. Ogasawara, M., Kotani, S., Nakajima, H., Furusho, H., Miyasaka, M., Shimoda, Y., Wu, W.-Y., Sugiura, M., Takahashi, T., Nakajima, M. *Angew. Chem. Int. Ed.* **2013**, *52*, 13798–13802.
35. Kotani, S., Ito, M., Nozaki, H., Sugiura, M., Ogasawara, M., Nakajima, M. *Tetrahedron Lett.* **2013**, *54*, 6430–6433.
36. Simonini, V., Benaglia, M., Benincori, T. *Adv. Synth. Catal.* **2008**, *350*, 561–564.
37. Rossi, S., Benaglia, M., Cozzi, F., Genoni, A., Benincori, T. *Adv. Synth. Catal.* **2011**, *353*, 848–854.
38. Rossi, S., Annunziata, R., Cozzi, F., Raimondi, L. M. *Synthesis* **2015**, *47*, 2113–2124.
39. Rossi, S., Benaglia, M., Genoni, A., Benincori, T., Celentano, G. *Tetrahedron* **2011**, *67*, 158–166.

40. Genoni, A., Benaglia, M., Rossi, S., Celentano, G. *Chirality* **2013**, *25*, 643–647.
41. Bonsignore, M., Benaglia, M., Cozzi, F., Genoni, A., Rossi, S., Raimondi, L. *Tetrahedron* **2012**, *68*, 8251–8256.
42. Cauteruccio S., Dova, D., Benaglia, M., Genoni, A., Orlandi, M., Licandro, E. *Eur. J. Org. Chem.* **2014**, 2694–2702.
43. Pu, X., Qi, X., Ready, J. M. *J. Am. Chem. Soc.* **2009**, *151*, 10364–10365.
44. Zhang, P., Han, Z., Wang, Z., Ding, K. *Angew. Chem. Int. Ed.* **2013**, *52*, 11054–11058.
45. Chen, J., Liu, D., Fan, D., Liu, Y., Zhang, W. *Tetrahedron* **2013**, *69*, 8161–8168.
46. Dogan, O., Bulut, A., Tecimer, M. A. *Tetrahedron: Asymmetry* **2015**, *26*, 966–969.
47. Sugiura, M., Sato, N., Kotani, S., Nakajima, M. *Chem. Commun.* **2008**, *44*, 4309–4311.
48. Sugiura, M., Sato, N., Sonoda, Y., Kotani, S., Nakajima, M. *Chem. Asian J.* **2010**, *5*, 478–481.
49. Sugiura, M., Kumahara, M., Nakajima, M. *Chem. Commun.* **2009**, *45*, 3585–3587.
50. Cao, J.-J., Zhou, F., Zhou, J. *Angew. Chem. Int. Ed.* **2010**, *49*, 4976–4980.
51. Chen, L., Shi, T.-D., Zhou, J. *Chem. Asian J.* **2013**, *8*, 556–559.
52. Chen, L., Du, Y., Zeng, X.-P., Shi, T.-D., Zhou, F., Zhou, J. *Org. Lett.* **2015**, *17*, 1557–1560.
53. Lu, J., Toy, P. H. *Chem. Asian J.* **2011**, *6*, 2251–2254.
54. Teng, Y., Lu, J., Toy, P. H. *Chem. Asian J.* **2012**, *7*, 351–359.
55. Zhang, P., Liu, J., Wang, Z., Ding, K. *Chin. J. Catal.* **2015**, *36*, 100–105.
56. Guizzetti, S., Benaglia, M. *Eur. J. Org. Chem.* **2010**, 5529–5541.
57. Jones, S., Warner, C. J. A. *Org. Biomol. Chem.* **2012**, *10*, 2189–2200.
58. Genoni, A., Benaglia, M., Massolo, E., Rossi, S. *Chem. Commun.* **2013**, *49*, 8365–8367.
59. Kurauchi, D., Hirano, K., Kato, H., Saito, T., Miyamoto, K., Uchiyama, M. *Tetrahedron* **2015**, *71*, 5849–5857.
60. Masaki, M., Fukui, K. *Chem. Lett.* **1977**, 151–152.
61. Xia, X., Toy, P. H. *Beilstein J. Org. Chem.* **2014**, *10*, 1397–1405.
62. van Kalkeren, H. A., van Delft, F. L., Rutjes, F. P. J. T. *ChemSusChem* **2013**, *6*, 1615–1624.
63. Denton, R. M., An, J., Adeniran, B. *Chem. Commun.* **2010**, *46*, 3025–3027.
64. Denton, R. M., An, J., Adeniran, B., Blake, A. J., Lewis, W., Poulton, A. M. *J. Org. Chem.* **2011**, *76*, 6749–6767.
65. Denton, R. M., Tang, X. P., Przeslak, A. *Org. Lett.* **2010**, *12*, 4678–4681.
66. Denton, R. M., An, J., Lindovska, P., Lewis, W. *Tetrahedron* **2012**, *68*, 2899–2905.
67. An, J., Tan, X. P., Moore, J., Lewis, W., Denton, R. M. *Tetrahedron* **2013**, *69*, 8769–8776.
68. Tang, X., An, J., Denton, R. M. *Tetrahedron Lett.* **2014**, *55*, 799–802.
69. Yu, T.-Y., Wang, Y., Xu, P. F. *Chem. Eur. J.* **2014**, *20*, 98–101.
70. Yu, T.-Y., Wang, Y., Hu, X.-Q., Xu, P.-F. *Chem. Commun.* **2014**, *50*, 7817–7820.
71. O'Brien, C. J., Tellez, J. L., Nixon, Z. S., Kang, L. J., Carter, A. L., Kunkel, S. R., Przeworski, K. C., Chass, G. A. *Angew. Chem. Int. Ed.* **2009**, *48*, 6836–6839.
72. O'Brien, C. J., Nixon, Z. S., Holohan, A. J., Kunkel, S. R., Tellez, J. L., Doonan, B. J., Coyle, E. J., Lavigne, F., Kang, L. J., Przeworski, K. C. *Chem. Eur. J.* **2013**, *19*, 15281–15289.
73. O'Brien, C. J., Lavigne, F., Coyle, E. E., Holohan, A. J., Doonan, B. J. *Chem. Eur. J.* **2013**, *19*, 5854–5858.
74. Coyle, E. E., Doonan, B. J., Holohan, A. J., Walsh, K. A., Lavigne, F., Krenske, E. H., O'Brien, C. J. *Angew. Chem. Int. Ed.* **2014**, *53*, 12907–12911.
75. Werner, T., Hoffmann, M., Deshmukh, S. *Eur. J. Org. Chem.* **2014**, 6630–6633.
76. Werner, T., Hoffmann, M., Deshmukh, S. *Eur. J. Org. Chem.* **2014**, 6873–6876.
77. Hoffmann, M., Deshmukh, S., Werner, T. *Eur. J. Org. Chem.* **2015**, 4532–4543.
78. Schirmer, M.-L., Adomeit, S., Werner, T. *Org. Lett.* **2015**, *17*, 3078–3081.
79. Werner, T., Hoffmann, M., Deshmukh, S. *Eur. J. Org. Chem.* **2015**, 3286–3295.

80. Tsai, Y.-L., Lin, W. *Asian J. Org. Chem.* **2015**, *4*, 1040–1043.
81. van Kalkeren, H. A., Leenders, S. H. A. M., Hommerson, C. R. A., Rutjes, F. P. J. T., van Delft, F. L. *Chem. Eur. J.* **2011**, *17*, 11290–11295.
82. van Kalkeren, H. A., Bruins, J. J., Ruitjes, F. P. J. T., van Delft, F. L. *Adv. Synth. Catal.* **2012**, *354*, 1417–1421.
83. van Kalkeren, H. A., te Grotenhuis, C., Haasjes, F. S., Hommerson, C. R. A., Rutjes, F. P. J. T., van Delft, F. L. *Eur. J. Org. Chem.* **2013**, 7059–7066.
84. Kosal, A. D., Wilson E. E., Ashfeld, B. L. *Angew. Chem. Int. Ed.* **2012**, *51*, 12036–12040.
85. Fourmy, K., Voituriez, A. *Org. Lett.* **2015**, *17*, 1537–1540.

7 *N*-Heterocyclic Carbene Catalysis

Homoenolate and Enolate Reactivity

Xinqiang Fang and Yonggui Robin Chi

CONTENTS

7.1 INTRODUCTION

During the past decade, *N*-heterocyclic carbene (NHC) catalysis has been the focus of global research interest.[1] The great substrate compatibility and versatile activation patterns make the utility of NHC catalysis particularly prominent. Through the activation of a large amount of compounds such as aldehydes, α,β-unsaturated aldehydes, ketenes, esters, α,β-unsaturated esters, acids, and α,β-unsaturated acids, NHC catalysis has realized many new transformation modes that were considered difficult or even impossible via other catalysis methods. Although plentiful reports have appeared in the field of NHC catalysis, most of them can be classified as several categories according to the different active intermediates formed during the transformations, such as acyl anion, homoenolate, enolate, acyl azolium, and α,β-unsaturated acyl azolium. In this article, transformations mediated by NHC-catalyzed homoenolate and enolate intermediates and their mechanistic details will be the focus of discussion.

7.2 HOMOENOLATE REACTIVITY IN *N*-HETEROCYCLIC CARBENE CATALYSIS

7.2.1 Reactivity of Homoenolate Derived from Enal Substrates

7.2.1.1 Reactions with Substrates Bearing C=X (X=O, N) Groups

Actually, before 2004, NHC catalysis mainly referred to the acyl anion-mediated benzoin and Stetter reactions.[2] In 2004, the research groups of Glorius and Bode reported simultaneously the formation of γ-butyrolactones through NHC-catalyzed reaction of α,β-unsaturated aldehydes with aromatic aldehydes.[3] This reaction stands for the first conceptual application of the homoenolate intermediate in NHC catalysis. Thus, cinnamaldehyde **1** reacted with catalytic amount of NHC **3** to form homoenolate equivalent **6**, which then underwent the [3 + 2] annulation with aromatic aldehyde **2** to produce lactone **4**, with *cis*-isomer as the major product (Figure 7.1).

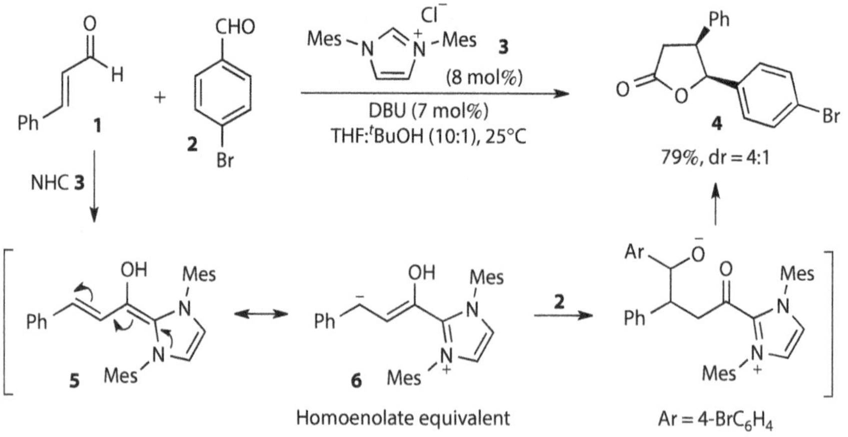

FIGURE 7.1 Homoenolate formation in NHC-catalyzed [3 + 2] annulation.

Sterically bulky imidazolium NHC **3** was proved to be necessary to preclude reactions from the acyl position and shepherd reactivity to the β-position. Although the reaction was a non-asymmetric version, it opened a new window of NHC catalysis.

Subsequently, a series of activated ketones such as trifluoroacetophenone were also found to be active toward the homoenolate intermediate in NHC catalysis.[4] The Nair group showed that multiple activated cyclic ketones such as 1,2-cyclohexanedione and isatin derivatives can participate in similar annulation reactions in the presence of NHC catalyst **3** to afford a series of spirolactones in good to excellent yields.[5] The enantioselective annulation reaction of enals with isatin **7** was completed by Ye et al., using chiral NHC **8** with a proximal hydroxyl substituent, and the corresponding product can be formed with 15:1 diastereomeric ratio (dr) and 99% enantiomeric excess (ee) (Figure 7.2).[6]

In 2005, the Bode group successfully expanded the application of NHC-mediated homoenolate activation to the formation of γ-lactams through the direct annulations of enals with aldimines.[7] Carefully screened *N*-4-methoxybenzenesulfonyl imines such as **10** were critical to prevent the electrophilic inhibition of catalyst by imines (Figure 7.3). And, later, the Bode group further expanded the substrate scope of this [3 + 2] annulation to ketimines that were derived from saccharin.[8]

The asymmetric version of the aforementioned γ-lactam formation was first reported by Chi et al., in 2012. With chiral NHC **13**, the highly enantioselective addition of enals to isatin-derived ketimines provided an efficient synthesis route to optically pure spirocyclic γ-lactams (Figure 7.4).[9]

For a long time, the issue of highly enantioselective formal [3 + 2] annulations of enals with noncyclic ketones through NHC-catalyzed homoenolate reactions was not addressed. Until 2013, with the aid of computational modeling of competing transition states, Scheidt and coworkers designed a new tailored C_1-symmetric biaryl-saturated

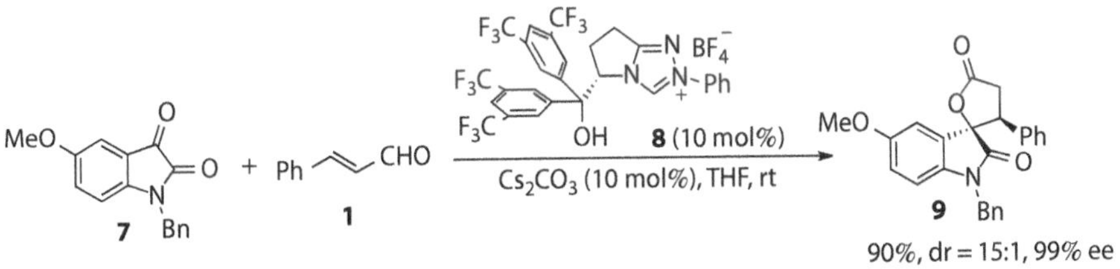

MeO; O; =O; N; Bn; 7; + Ph; CHO; 1; F_3C; CF_3; F_3C; F_3C; =N; BF_4^-; N; N-Ph; +; OH; **8** (10 mol%); Cs_2CO_3 (10 mol%), THF, rt; MeO; O; O; Ph; =O; N; **9** Bn; 90%, dr = 15:1, 99% ee

FIGURE 7.2 NHC-catalyzed spirolactone formation.

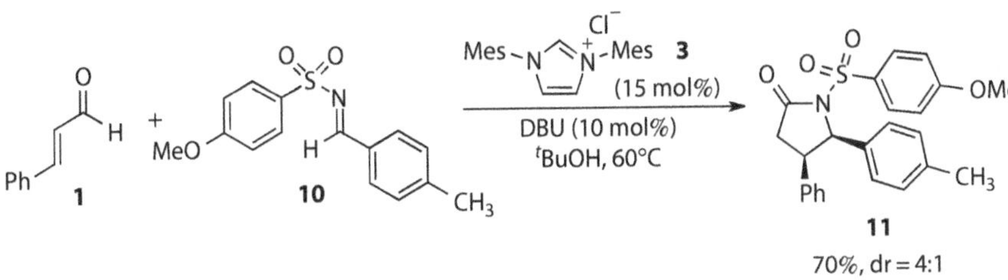

O; H; Ph; **1**; +; MeO; O; O; S; N; H; **10**; CH_3; Cl^-; Mes–N; +N–Mes **3**; (15 mol%); DBU (10 mol%); ^{t}BuOH, 60°C; O; O=S; O; N; OMe; Ph; CH_3; **11**; 70%, dr = 4:1

FIGURE 7.3 γ-Lactam formation in the presence of carbene catalyst.

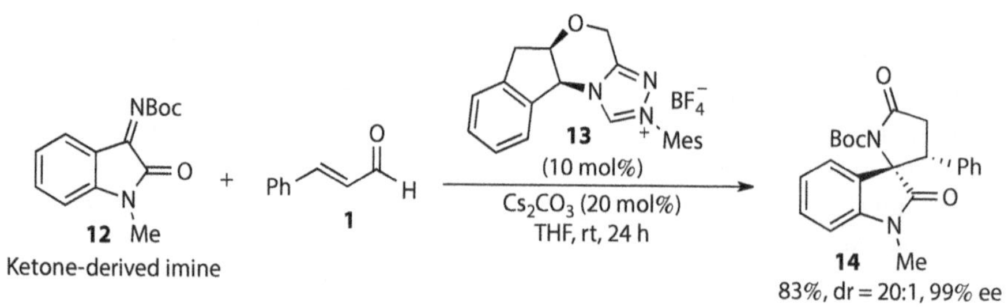

FIGURE 7.4 Enantioselective annulation of cinnamaldehyde with isatin-derived imine.

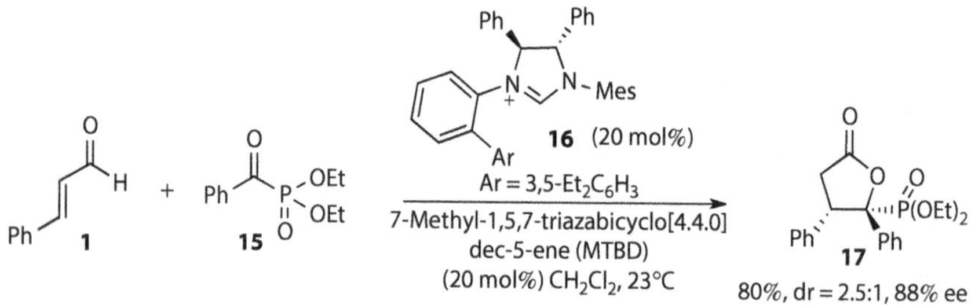

FIGURE 7.5 Enantioselective γ-lactone formation from activated ketone.

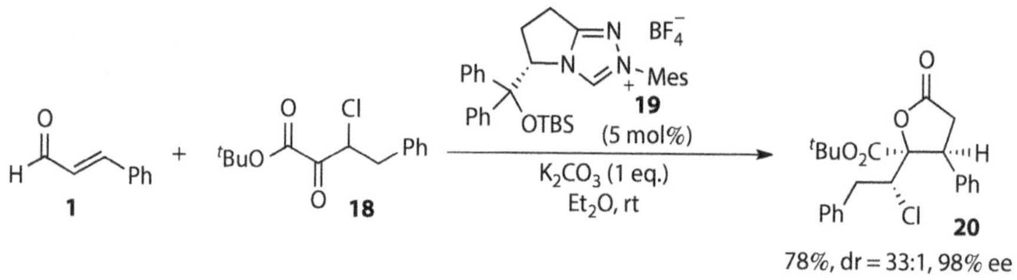

FIGURE 7.6 Dynamic kinetic resolution of β-chloro-α-keto ester.

imidazolium-derived NHC catalyst **16**, which can catalyze the highly selective synthesis of γ-butyrolactones **17** from enals and α-ketophosphonates **15**. The products can be gotten with high enantioselectivity and in good yields, albeit the diastereoselectivity was not higher than 3:1 (Figure 7.5).[10]

The Johnson group very recently reported an elegant work of NHC-catalyzed reaction between enals and β-chloro-α-keto esters.[11] Through dynamic kinetic resolution (DKR), three chiral centers can be formed in the enantiodetermining step, which has not been realized previously. Chiral catalyst **19**[14] was proved to be optimal and product **20**, containing three consecutive chiral centers, can be generated with excellent diastereoselectivity and enantioselectivity (Figure 7.6).

7.2.1.2 Reactions with α,β-Unsaturated Ketones

In 2006, Nair et al. reported a novel cyclopentene formation through NHC-catalyzed annulation–decarboxylation sequence of enals and chalcones.[12] This reaction type has

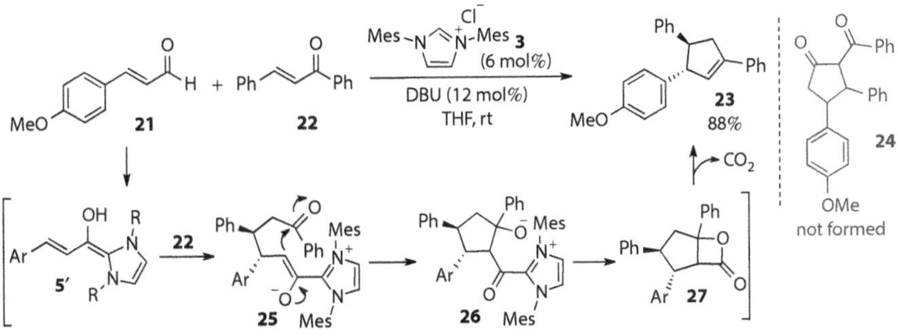

FIGURE 7.7 NHC-catalyzed cyclopentene formation.

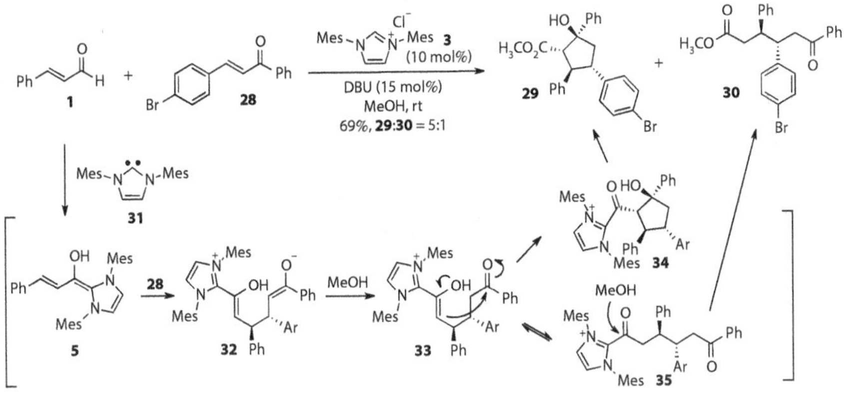

FIGURE 7.8 Reaction of cinnamaldehyde and chalcone in the presence of MeOH.

become one of the basic reaction modes in NHC catalysis. In spite of the proposed C-acylation product **24**, the reaction first underwent a proton transfer to form acyl azolium enolate **25**, and then after intramolecular aldol reaction and decarboxylation, *trans*-cyclopentene **23** was obtained in good yield and with excellent diastereoselectivity (Figure 7.7). The intermediate of β-lactone **34** was observed by Fourier transform infrared spectroscopy (FTIR).

A following work by the Nair group showed that when the same reaction of enal and enone was conducted in MeOH with the same catalyst **3**, the corresponding intermediate **32** formed after the homoenolate addition can be trapped by MeOH to finally afford the mixture of substituted cyclopentane **29** and 6-ketoester **30**, with the former dominated (Figure 7.8).[13]

In 2011, Chi and Fang et al. reported a novel a3–d3 umpolung/Michael/Michael/esterification domino reaction sequence between enals and benzodienones.[14] In the presence of chiral NHC catalyst **13′**, benzobicyclic compound **37** with four contiguous stereogenic centers was formed in moderate to good yields and with excellent diastereo- and enantioselectivities. No cyclopentene products were observed as reported by Nair et al.[12] For unsymmetric benzodienones **36**, the regioselectivity was excellent as long as the substituents had different electronic properties (Figure 7.9).

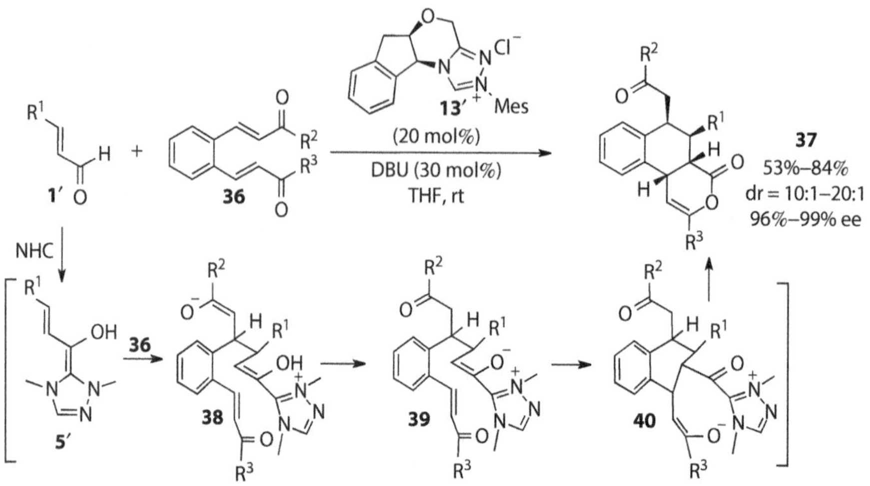

FIGURE 7.9 Cascade annulations of enals with benzodienones.

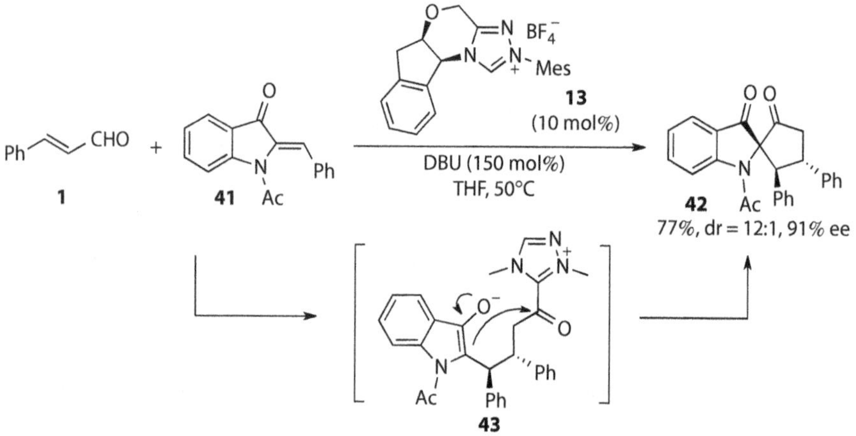

FIGURE 7.10 NHC-catalyzed annulation of enal with azaaurone.

This work provides a rapid entry to relatively complex structures from simple starting materials in a highly selective manner.

In 2015, Glorius et al. showed that with structurally well-defined substrate **41**, the C-acylation can happen in the last step of the mechanism to generate spiroheterocycle **42** that bears a quaternary stereogenic center with high optical purity (Figure 7.10).[15]

The formal [3 + 4] annulation of enal with enone catalyzed by NHC has been a challenge for a long time. But using *o*-quinonemethides as substrates and taking advantage of the driving force of rearomatization, Ye et al. showed that benzo-ε-lactone **46** can be obtained in high yields and with high ee values.[16] Catalyst **45** with free hydroxyl group showed high reactivity, possibly because the hydrogen bonding

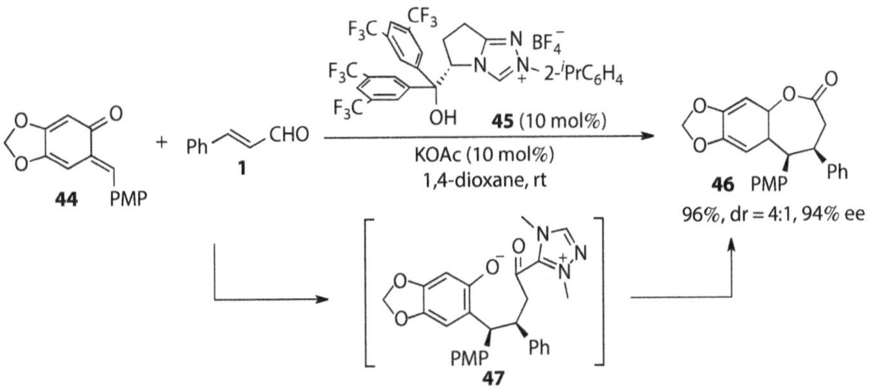

FIGURE 7.11 Enantioselective benzo-ε-lactone formation.

between the OH and *o*-quinonemethide could enhance the substrate reactivity and could improve the enantioselectivity (Figure 7.11).

7.2.1.3 Heterocyclic Compounds Formation

Inspired by the pioneering work of NHC-catalyzed reactions of enals with aldehydes and activated ketones, a series of homoenolate equivalent-mediated annulations for the synthesis of heterocyclic compounds were discovered. As shown in Figure 7.12, Scheidt et al. reported the first formal [3 + 3] cycloaddition of enal with azomethine imine **48** to afford bicyclic heterocycle of pyridazinone **50** with excellent diastereoselectivity.[17]

Following this work, the enantioselective homoenolate additions to nitrone **51** and diazene **55** were able to deliver multiple-substituted chiral γ-amino acid **54** and racemic pyrazolidinone **57** under the catalysis of chiral NHC **52** and **56**, respectively (Figure 7.13).[18]

In 2015, Glorius and coworkers developed the first NHC-catalyzed asymmetric formal [4 + 3] reaction of enal with *in situ* formed azoalkene **60** to generate a diverse set of 1,2-diazepine heterocycle **59** in good yield and with excellent enantioselectivity (Figure 7.14).[19]

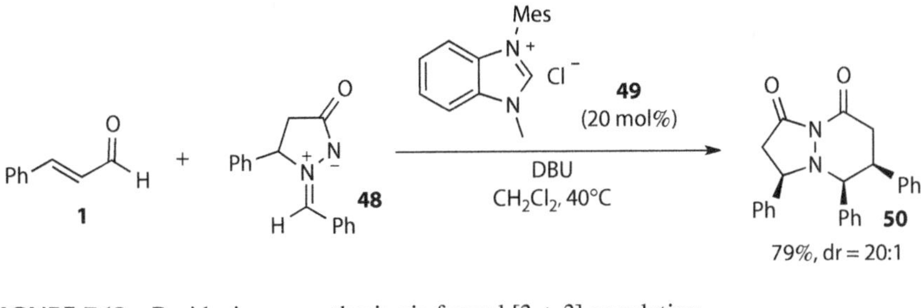

FIGURE 7.12 Pyridazinone synthesis via formal [3 + 3] annulation.

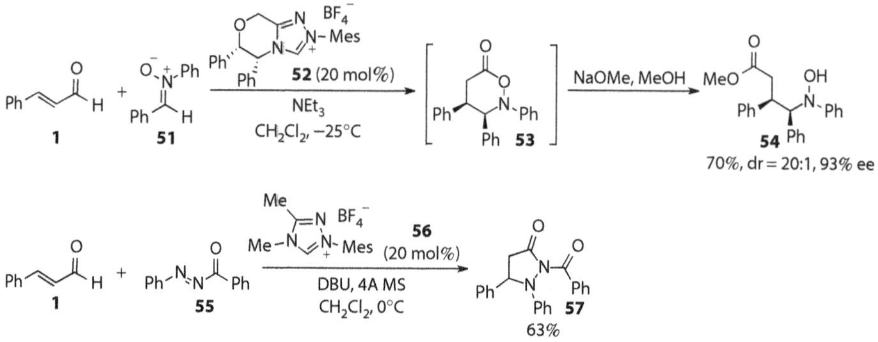

FIGURE 7.13 NHC-catalyzed γ-amino acid and pyrazolidinone synthesis.

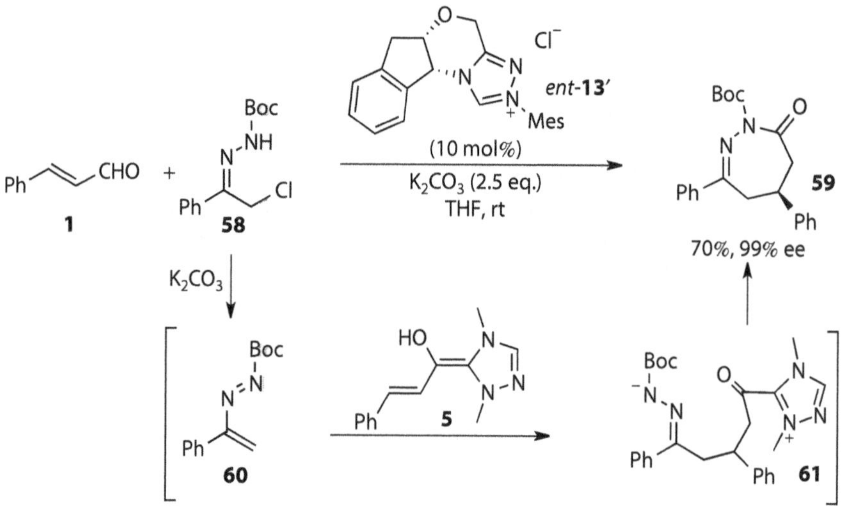

FIGURE 7.14 NHC-catalyzed generation of 1,2-diazepine heterocycle.

7.2.1.4 Addition to Nitroalkenes

Nitroalkene is one of the most useful Michael acceptors, because transformations of the unique nitro group in the resulting products can facilitate further structural elaboration.[20] In 2009, Nair et al. reported the first imidazolium **3**-catalyzed homoenolate addition to β-nitrostyrenes, affording racemic 5-nitroesters in moderate yields, and both the enals and nitroalkenes were limited to aromatic substituents.[21] In 2012, Liu et al. developed an enantioselective version of this reaction. A variety of nitroalkenes, such as nitrodienes, nitroenynes, and nitrostyrenes all tolerated well in the standard conditions, and the *anti*-products were generated with good to excellent enantioselectivities (Figure 7.15).[22]

Rovis and coworkers subsequently developed a highly effective catalytic system for the asymmetric and diastereoselective generation of a diverse array of δ-nitroesters.[23] This work was highlighted by the generation of previously

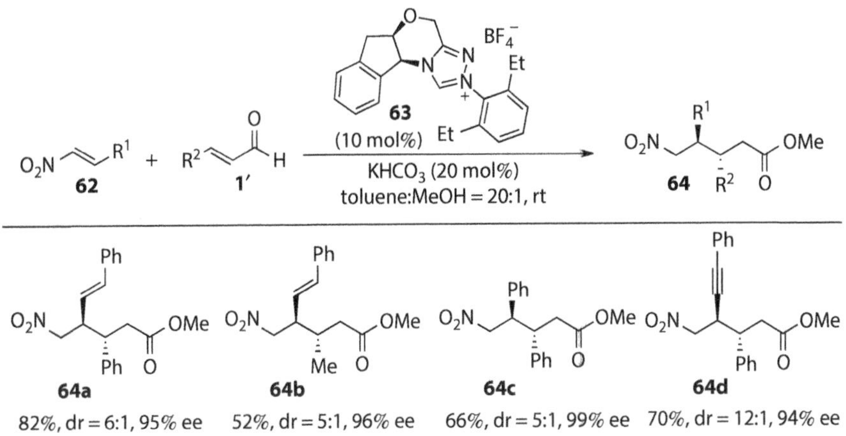

FIGURE 7.15 Enantioselective homoenolate addition to nitroalkenes.

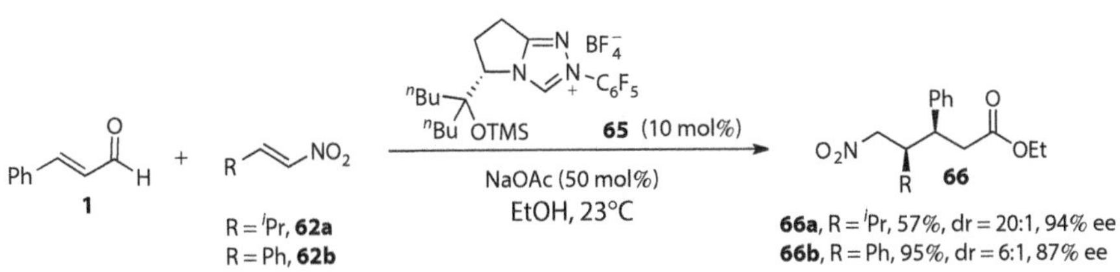

FIGURE 7.16 Enantioselcetive synthesis of *syn*-δ-nitroesters.

inaccessible *syn*-products. Key to the success was the development of NHC catalyst **65** that favored the homoenolate addition pathway instead of the Stetter reaction. This work also shed light on the successful use of aliphatic nitroalkenes as substrates, which had not been realized previously (Figure 7.16).

7.2.1.5 Reactions Mediated by Dual Activation/Catalysis

Although carbene catalysts are highly nucleophilic, in certain cases, they can coexist with other types of catalysts or activators, such as Lewis acids, Brønsted acids, Lewis bases, Brønsted bases, and even transition metals. So, the combination of NHC catalysis with other types of activation modes further expands the scope of influence of NHC catalysis, and new reactivity and selectivity patterns can be reached.

In 2010, Scheidt et al., reported the first path-breaking work of cooperative NHC/Lewis acid catalysis.[24] NHCs, traditionally, are good ligands and will combine with most of the transition metals irreversibly, but Scheidt et al. showed that early metals such as Mg (II) can combine reversibly with NHCs, which provided the opportunity of activating materials simultaneously. Using this strategy, the combination of NHC **63** and the Lewis acid of $Mg(O^tBu)_2$ achieved the synthesis of highly substituted γ-lactams **68** in moderate to good yields and with high levels of diastereo- and enantioselectivity. In contrast, lower yield, dr, and ee were observed in the absence of Lewis acid (Figure 7.17).

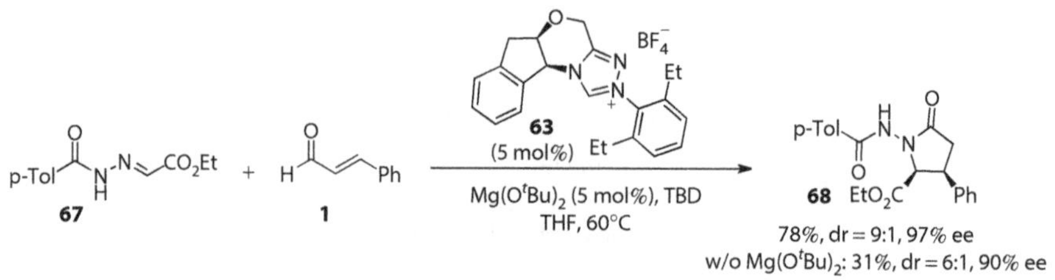

FIGURE 7.17 NHC/Lewis acid cocatalyzed synthesis of γ-lactam.

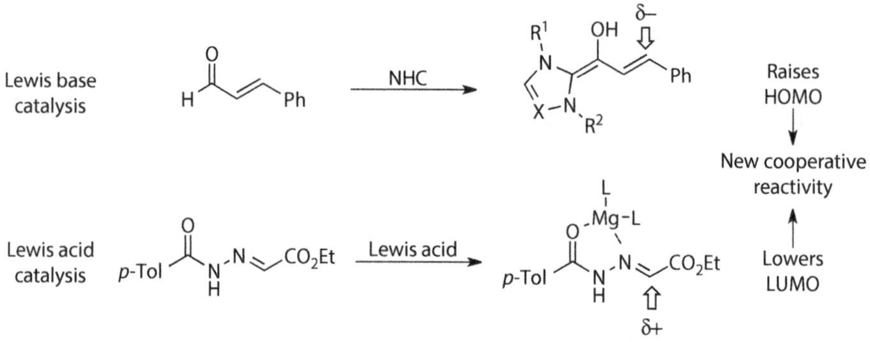

FIGURE 7.18 Cooperative NHC/Lewis acid catalysis strategy.

The dual activation strategy can be illustrated in Figure 7.18. NHC reacts with aldehyde to generate an unusual nucleophile, raising the energy of the highest occupied molecular orbital (HOMO); simultaneously, the Lewis acid activates the hydrazone by lowering the energy of the lowest unoccupied molecular orbital (LUMO) for the overall productive reaction.

Almost at the same time, Scheidt et al., disclosed the annulations of enals with enones, affording *cis*-cyclopentenes **69** with high diastereoselectivity and excellent enantioselectivity in the presence of $Ti(O^iPr)_4$ and chiral NHC **63**.[25] Ti (IV) affects the reaction pathway by coordinating with oxo-intermediates such as **70–72**. To be noted, *trans*-product was produced without Ti (IV) Lewis acid, which further demonstrated the priority of this cocatalysis system (Figure 7.19).

Rovis and coworkers envisioned that if an acid with low pK_a value does not neutralize a carbene, then the carbene and the acid could play different roles in the same reaction system, leading to new reaction discovery. They found that with the catalysis of chiral NHC **76** and *in situ* generated 2-ClPhCO$_2$H (derived from the sodium salt **77**), lactam **78** can be formed in high yield and with good stereocontrol.[26] To be noted, electron-poor NHC catalyst **76** worked well in this homoenolate-mediated transformation, and the product was *trans*-lactam, which was not observed in the previous reports.[7,24] The postulated reaction mechanism for *trans*-lactam formation is shown in Figure 7.20. The hydrogen bonding between aldimine, acid, and the homoenolate intermediate was considered critical in this transformation.

In 2013, Scheidt group reported the first application of NHC/Lewis base dual activation strategy in the synthesis of chiral 2-benzoxopinone **82**. This method further expanded the substrate scope from reactive electrophiles to some less-reactive potential partners. For example, substrate **81** can produce high-reactive *o*-quinone-methide **84** under fluoride conditions through a desilylation/elimination cascade, and the latter can undergo formal [4 + 3] annulation with homoenolate equivalents under the catalysis of NHC.[27] Both of the two reactive species were formed *in situ*, and many different functional groups were tolerated in this relatively complex system (Figure 7.21). This new strategy further showed the great potential of NHC catalysis based on different dual activation modes.

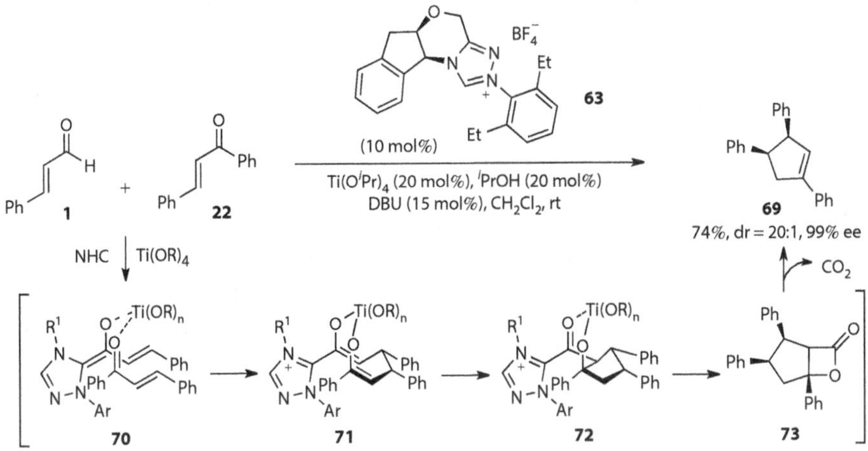

FIGURE 7.19 NHC/Lewis acid cocatalyzed synthesis of *cis*-cyclopentene.

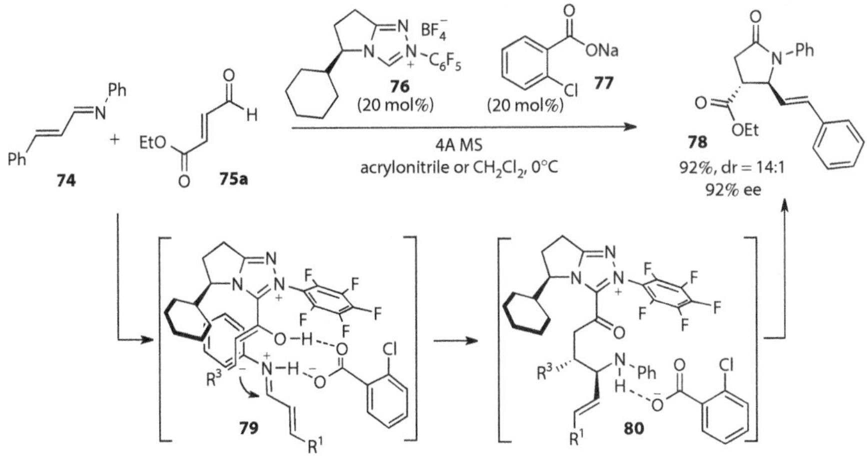

FIGURE 7.20 *trans*-Lactam formation via NHC/Brønsted acid cocatalysis.

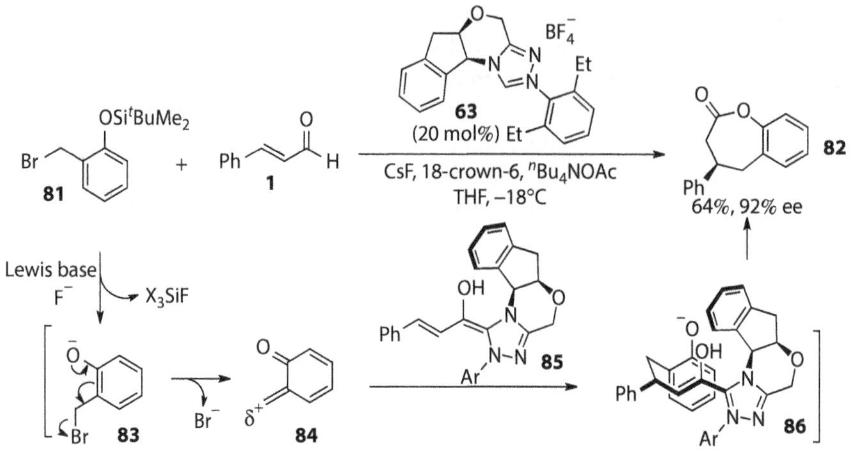

FIGURE 7.21 NHC/Lewis base dual activation for the synthesis of 2-benzoxopinone.

7.2.1.6 Redox Reaction of Enals via Homoenolate Pathway

In 2005, Scheidt et al. reported the first internal redox esterification of enals under the catalysis of NHC.[28] The catalytically formed nucleophilic homoenolate can be protonated to produce enol **89**, and activated ester **90** was then formed after tautomerization. The attack of external electrophiles such as ethanol afforded saturated ester **88** as the final product. PhOH was used as the proton source to improve the yield (Figure 7.22). A variety of enals, including di- and trisubstituted ones, were good candidates under these conditions.

Both the Glorius and Bode groups have attempted to achieve the enantioselective protonation of β-methyl cinnamaldehyde with chiral NHC catalysts, but only moderate ee of 53% was realized.[27] This issue was finally addressed by Scheidt et al., using a cooperative catalysis strategy. In this work, cocatalyst **93** coordinates to the homoenolate intermediate through H-bonding, providing additional steric interactions near

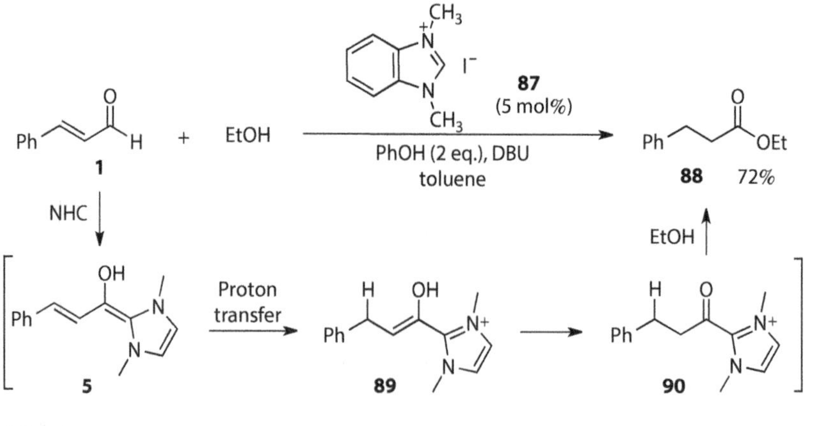

FIGURE 7.22 Transformation of enal to ester.

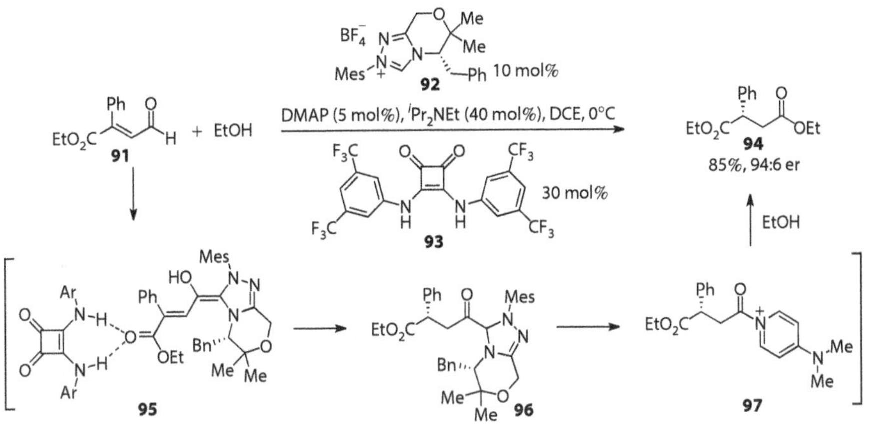

FIGURE 7.23 Enantioselective protonation of β,β-disubstituted enal.

the β-position, and enhanced facial selectivity. 4-Dimethylaminopyridine (DMAP) can enhance the yield by accelerating the carbene catalyst turnover (Figure 7.23).

7.2.1.7 Oxidative *N*-Heterocyclic Carbene Catalysis of Homoenolate Intermediate

In 2010, Studer et al., first reported the two electron oxidation of the homoenolate equivalent **5′** that derived from enal **1′**.[28] The *in situ* formed α,β-unsaturated acyl azolium **102** can undergo 1,4-Michael addition with soft carbon nucleophile such as acetylacetone. After the proton transfer and lactonization, dihydropyranone **101** can be produced in high yields generally (Figure 7.24). Sterically bulky oxidant **100** was a good choice by avoiding the O-acylation of the corresponding acyl azolium. Following this pioneering report, Studer and many other groups have made great contributions to expand this new NHC catalysis strategy.[29] As the real reactive intermediate in this reaction is α,β-unsaturated acyl azolium **102**, detailed demonstration of the application of this intermediate is not the emphasis of this chapter.

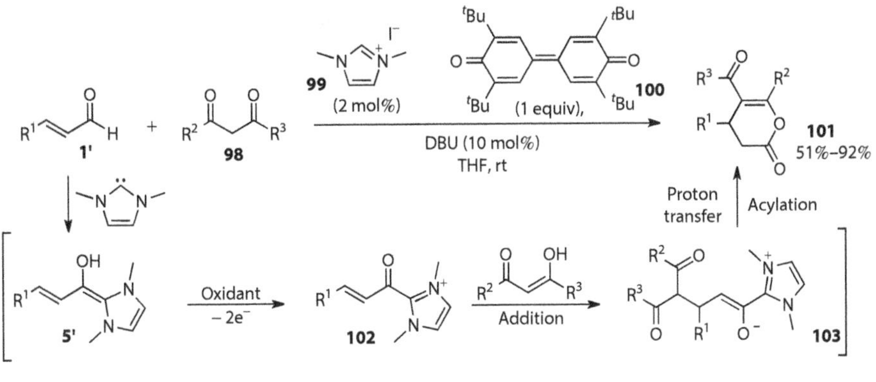

FIGURE 7.24 Dihydropyranone formation via oxidative NHC catalysis.

In 2014, Rovis and coworkers reported a novel work illustrating that alkyl and aryl enals can undergo β-hydroxylation via oxygen atom transfer from electron-deficient nitrobenzenes.[30] To be noted, product **105a** can be afforded in moderate yields and with high enantioselectivity under the catalysis of NHC **65**. The proposed mechanism involves a single electron oxidation of the homoenolate equivalent by nitrobenzene to furnish the radical anion **106** and radical cation **107**. The combination of the two radicals of **106** and **107** affords intermediate **108**, which finally produces β-hydroxy ester **105a** (Figure 7.25). The nitrobenzene derivative with the reduction potential of bigger than −0.49 V is required to oxidize Breslow intermediate **5'**.

Chi and coworkers developed a similar strategy for the synthesis of optically pure β-hydroxyesters.[31] Compared to Rovis' work,[30] a different oxidant of **109** was used, and chiral catalyst *ent*-**13'** was the best choice for both the yields and ee. Two equivalents of enals were necessary to get higher yields. Excellent enantioselectivity (~95% ee for most products) was observed (Figure 7.26). To be noted, the radical clock

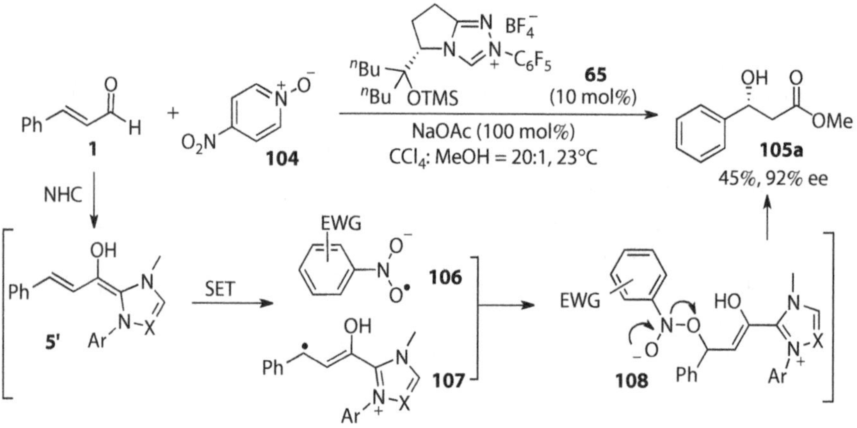

FIGURE 7.25 Rovis' enantioselective β-hydroxylation of cinnamaldehyde.

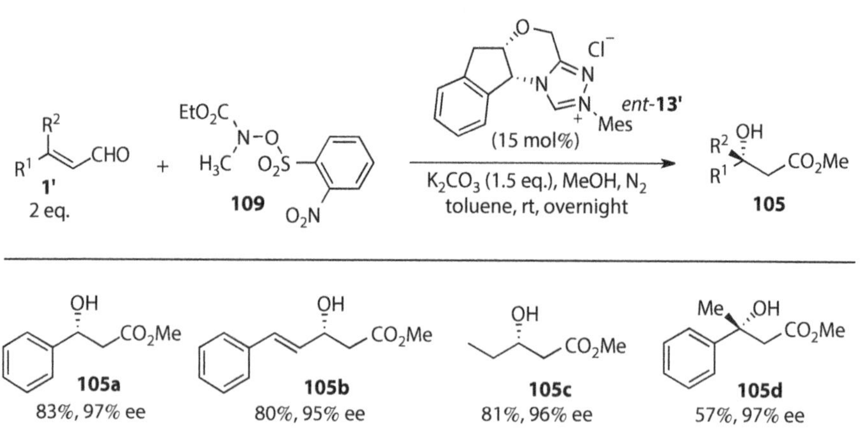

FIGURE 7.26 Chi's enantioselective β-hydroxylation of cinnamaldehyde.

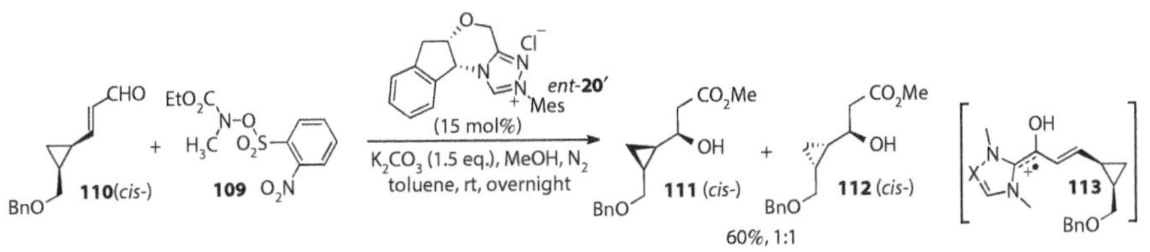

FIGURE 7.27 Radical clock experiment in β-hydroxylation of cinnamaldehyde.

experiment (Figure 7.27) supported the conclusion that the radical spin is more likely delocalized between the enal formyl carbon and the triazolium NHC unit (as showed in **113**, Figure 7.27), because no ring-opening products were detected.

7.2.2 Reactivity of Homoenolate Derived from α′-Hydroxyenones

Although enals have been widely used as homoenolate precursors in NHC catalysis, their drawbacks are also notable: some enals are not stable; usually several steps are needed to make them; and side reaction of dimerization occurs under certain conditions. The Bode group found that α′-hydroxyenones **114**, which are easy for preparation and handling, can be used as enal surrogates in a number of NHC-catalyzed annulation reactions.[32] Side reactions such as decomposition or dimerization can be avoided or diminished with this substrate. A series of transformations have been realized by Bode et al. using α′-hydroxyenones (Figure 7.28). The major disadvantage is that this method is presently limited to prepare racemic products.

By elaborate design of the reaction process, Bode and coworkers developed an elegant relay catalysis approach for the catalytically kinetic resolution of cyclic secondary amines.[33] Chiral hydroxamic acid **120** was the key point of the reaction, but the work also took advantage of the reactivity of acyl azolium **122**, which was produced from the reaction of NHC **115′** with α′-hydroxyenone **118**. The catalytically generated azylazolium **122** was inert to amine **119** but active to hydroxamic acid **120**, and the corresponding ester derivative **123** was the active acylating agent. The *s* factor ranged from 8 to 74 in the kinetic resolution of cyclic secondary amines (Figure 7.29).

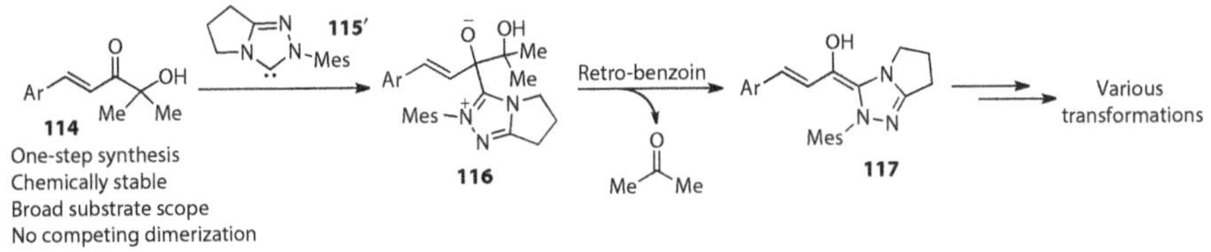

FIGURE 7.28 NHC-catalyzed homoenolate formation from α′-hydroxyenones.

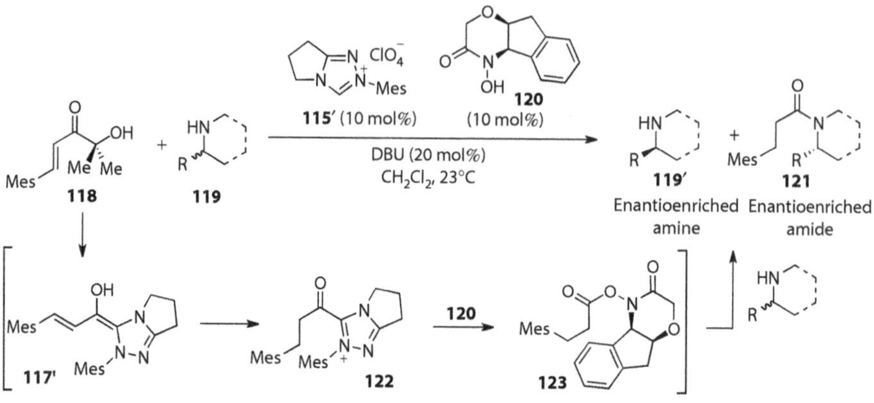

FIGURE 7.29 Dynamic resolution of secondary amines from α′-hydroxyenone.

7.2.3 Reactivity of Homoenolate Derived from Esters

In 2013, Chi and coworkers developed a creative activation strategy for activating β-carbons to form homoenolate equivalent from saturated ester **124**.[34] This work expanded the substrate scope of NHC-mediated homoenolate reactions from enals to saturated substrates. *trans*-Cyclopentene **129**, previously the product from NHC-catalyzed enal reaction, is now available from the saturated ester. The electrophilic substrates were not limited to chalcone **22**. Activated ketone **125** and aldimine **126** also worked well in this reaction system, affording corresponding lactone **128** and lactam **130** with good to excellent stereoselectivities (Figure 7.30). The mechanism is relatively complicated, and the transformation of NHC-mediated enolate **132** to homoenolate equivalent **5′** through β-carbon deprotonation is one of the key steps (Figure 7.31).

The strategy of NHC-catalyzed ester β-carbon activation was further utilized in the reaction of amino enone **136** as electrophile.[35] After the catalytically formed homoenolate intermediate **5′** was added to the enone **136**, an aldol–lactamization–dehydration sequence occurred to afford the multicyclic oxoquinoline-type heterocycle **137** with high enantiomeric ratio (Figure 7.32). The normal lactonization process to form the intermediate of **141** was interrupted by the lactamization step, so cyclopentene **138** was not observed.

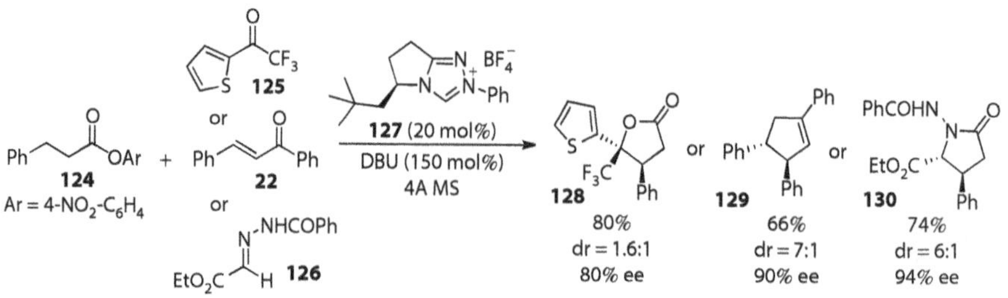

FIGURE 7.30 Chi's homoenolate reaction from saturated esters.

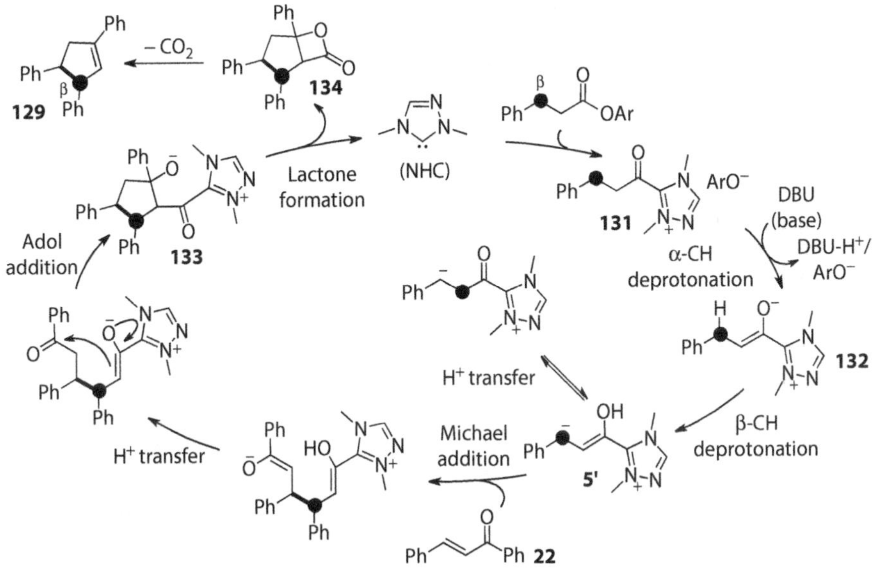

FIGURE 7.31 Organocatalytic β-sp^3-CH activation of saturated ester.

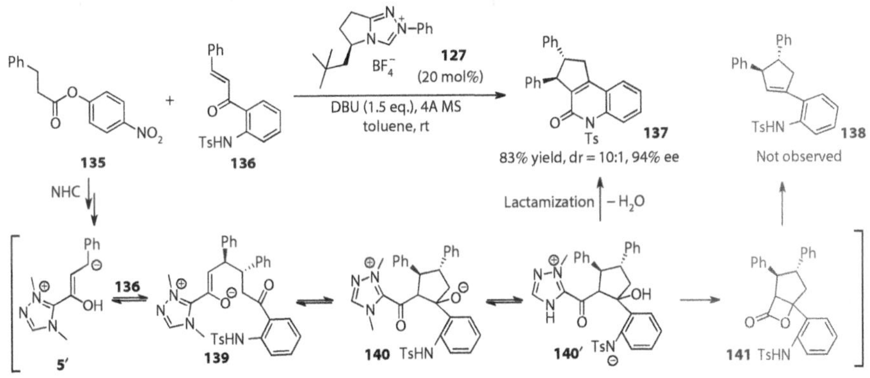

FIGURE 7.32 Oxoquinoline heterocycle formation from ester and amino enone.

7.3 ENOLATE REACTIVITY IN *N*-HETEROCYCLIC CARBENE CATALYSIS

NHC-catalyzed transformations through enolate azoliums have been another important component in the family of NHC catalysis. Enals, α-functionalized aldehydes, ketenes, esters, and acids are all suitable substrates for this fundamental reactivity.[36] A variety of reaction types, such as annulation reactions, redox esterifications, and ring-opening/expansion reactions have been developed. Non-NHC bonded enolate-related reactions, such as the enolates formed from the conjugate addition of carbenes to olefins with electron-withdrawing groups,[37] will not be the subject of discussion in this chapter.

7.3.1 Reactivity of Enolate Derived from Enals (or Enones)

Mechanistically, the homoenolate equivalents derived from the reactions of enals and carbene catalysts can be protonated and can lead to the formation of enolate intermediates. The latter can undergo a series of C–C and C–X (X = F, N, O, S, Cl, etc.) bond formation-related transformations and thus is synthetically very useful (Figure 7.33).

Early in 2006, the Glorius group found the formation of β-lactone **143** and **143′** in carbene **3**-catalyzed reaction of enal **142** and trifluoroacetophenone. The mechanism involves the transformation of the initially formed homoenolate to enolate intermediate **144**, and then a formal [2 + 2] annulation affords the products. Less polar solvent such as toluene, higher temperature, and the base of triethylamine favored the β-lactone formation (Figure 7.34).[38]

Almost at the same time, Bode et al. reported the NHC-catalyzed generation of enolate that participated in LUMO-controlled Diels–Alder cyclization with α,β-unsaturated imines.[39] The corresponding dihydropyridinone product **146** was obtained in >99% ee, and only a single diastereomer was observed. Enals bearing electron-withdrawing groups such as **75** are suitable substrates. It is possibly because the enhanced reactivity of **75** with nucleophilic carbenes favors the protonation of the homoenolate intermediate to generate the enolate (Figure 7.35).

Later, Bode et al.[40] successfully expanded the substrate scope of the aforementioned enantioselective Diels–Alder reactions to a variety of differently substituted enones **22′**. Critical to the success of the enolate generation was the strength of the catalytic base used; with weaker bases such as *N*-methylmorpholine, it favors the pathway of Diels–Alder reaction (Figure 7.36).

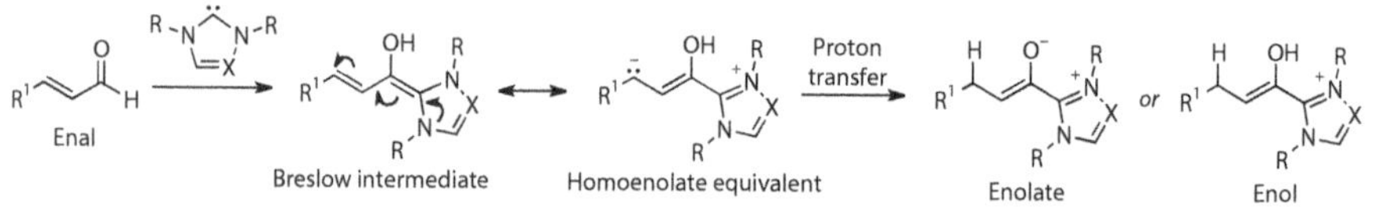

FIGURE 7.33 Mechanism of transformation from homoenolate to enolate intermediate.

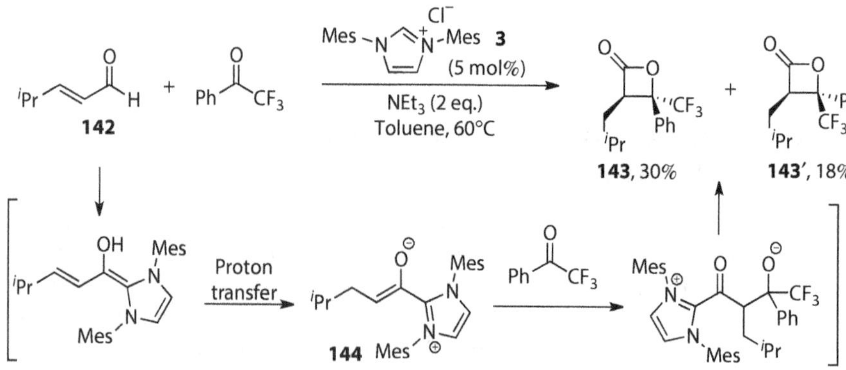

FIGURE 7.34 β-Lactone formation by conjugate umpolung.

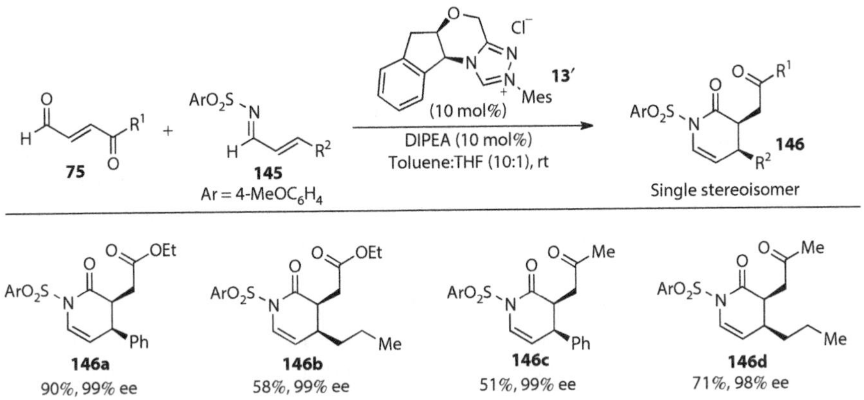

FIGURE 7.35 NHC-catalyzed Diels–Alder reaction of enals with α,β-unsaturated imines.

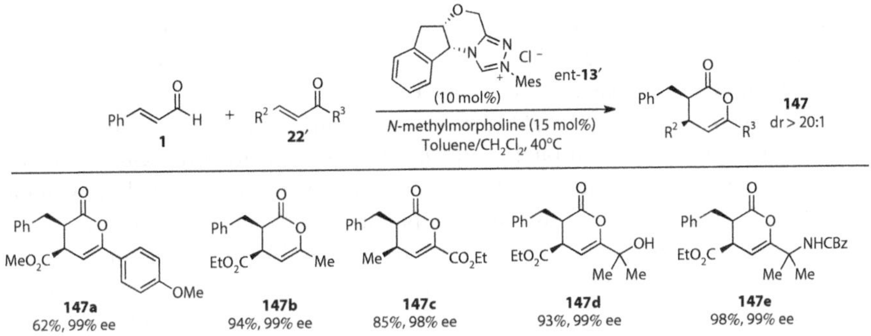

FIGURE 7.36 NHC-catalyzed Diels–Alder reaction of enals with enones.

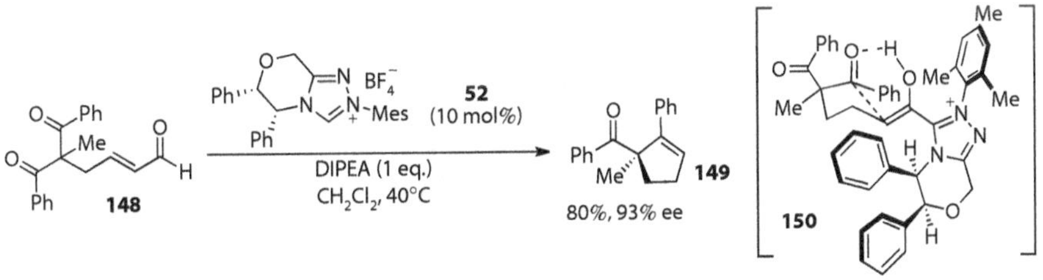

FIGURE 7.37 Desymmetrization of 1,3-diketone.

In the same year, NHC-catalyzed intramolecular aldol reaction that achieved desymmetrization of 1,3-diketone **148** was developed by the Scheidt group.[41] The reaction proceeded with the sequence of enolate formation–aldol reaction–lactonization–decarboxylation to afford the final product of cyclopentene **149** with one quaternary carbon center (Figure 7.37). The proposed mechanistic mode for the reaction involves the *Z(O)*-enol intermediate **150**. The six-membered hydrogen-bonding transition state minimizes the nonbonding interactions between the catalyst, with the phenyl ketone not being attacked.

If there is a substituent at the α-position of the enal, then it is possible that the enolate formed from the protonation of homoenolate intermediate can be protonated enantioselectively under the catalysis of chiral NHC. In 2010, Rovis et al., successfully developed a method for the enantioselective synthesis of α-fluoroacid **153** from α-fluoroenal **151** through chiral NHC **152**-catalyzed enal redox reaction. The key step is the protonation of intermediate **155**. Buffer solution was used to overcome the epimerization of the products (Figure 7.38).[42]

This strategy can also be used in NHC and Brønsted acid cocatalyzed aminomethylation of α,β-unsaturated aldehyde to afford β^2-amino ester **157** in good yield and with high enantioselectivity.[43] The catalytically generated conjugated acid cocatalyst plays dual roles in promoting the generation of azolium enolate intermediate **158** and the formation of formaldehyde-derived iminium ion **159**. The β^2-amino ester **157** can be easily converted into the corresponding N-protected β^2-amino acid by simple operations (Figure 7.39).

In 2012, Chi group disclosed the first oxidative generation of vinyl enolates from **160** to undergo γ-functionalization of enals under NHC catalysis.[44] The corresponding vinyl enolates (dienoate) **164** can undergo formal [4 + 2] annulation to release δ-lactone **161**. Lewis acid proved to be critical to improve both the yield and enantioselectivity, presumably through multisite coordination to bring close both the ketone and the dienolate intermediate (Figure 7.40). This work opened the window of NHC-catalyzed remote chiral control of previously challenging transformations.

NHC-catalyzed highly diastereo- and enantioselective [3 + 4] annulation of azomethine imines and enals was also revealed by Chi group.[45] Racemic azomethine imine **48** can be used as 1,3-dipolar substrate to react with enal-derived dienolate to afford dinitrogen-fused cyclic compound **167** with high ee. Highly effective kinetic resolution of azomethine imine **48** can also be achieved. It is impressive that the stereocontrol of the remote chiral center in the substrate **48** was still very well (Figure 7.41).

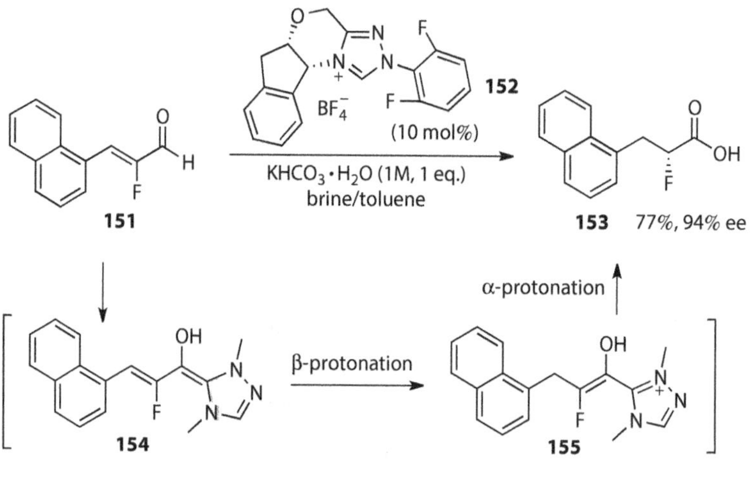

FIGURE 7.38 Enantioselective synthesis of α-fluoroacid.

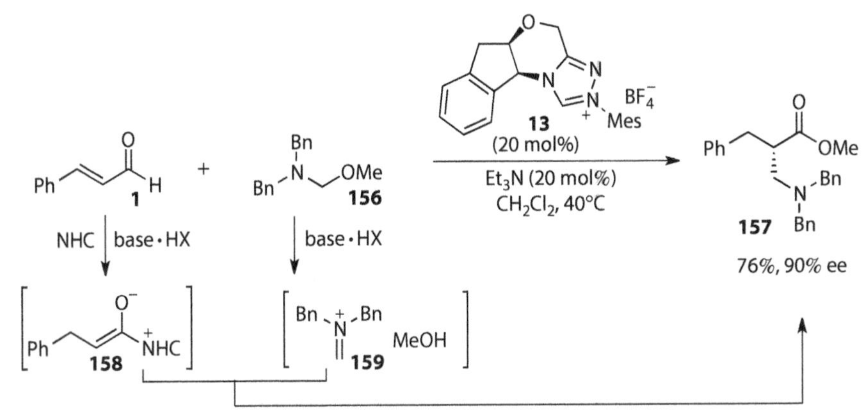

FIGURE 7.39 Enantioselective α-aminomethylation of enal.

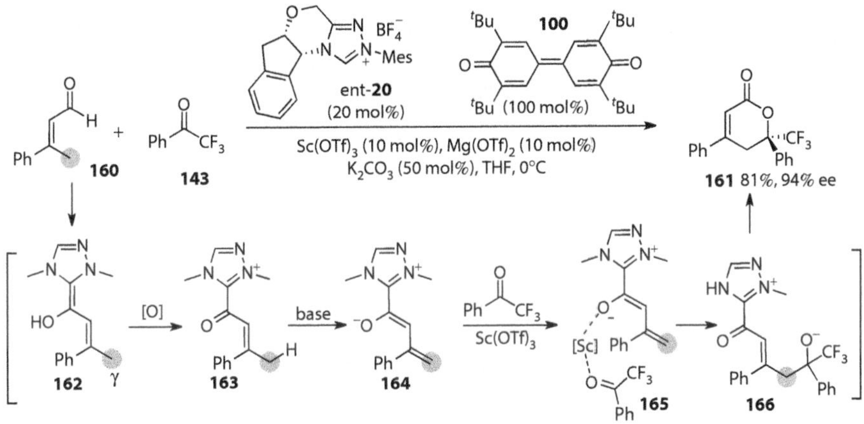

FIGURE 7.40 NHC-catalyzed γ-functionalization of enal.

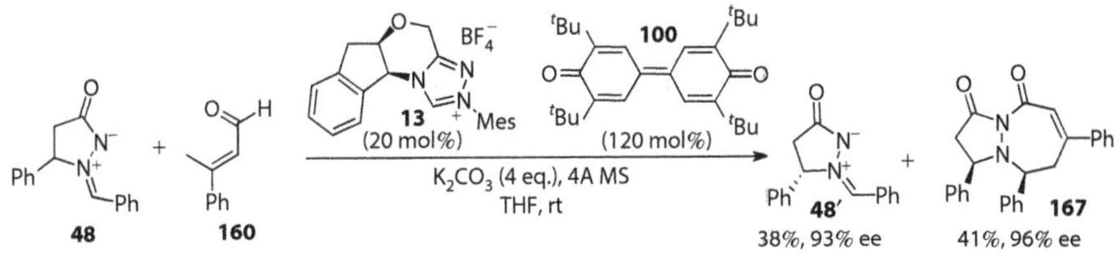

FIGURE 7.41 NHC-catalyzed kinetic resolution of azomethine imine.

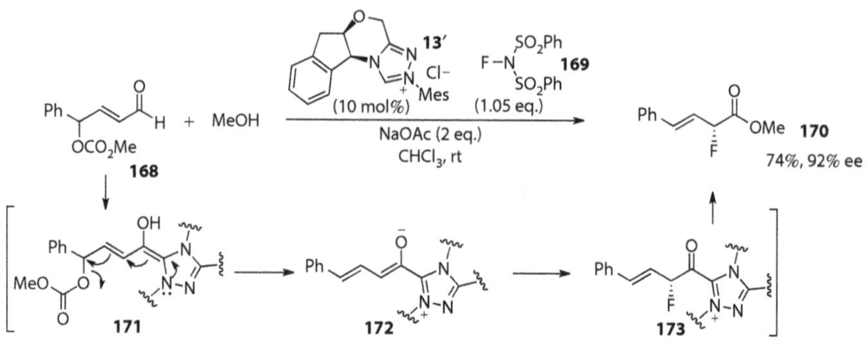

FIGURE 7.42 Enantioselective α-fluoroester formation.

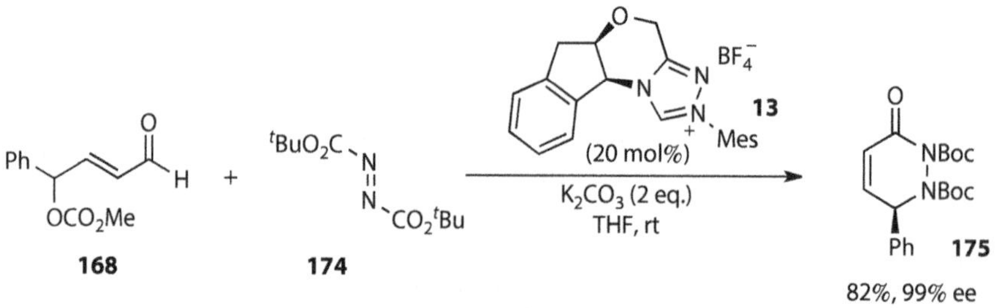

FIGURE 7.43 Stereoselective dihydropyridazinone synthesis.

Sun and coworkers showed that, under NHC catalysis, enals bearing a leaving group in the γ-position generate NHC-bound dienolate **172**, which can react with electrophilic fluorinating reagent **169** N-fluorobenzenesulfonimide (NSFI) to afford fluorinated aster **170** in good yield and with high ee.[46] α-Position of the dienolate was proved to be more reactive than the γ-position. Mechanistic study revealed that free NHC reacts faster with aldehydes than NFSI, which guarantees the smooth turnover of the catalyst (Figure 7.42).

Different to the Sun group's work, Ye et al. reported the formal [4 + 2] annulation of enal **168** bearing γ-leaving group with azodicarboxylate **174**, giving dihydropyridazinone **175** in good yield and with excellent enantioselectivity (Figure 7.43).[47]

Dienolate formation from the reaction of NHC with cyclobutenone **176** was disclosed by Chi group.[48] Ring opening of the adduct **179** generates dienolate **180**, and after formal [4 + 2] annulation with imine **177**, lactam **178** can be afforded with good enantioselectivity (Figure 7.44).

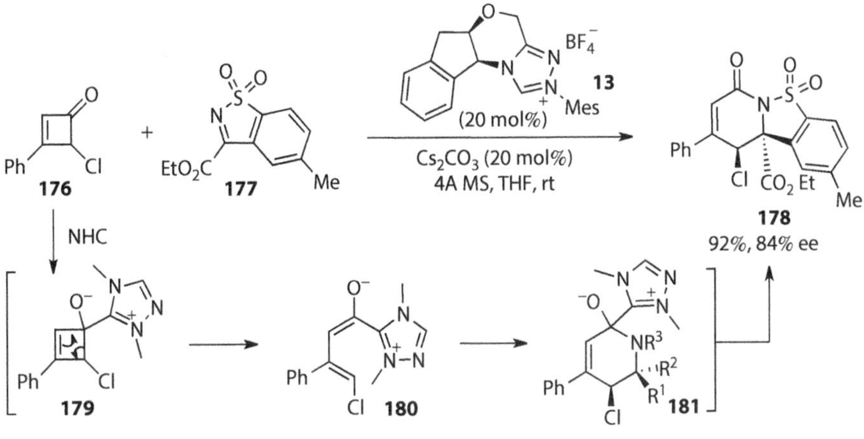

FIGURE 7.44 NHC-catalyzed reaction with cyclobutenone as substrate.

7.3.2 Reactivity of Enolate Derived from α-Functionalized Aldehydes

In 2004, Rovis group reported the conversion of α-haloaldehydes into esters by an internal redox reaction catalyzed by nucleophilic carbenes.[49] The mechanism involves enol/enolate **186** formation from the adduct of aldehyde **182** with carbene, and after tautomerization or protonation, activated acyl azolium **187** is formed. Nucleophilic attack of this intermediate by alcohol affords ester product and regenerates carbene (Figure 7.45).

Using similar strategy, Rovis and coworkers were able to demonstrate a unique synthesis of chiral α-chloroester **190** based on an enantioselective protonation of *in situ* generated chiral α-chloroenolate **192** (Figure 7.46).[50] Later, this strategy was

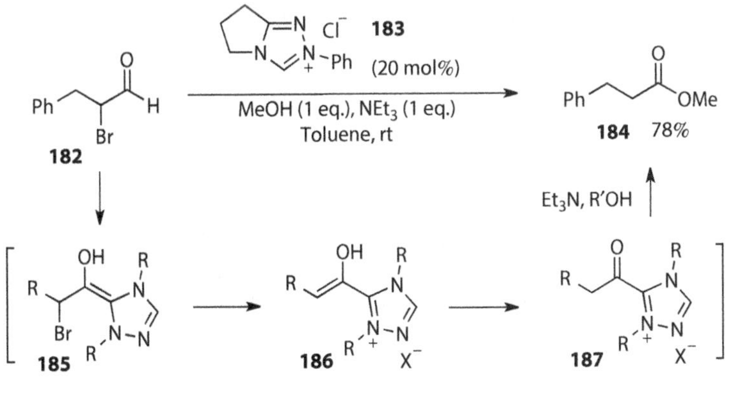

FIGURE 7.45 Internal redox reaction of α-haloaldehydes.

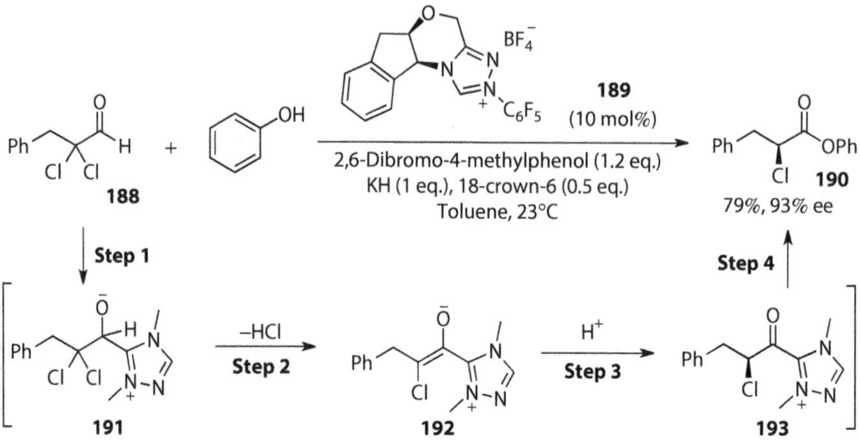

FIGURE 7.46 Enantioselective α-chloroester formation.

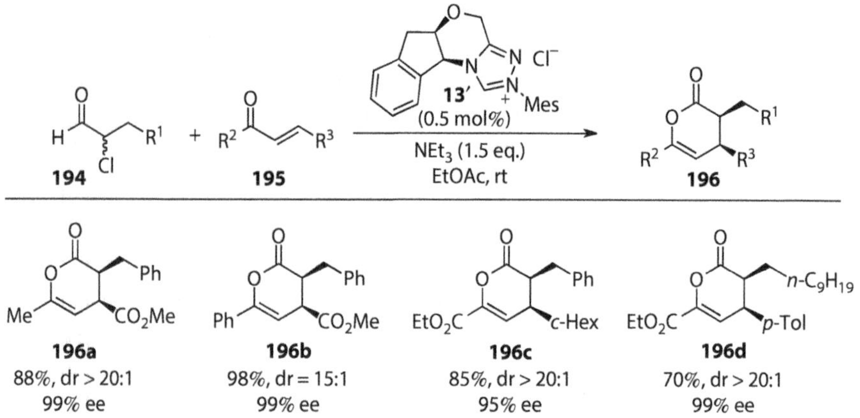

FIGURE 7.47 Diels–Alder reaction with α-chloroaldehydes as substrates.

employed to synthesize α-chloroamides and α-chloroacids under the asymmetric catalysis of NHC.[42,51]

Bode and coworkers showed that α-chloroaldehydes **194** can be used as excellent enolate precursor in NHC-catalyzed Diels–Alder reaction with enones **195**.[52] Only 0.5 mol% of catalyst loading is necessary to achieve high yields and excellent stereoselectivities (Figure 7.47).

Scheidt group developed a highly selective and versatile Mannich reaction using α-aryloxyacetaldehyde **197** and carbene catalyst **199** to initiate the enol/enolate formation. In the presence of activated imine **198**, a Mannich reaction occurs to afford β-amino acyl azolium intermediate. After acylation, the corresponding ester can be intercepted *in situ* by amine to yield valuable nitrogen-containing compound **200** (Figure 7.48).[53]

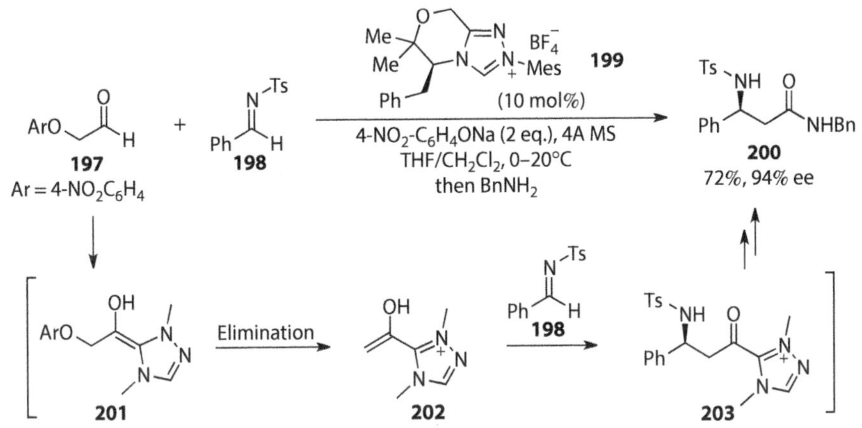

FIGURE 7.48 NHC-catalyzed enantioselective Mannich reaction.

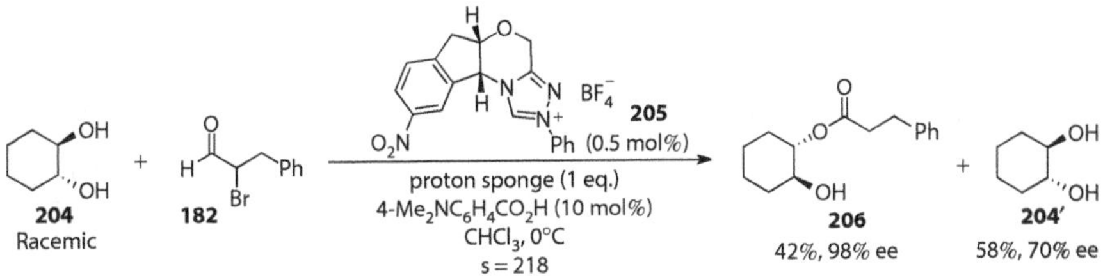

FIGURE 7.49 Kinetic resolution and desymmetrization of secondary alcohol.

In 2013, Yamada et al. developed a highly enantioselective kinetic resolution of secondary alcohols that possess an adjacent hydrogen-bond donor such as **204**, using a newly developed NHC **205** that bears a nitro group on the indane moiety.[54] The presence of a carboxylate cocatalyst can markedly enhance the reaction rate and selectivity, which likely works as a base to facilitate the C–O bond-forming step (Figure 7.49).

7.3.3 Reactivity of Enolate Derived from Ketenes

In 2008, Ye and Smith et al. independently reported the NHC-catalyzed Staudinger reaction between imine **207** and ketene **208**.[55] Using triazolium catalyst **209**, Ye and coworkers were able to develop the highly stereoselective [2 + 2] annulation for the synthesis of β-lactam **210**. Azolium enolate **211** is the key intermediate derived from the addition of carbene to ketene (Figure 7.50).

Since then, Ye group has been devoted to the NHC-catalyzed annulations between ketenes and a series of alternate electrophiles.[56] Some selected products are summarized in Figure 7.51.

The enolates generated from ketenes and carbenes can also be stereoselectively protonated to produce chiral esters. Ye and Smith et al. indicated that the

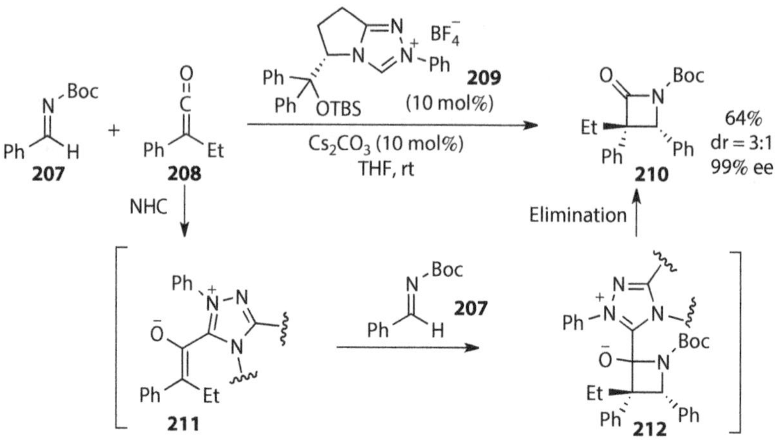

FIGURE 7.50 [2 + 2] Annulation of ketene with imine.

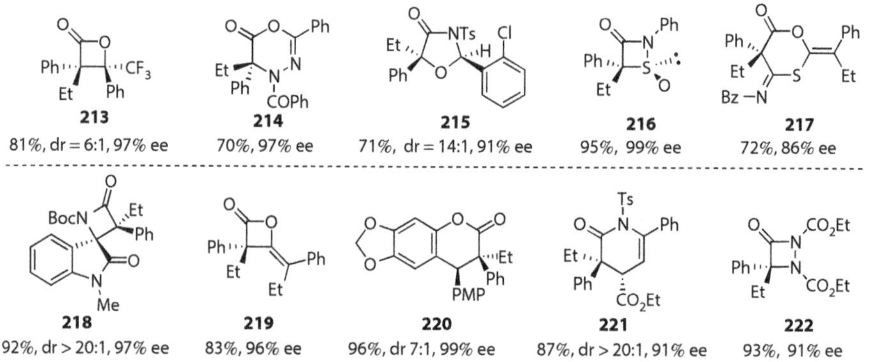

FIGURE 7.51 Selected products of NHC-catalyzed reactions with ketenes as substrates.

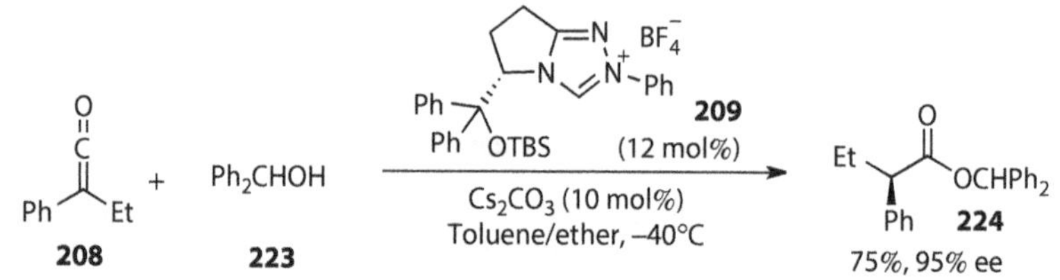

FIGURE 7.52 Stereoselective α-protonation of ketene.

corresponding esters can be obtained with up to 95% ee under the catalysis of chiral NHCs (Figure 7.52).[57]

Enantioselective chlorination of azolium enolates from ketenes was reported by Smith and coworkers.[58] Chiral triazolium **209** resulted in the highest asymmetric induction by using polyhalogenated quinines such as **225** (Figure 7.53).

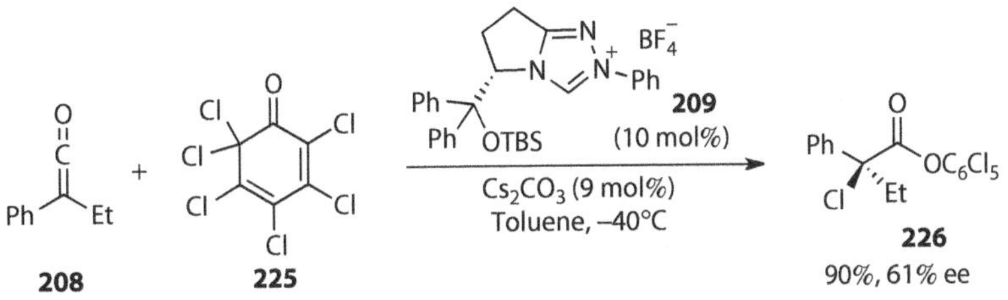

FIGURE 7.53 Stereoselective α-chlorination of ketene.

7.3.4 Reactivity of Enolate Derived from Saturated Aldehydes/Esters/Acids

The Rovis and Chi groups independently developed the oxidative NHC catalysis of simple saturated aldehydes for the formation of enolate intermediates that participated in Diels–Alder reactions with enones or α,β-unsaturated imines.[59] The mechanism involves the generation of acyl azolium **230** from the oxidation of **229** first and then the formation of enolate **231** from the deprotonation of **230**. Under the catalysis of NHC *ent*-**13**, product **228** was generated in very high yield and with excellent ee (Figure 7.54).

Based on this strategy, the Wang and Sun groups simultaneously reported the enantioselective α-fluoronation of enolate azoliums generated from the oxidative NHC catalysis of aliphatic aldehydes **232**.[60] The process provided facile access to a wide range of α-fluoro esters **233** with excellent enantioselectivity. Both work highlighted on the reagent of NSFI that acted as both *F* source and an oxidant (Figure 7.55).

Since 2012, the Chi group has been devoted to the activation of stable carboxylate esters with NHC catalysts.[61] The suitable ester needs to be both stable and reactive

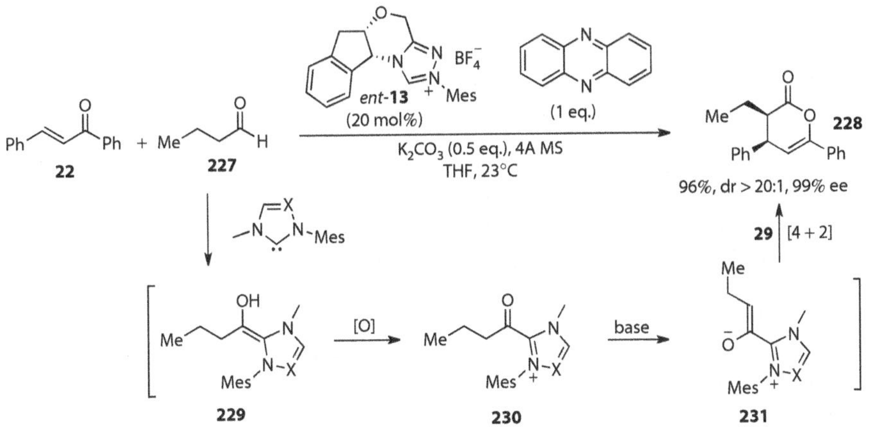

FIGURE 7.54 Oxidative enolate formation from simple aldehyde.

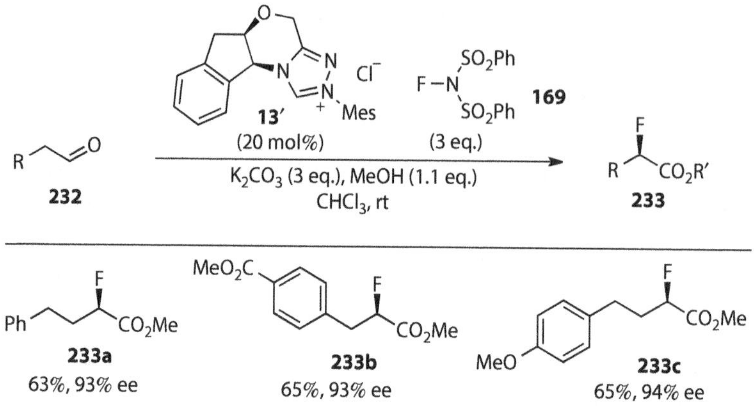

FIGURE 7.55 Stereoselective α-fluorination of simple aldehyde.

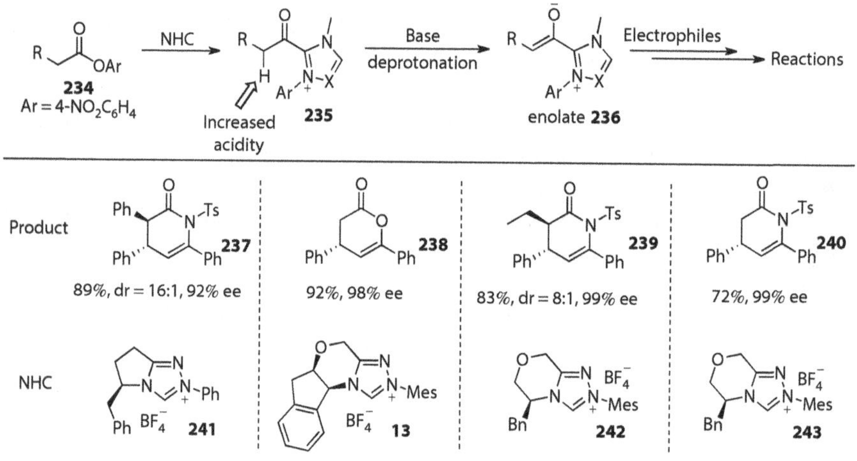

FIGURE 7.56 Carbene-catalyzed enolate formation from stable esters.

enough toward a nucleophilic carbene. 4-Nitrophenyl esters **234** proved to be a good choice. The typical utility of this study is the formation of a series of enolate equivalents that participate in the Diels–Alder reactions with enones or α,β-unsaturated imines (Figure 7.56).

In 2014, Scheidt et al., developed a convergent, catalytic asymmetric formal [4 + 2] annulation for the synthesis of dihydroquinolone **247** (Figures 7.57).[62] Carboxylic acid **244** was employed as the precursor of enolate **249** through an *in situ* activation strategy. At the same time, *in situ* generated aza-*o*-quinonemethide **250** can react with enolate **249** to provide dihydroquinolone **247** with good to excellent ee (Figure 7.58).

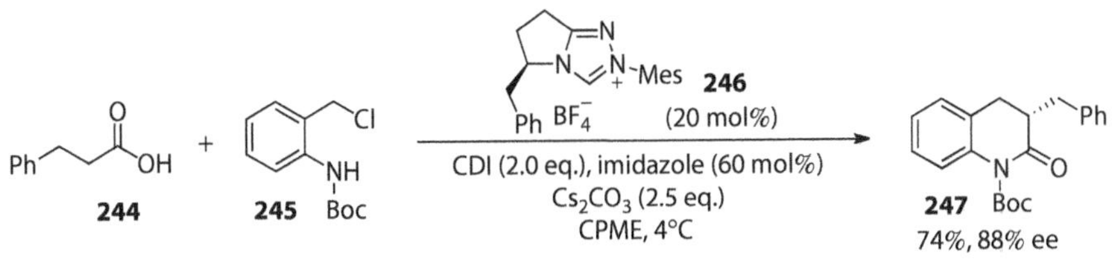

FIGURE 7.57 Synthesis of dihydroquinolone from saturated acid.

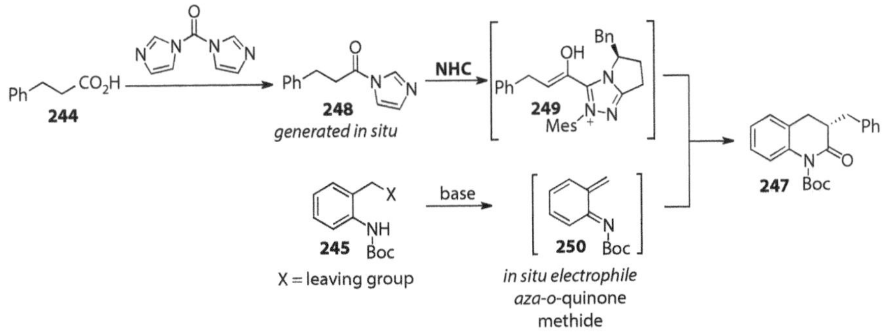

FIGURE 7.58 Mechanism of the formation of dihydroquinolone.

7.3.5 Reactivity of Enolate Derived from Small Ring Opening

Aldehydes bearing small ring structures adjacent to the carbonyl groups are prone to undergo ring opening in the presence of NHC catalysts; enol/enolate structures are thus formed and can be used in further transformations. In 2004, Bode et al. reported the direct and stereoselective synthesis of β-hydroxyesters **253** from epoxyaldehydes **251** with thiazolium **252** as the catalyst.[63] Mechanistically, the catalytically formed enolate intermediate **256** is protonated to form activated carboxylate **257**, and further acylation affords the ester **253** (Figure 7.59). Following this work, Bode et al. later documented the NHC-catalyzed redox amidation of α-cyclopropyl aldehydes with amines.[64]

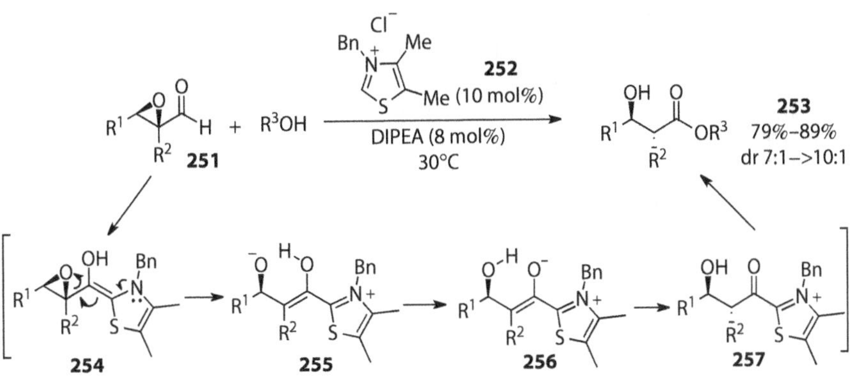

FIGURE 7.59 NHC-catalyzed synthesis of β-hydroxyesters from epoxyaldehydes.

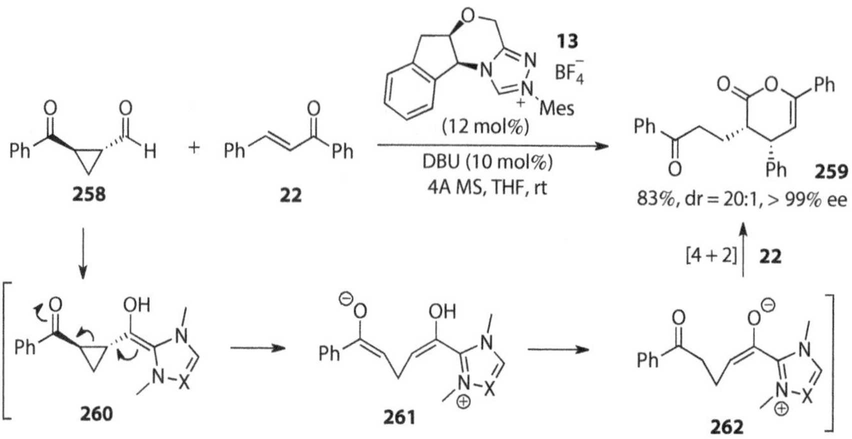

FIGURE 7.60 δ-Lactone formation from α-cyclopropyl aldehyde.

The enolate formed from ring opening can also be used in annulation reactions. Chi and coworkers found that enolate intermediate **262** derived from the NHC-catalyzed ring opening of α-cyclopropyl aldehyde **258** can participate in the formal Diels–Alder reaction with enone **22**.[65] The resulting functionalized δ-lactone product **259** can be easily obtained with high yield and excellent diastereo- and enantioselectivity (Figure 7.60).

7.4 CONCLUSION AND OUTLOOK

The application of NHCs as catalysts for reactions that are mediated by homoenolate and enolate intermediates has grown dramatically over the past decade. The unique *umpolung* feature of NHC catalysis constitutes the foundation of its novel reactivity. The combination of these two reaction patterns with other activation methods such as Lewis acid/base activation and Brønsted acid/base activation has widely broadened their domain of influence. The integration of experimental data with computational analysis has also provided additional insights into the mechanistic details of these powerful reactions that will drive further development in the field of NHC catalysis. We have also seen the promising potential for the application of carbene catalysis in the field of total synthesis. It can be foreseen that more new types of molecules activated by NHCs will be disclosed, new bond cleavage/formation modes will be established, and new synthetic strategies will be developed in the future, which will finally be of great value in the fields of total synthesis of natural products, discovery of medicine, and finally the development of organic chemistry.

REFERENCES

1. For selected recent reviews of NHC catalysis, see: (a) Enders, D., Balensiefer, T. *Acc. Chem. Res.* **2004**, *37*, 534–541. (b) Enders, D., Niemeier, O., Henseler, A. *Chem. Rev.* **2007**, *107*, 5606–5655. (c) Nair, V., Vellalath, S., Babu, B. P. *Chem. Soc. Rev.* **2008**, *37*, 2691–2698. (d) Phillips, E. M., Chan, A., Scheidt, K. A. *Aldrichimica Acta* **2009**,

42, 55–56. (e) Nair, V., Menon, R. S., Biju, A. T., Sinu, C. R., Paul, R. R., Jose, A., Sreekumar, V. *Chem. Soc. Rev.* **2011**, *40*, 5336–5346. (f) Ryan, S. J., Candish, L., Lupton, D. W. *Chem. Soc. Rev.* **2013**, *42*, 4906–4917. (g) Mahatthananchai, J., Bode, J. W. *Acc. Chem. Res.* **2014**, *47*, 696–707. (h) Hopkinson, M. N., Richter, C., Schedler, M., Glorius, F. *Nature* **2014**, *510*, 485–496.

2. For selected representative reports of benzoin reactions before 2004, see: (a) Enders, D., Kallfass, U. *Angew. Chem., Int. Ed.* **2002**, *41*, 1743–1745. (b) Kerr, M. S., de Alaniz, J. R., Rovis, T. *J. Am. Chem. Soc.* **2002**, *124*, 10298–10299. (c) Ugai, T., Tanaka, S., Dokawa, S. *J. Pharm. Soc. Jpn.* **1943**, *63*, 296–300. (d) Breslow, R. *J. Am. Chem. Soc.* **1958**, *80*, 3719–3726. (e) Stetter, H., Rämsch, R. Y., Kuhlmann, H. *Synthesis* **1976**, 733–735. (f) Stetter, H., Kuhlmann, H. *Org. React.* **1991**, *40*, 407–496. (g) Sheehan, J., Hunneman, D. H. *J. Am. Chem. Soc.* **1966**, *88*, 3666–3667. (h) Sheehan, J., Hara, T. *J. Org. Chem.* **1974**, *39*, 1196–1199. (i) Knight, R. L., Leeper, F. J. *J. Chem. Soc., Perkin Trans.* **1998**, *1*, 1891–1893. (j) Hachisu, Y., Bode, J. W., Suzuki, K. *J. Am. Chem. Soc.* **2003**, *125*, 8432–8433. (k) For selected representative reports of Stetter reactions before 2004, see: (k) Stetter, H. *Angew. Chem., Int. Ed.* **1976**, *15*, 639–712. (l) Enders, D. *In Stereoselective Synthesis*. Springer-Verlag: Heidelberg, Germany, 1993, p. 63. (m) Ciganek, E. *Synthesis* **1995**, 1311–1315, (n) Enders, D., Breuer, K., Runsink, J., Teles, J. H. *Helv. Chim. Acta* **1996**, *79*, 1899–1902.
3. (a) Burstein, C., Glorius, F. *Angew. Chem.,Int. Ed.* **2004**, *43*, 6205–6208. (b) Sohn, S. S., Rosen, E. L., Bode, J. W. *J. Am. Chem. Soc.* **2004**, *126*, 14370–14371.
4. (a) Hirano, K., Piel, I., Glorius, F. *Adv. Syn. Catal.* **2008**, *350*, 984–988. (b) Li, Y., Zhao, Z.-A., He, H., You, S.-L. *Adv. Syn. Catal.* **2008**, *350*, 1885–1890.
5. Nair, V., Vellalath, S., Poonoth, M., Mohan, R., Suresh, E. *Org. Lett.* **2006**, *8*, 507–509.
6. Sun, L.-H., Shen, L.-T., Ye, S. *Chem. Commun.* **2011**, *47*, 10136–10138.
7. He, M., Bode, J. W. *Org. Lett.* **2005**, *7*, 3131–3134.
8. Rommel, M., Fukuzumi, T., Bode, J. W. *J. Am. Chem. Soc.* **2008**, *130*, 17266–17267.
9. Lv, H., Tiwari, B., Mo, J., Xing, C., Chi, Y. R. *Org. Lett.* **2012**, *14*, 5412–5415.
10. Jang, K. P., Hutson, G. E., Johnston, R. C., McCusker, E. O., Cheong, P. H.-Y., Scheidt, K. A. *J. Am. Chem. Soc.* **2014**, *136*, 76–79.
11. Goodman, C. G., Walker, M. M., Johnson, J. S. *J. Am. Chem. Soc.* **2015**, *137*, 122–125.
12. Nair, V., Vellalath, S., Poonoth, M., Suresh, E. *J. Am. Chem. Soc.* **2006**, *128*, 8736–8737.
13. Nair, V., Babu, B. P., Vellalath, S., Varghese, V., Raveendran, A. E., Suresh, E. *Org. Lett.* **2009**, *11*, 2507–2510.
14. Fang, X., Jiang, K., Xing, C., Hao, L., Chi, Y. R. *Angew. Chem. Int. Ed.* **2011**, *50*, 1910–1913.
15. Guo, C., Schedler, M., Daniliuc, C. G., Glorius, F. *Angew. Chem. Int. Ed.* **2014**, *53*, 10232–10236.
16. Lv, H., Jia, W.-Q., Sun, L.-H., Ye, S. *Angew. Chem. Int. Ed.* **2013**, *52*, 8607–8610.
17. Chan, A., Scheidt, K. A. *J. Am. Chem. Soc.* **2007**, *129*, 5334–5335.
18. (a) Phillips, E. M., Reynolds, T. E., Scheidt, K. A. *J. Am. Chem. Soc.* **2008**, *130*, 2416–2417. (b) Chan, A., Scheidt, K. A. *J. Am. Chem. Soc.* **2008**, *130*, 2740–2741.
19. Guo, C., Sahoo, B., Daniliuc, C. G., Glorius, F. *J. Am. Chem. Soc.* **2014**, *136*, 17402–17405.
20. Ono, N. *The Nitro Group in Organic Synthesis*. Wiley-VCH: New York, 2001.
21. Nair, V., Sinu, C. R., Babu, B. P., Varghese, V., Jose, A., Suresh, E. *Org. Lett.* **2009**, *11*, 5570–5573.
22. Maji, B., Ji, L., Wang, S., Vedachalam, S., Ganguly, R., Liu, X.-W. *Angew. Chem. Int. Ed.* **2012**, *51*, 8276–8280.
23. White, N. A., DiRocco, D. A., Rovis, T. *J. Am. Chem. Soc.* **2013**, *135*, 8504–8507.
24. Raup, D. E. A., Cardinal-David, B., Holte, D., Scheidt, K. A. *Nat. Chem.* **2010**, *2*, 766–771.

25. Cardinal-David, B., Raup, E. E. A., Scheidt, K. A. *J. Am. Chem. Soc.* **2010**, *132*, 5345–5347.
26. Zhao, X., DiRocco, D. A., Rovis, T. *J. Am. Chem. Soc.* **2011**, *133*, 12466–12469.
27. Izquierdo, J., Orue, A., Scheidt, K. A. *J. Am. Chem. Soc.* **2013**, *135*, 10634–10637.
28. Sarkar, S. D., Studer, A. *Angew. Chem. Int. Ed.* **2010**, *49*, 9266–9269.
29. Sarkar, S. D., Biswas, A., Samanta, R. C., Studer, A. *Chem. Eur. J.* **2013**, *19*, 4664–4678.
30. White, N. A., Rovis, T. *J. Am. Chem. Soc.* **2014**, *136*, 14674–14677.
31. Zhang, Y., Du, Y., Huang, Z., Xu, J., Wu, X., Wang, Y., Wang, M., Yang, S., Webster, R. D., Chi, Y. R. *J. Am. Chem. Soc.* **2015**, *137*, 2416–2419.
32. Chiang, P.-C., Rommel, M., Bode, J. W. *J. Am. Chem. Soc.* **2009**, *131*, 8714–8718.
33. Binanzer, M., Hsieh, S.-Y., Bode, J. W. *J. Am. Chem. Soc.* **2011**, *133*, 19698–19701.
34. Fu, Z., Xu, J., Zhu, T., Leong, W. W. Y., Chi, Y. R. *Nat. Chem.* **2013**, *5*, 835–839.
35. Fu, Z., Jiang, K., Zhu, T., Torres, J., Chi, Y. R. *Angew. Chem. Int. Ed.* **2014**, *53*, 6506–6510.
36. For a recent review, see: Flanigan, D. M., Romanov-Michailidis, F., White, N. A., Rovis, T. *Chem. Rev.* **2015**, *115*, 9307–9387.
37. For selected examples, see: (a) He, L., Jian, T.-Y., Ye, S. *J. Org. Chem.* **2007**, *72*, 7466–7468. (b) Chen, X.-Y., Sun, L.-H., Ye, S. *Chem. Eur. J.* **2013**, *19*, 4441–4445. (c) Matsuoka, S.-I., Ota, Y., Washio, A., Katada, A., Ichioka, K., Takagi, K., Suzuki, M. *Org. Lett.* **2011**, *13*, 3722–3725. (d) Biju, A. T., Padmanaban, M., Wurz, N. E., Glorius, F. *Angew. Chem. Int. Ed.* **2011**, *50*, 8412–8415.
38. Burstein, C., Tschan, S., Xie, X., Glorius, F. *Synthesis* **2006**, *14*, 2418–2439.
39. He, M., Struble, J. R., Bode, J. W. *J. Am. Chem. Soc.* **2006**, *128*, 8418–8420.
40. Kaeobamrung, J., Kozlowski, M. C., Bode, J. W. *Proc. Natl. Acad. Sci.* **2010**, *107*, 20661–20665.
41. Wadamoto, M., Phillips, E. M., Reynolds, T. E., Scheidt, K. A. *J. Am. Chem. Soc.* **2007**, *129*, 10098–10099.
42. Vora, H. U., Rovis, T. *J. Am. Chem. Soc.* **2010**, *132*, 2860–2861.
43. Xu, J., Chen, X., Wang, M., Zheng, P., Song, B. A., Chi, Y. R. *Angew. Chem. Int. Ed.* **2015**. doi:10.1002/anie.201412132.
44. Mo, J., Chen, X., Chi, Y. R. *J. Am. Chem. Soc.* **2012**, *134*, 8810–8813.
45. Wang, M., Huang, Z., Xu, J., Chi, Y. R. *J. Am. Chem. Soc.* **2014**, *136*, 1214–1217.
46. Zhao, Y. M., Cheung, M. S., Lin, Z., Sun, J. *Angew. Chem. Int. Ed.* **2012**, *51*, 10359–10363.
47. Chen, X. Y., Xia, F., Cheng, J. T., Ye, S. *Angew. Chem. Int. Ed.* **2013**, *52*, 10644–10647.
48. Li, B. S., Wang, Y., Jin, Z., Zheng, P., Ganguly, R., Chi, Y. R. *Nat. Commun.* **2015**, *6*, 6207–6212.
49. Reynolds, N. T., de Alaniz, J. R., Rovis, T. *J. Am. Chem. Soc.* **2004**, *126*, 9518–9519.
50. Reynolds, N. T., Rovis, T. *J. Am. Chem. Soc.* **2005**, *127*, 16406–16407.
51. Vora, H. U., Rovis, T. *J. Am. Chem. Soc.* **2007**, *129*, 13796–13797.
52. He, M., Uc, G. J., Bode, J. W. *J. Am. Chem. Soc.* **2006**, *128*, 15088–15089.
53. Kawanaka, Y., Phillips, E. M., Scheidt, K. A. *J. Am. Chem. Soc.* **2009**, *131*, 18028–18029.
54. Kuwano, S., Harada, S., Kang, B., Oriez, R., Yamaoka, Y., Takasu, K., Yamada, K. *J. Am. Chem. Soc.* **2013**, *135*, 11485–11488.
55. Zhang, Y.-R., He, L., Wu, X., Shao, P.-L., Ye, S. *Org. Lett.* **2008**, *10*, 277–280. (b) Duguet, N., Campbell, C. D., Slawin, A. M., Smith, A. D. *Org. Biomol. Chem.* **2008**, *6*, 1108–1113.
56. (a) Lv, H., Zhang, Y.-R., Huang, X-L., Ye, S. *Adv. Synth. Catal.* **2008**, *350*, 2715–2718. (b) Lv, H., You, L., Ye, S. *Adv. Synth. Catal.* **2009**, *351*, 2822–2826. (c) Shao, P.-L., Chen, X.-Y., Ye, S. *Angew. Chem. Int. Ed.* **2010**, *49*, 8412–8416. (d) Jian, T.-Y., He, L., Tang, C., Ye, S. *Angew. Chem., Int. Ed.* **2011**, *50*, 9104–9107. (e) He, L., Lv, H., Zhang, Y.-R., Ye, S. *J. Org. Chem.* **2008**, *73*, 8101–8103. (f) Huang, X.-L., Chen, X.-Y.,

Ye, S. *J. Org. Chem.* **2009**, *74*, 7585–7587. (g) Wang, X.-N., Shen, L.-T., Ye, S. *Org. Lett.* **2011**, *13*, 6382–6385. (h) Wang, X.-N., Shao, P.-L., Lv, H., Ye, S. *Org. Lett.* **2009**, *11*, 4029–4031. (i) Zhang, H.-M., Gao, Z.-H., Ye, S. *Org. Lett.* **2014**, *16*, 3079–3081.
57. (a) Wang, X. N., Lv, H., Huang, X. L., Ye, S. *Org. Biomol. Chem.* **2009**, *7*, 346–350. (b) Concellón, C., Duguet, N., Smith, A. D. *Adv. Synth. Catal.* **2009**, *351*, 3001–3009.
58. Douglas, J., Ling, K. B., Concellón, C., Churchill, G., Slawin, A. M. Z., Smith, A. D. *Eur. J. Org. Chem.* **2010**, 5863–5869.
59. (a) Zhao, X., Ruhl, K. E., Rovis, T. *Angew. Chem., Int. Ed.* **2012**, *51*, 12330–12333. (b) Mo, J., Yang, R., Chen, X., Tiwari, B., Chi Y. R. *Org. Lett.* **2013**, *15*, 50–53.
60. (a) Li, F., Wu, Z., Wang, J. *Angew. Chem., Int. Ed.* **2014**, *53*, 656–659. (b) Dong, X., Yang, W., Hu, W., Sun, J. *Angew. Chem., Int. Ed.* **2014**, *53*, 660–663.
61. (a) Hao, L., Du, Y., Lv, H., Chen, X., Jiang, H., Shao, Y., Chi, Y. R. *Org. Lett.* **2012**, *14*, 2154–2157. (b) Hao, L., Chen, S., Xu, J., Tiwari, B., Fu, Z., Li, T., Lim, J., Chi, Y. R. *Org. Lett.* **2013**, *15*, 4956–4959. (c) Chen, S., Hao, L., Zhang, Y., Tiwari, B., Chi, Y. R. *Org. Lett.* **2013**, *15*, 5822–5825.
62. Lee, A., Younai, A., Price, C. K., Izquierdo, J., Mishra, R. K., Scheidt, K. A. *J. Am. Chem. Soc.* **2014**, *136*, 10589–10592.
63. Chow, K. Y.-K., Bode, J. W. *J. Am. Chem. Soc.* **2004**, *126*, 8126–8127.
64. Bode, J. W., Sohn, S. S. *J. Am. Chem. Soc.* **2007**, *129*, 13798–13799.
65. Lv, H., Mo, J., Fang, X., Chi, Y. R. *Org. Lett.* **2011**, *13*, 5366–5369.

8 Other Nonnitrogenous Organocatalysts

Andrew M. Harned

CONTENTS

This chapter will discuss several important classes of nonnitrogenous organocatalysts that, for one reason or another, were not able to be included as their own chapter. Even though this chapter may be less comprehensive than the others, this should not be taken as a commentary on the importance of these catalysts. To the contrary, some of these catalysts have served as the basis for the renaissance of organocatalysis that we are currently experiencing. As such, this resource would feel incomplete without their presence. It should be noted that many of these have been the subject of recent review articles. These have been referenced where appropriate, and the reader is encouraged to seek those out for a more in-depth discussion of these topics.

8.1 KETONE CATALYSTS FOR OXIDATION REACTIONS

8.1.1 Historical and Fundamental Studies

Oxidation reactions are of fundamental importance to organic synthesis, and our ability to carry out selective, in all senses of the word, oxidations is critical for achieving efficient synthetic processes.[1] Alkenes are particularly useful substrates for oxidation, and many reagents and catalysts have been developed for converting alkenes into a variety of oxidized products (epoxides, 1,2-diols, ketones, aldehydes, aziridines, etc.). Most of these reagents are based on transition metals; or, if they are "non-metallic" in nature, would be difficult to render catalytic. However, one class of reagent stands out in this regard: dioxiranes. The discussion here will, by necessity, be somewhat selective. Several excellent reviews by Denmark,[2] Yang,[3] and Shi[4] provide a more complete discussion on the development of enantioselective ketone catalysts for oxidation reactions.

Dioxiranes, structurally, are peroxides that have been constrained to a three-membered ring. The strained nature of these compounds makes them powerful oxygen-atom transfer agents.[5] Dioxiranes can trace their history back to 1899 when Baeyer and Villiger proposed the intermediacy of dioxirane **2** during the conversion of menthone (**1**) into lactone **3** by monoperoxysulfonic acid (Figure 8.1).[6] It would be another 70 years before the first documented case of a dioxirane preparation would appear. In 1972, Talbot and Thompson reported that 3,3-bis(trifluoromethyl)dioxirane (**6**) and 3-(chlorodifluoromethyl)-3-(trifluoromethyl)dioxirane (**7**) could be prepared by F_2 oxidation of bisalkoxides **4** and **5**, respectively (Figure 8.2).[7] Dioxirane **7** was reported to be explosive, but both were characterized by UV, ^{19}F NMR, and MS techniques. Although this represented a major step forward in the study of this class of compounds, this method is clearly not going to serve as an everyday preparation of dioxiranes.

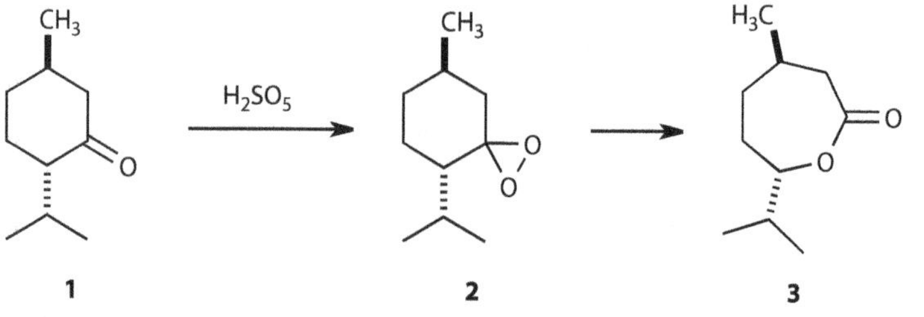

FIGURE 8.1 Baeyer and Villiger's proposal for the formation of a dioxirane from menthone.

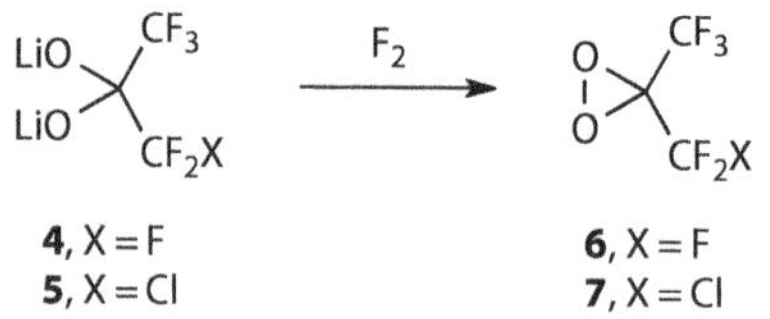

FIGURE 8.2 First preparation of a dioxirane.

The first breakthrough in using dioxiranes as *in situ* oxidants for organic molecules came in 1974 when Montgomery reported that certain ketones catalyzed the decomposition of monoperoxysulfate ion (HSO_5^-) as shown in Figure 8.3a.[8] Acetone, cyclohexanone, and the oxopiperidinium salt **8** were particularly found to be effective catalysts. Montgomery proposed that dioxiranes might be the key intermediates in this decomposition reaction and that they could be formed by the addition of HSO_5^- to acetone (Figure 8.3b). Thus, an initial attack of HSO_5^- onto the carbonyl carbon of the ketone would give **9**. This intermediate is analogous to the Crigee intermediate proposed for the Baeyer–Villiger reaction. However, Montgomery recognized that Baeyer–Villiger reactions using monoperoxysulfate are typically carried out under acidic conditions, whereas the ketone-promoted decompositions he studied were performed under mild alkaline conditions (pH 9.0). One might imagine that deprotonation of **9** would give alkoxide **10**, which could then suffer intramolecular displacement of sulfate in order to arrive at the dioxirane. Importantly, Montgomery observed little competing Baeyer–Villiger oxidation when these reactions were run under alkaline conditions. A few years later, Edwards and Curci presented a series of kinetics experiments and ^{18}O-labeling studies that essentially confirmed the mechanistic hypothesis proposed by Montgomery.[9]

Several details of the mechanistic work performed by Curci and Edwards would prove to be critical in developing a catalytic asymmetric variant. Foremost among these was the finding that Oxone® (caroate) can be used as the source of HSO_5^-. Oxone® is a so-called triple salt with the formula $2KHSO_5 \cdot KHSO_4 \cdot K_2SO_4$. Control of pH is also important. If the pH is too low, then decomposition of the ketone by a Baeyer–Villiger reaction becomes a problem. If the pH is too high, then decomposition of the dioxirane by SO_5^{2-} becomes a problem. Consequently, running the reaction with water that is buffered to pH 7.5 was found to be optimal. Also important was the addition of small amounts of Na_2EDTA as a scavenger of trace metals that may also decompose peroxide intermediates. By following these precautions, Curci and Edwards were able to show that caroate and acetone could function as a useful combination for alkene epoxidation reactions (Figure 8.4), presumably through

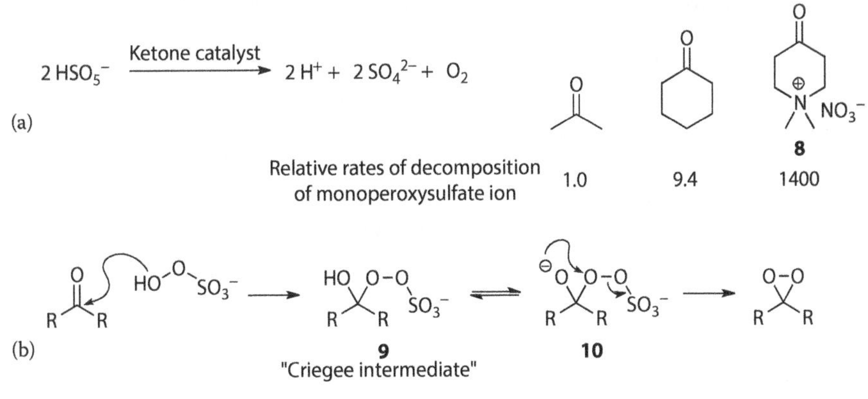

FIGURE 8.3 (a) Catalytic decomposition of monoperoxysulfate by various ketones. (b) Montgomery's proposal for the formation of dioxiranes from ketones.

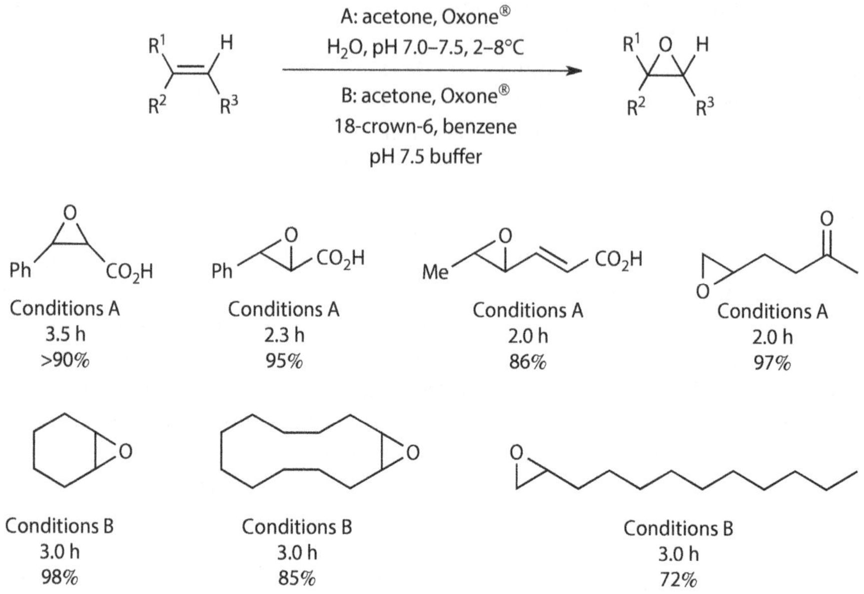

A: acetone, Oxone®
H_2O, pH 7.0–7.5, 2–8°C
B: acetone, Oxone®
18-crown-6, benzene
pH 7.5 buffer

Conditions A
3.5 h
>90%

Conditions A
2.3 h
95%

Conditions A
2.0 h
86%

Conditions A
2.0 h
97%

Conditions B
3.0 h
98%

Conditions B
3.0 h
85%

Conditions B
3.0 h
72%

FIGURE 8.4 Curci's epoxidation of alkenes using DMDO formed *in situ*.

in situ formation of dimethyldioxirane (DMDO).[10] Also important was Curci's finding that the use of fluorinated ketones such as trifluoroacetone had a pronounced impact on the reactivity and stability of the corresponding dioxirane.[11,12] Yang would show that this reactivity could be harnessed for practical alkene epoxidations by adapting Curci's procedure for *in situ* formation of DMDO to work with trifluoroacetone (Figure 8.5).[13] Especially important was the finding that $NaHCO_3$ could effectively regulate the pH of the reaction without requiring constant monitoring.

The *in situ* formation of DMDO and methyl(trifluoromethyl)dioxirane (TFDO) is nice; in that, it obviates the need for isolating the unstable dioxirane. But, the low molecular weight and volatility of acetone and trifluoroacetone make it difficult to carry out epoxidation reactions with substoichiometric amounts of these catalysts. Fortunately, their cost is sufficiently low, which is not a significant issue. But, if one is interested in using a chiral ketone to generate an enantioenriched epoxide that is being able to use a substoichiometric amount of the ketone becomes a priority. In the mid-1990s, Denmark demonstrated that higher molecular weight ketones, in particular 4-oxopiperidinium salts (e.g., **11**), could indeed serve as effective catalysts for nonasymmetric epoxidation reactions (Figure 8.6).[14] Certain fluorinated ketones were also effective catalysts.[15]

8.1.2 Transition-State Considerations

Dioxirane-mediated epoxidations are stereospecific and appear to pass through a concerted, though possibly asynchronous,[16,17] transition state. Two transition states have been proposed: planar transition state **12P** and spiro transition state **12S** (Figure 8.7). By analyzing kinetic data of various substituted alkenes, Baumstark and coworkers

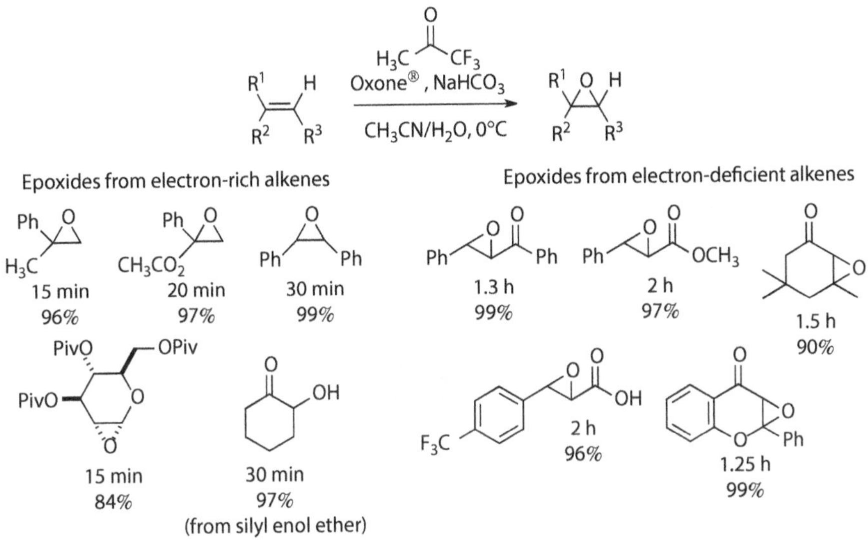

FIGURE 8.5 Yang's epoxidation of alkenes using TFDO formed *in situ*.

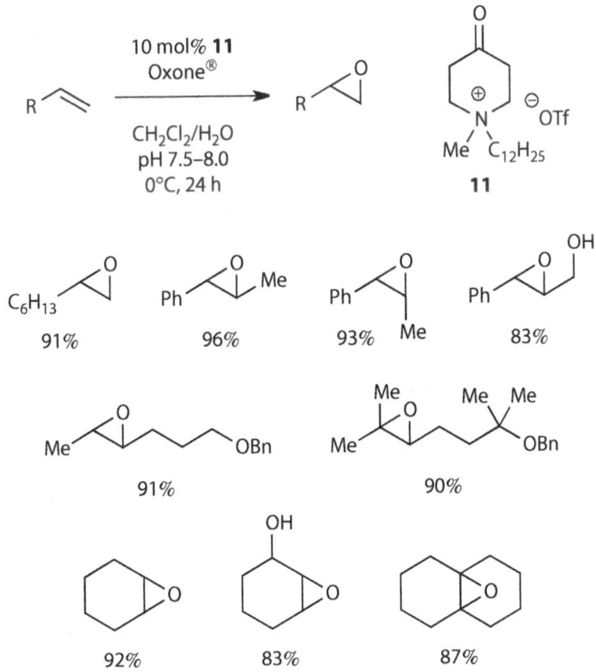

FIGURE 8.6 Denmark's catalytic epoxidation reaction.

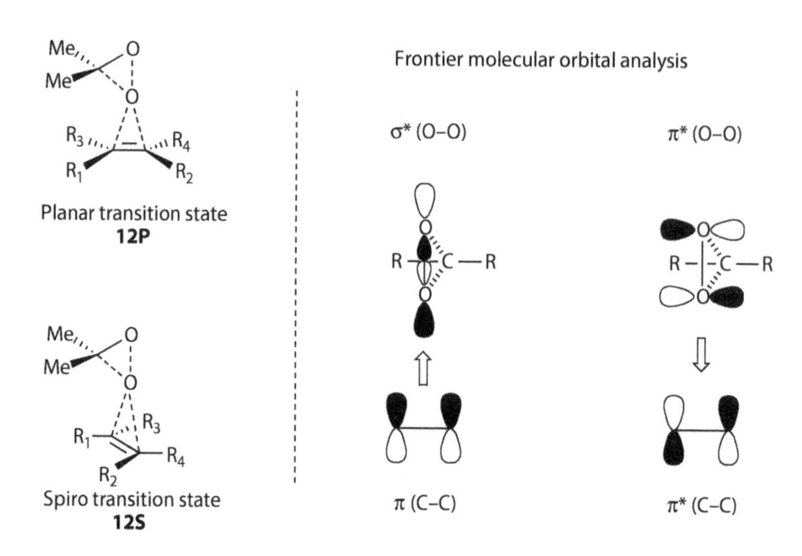

FIGURE 8.7 Possible transition states and orbital analysis for dioxirane epoxidation.

proposed that DMDO prefers to react through a spiro transition state.[18] This is due to the observation that some *cis*-alkenes (R^1, R^2 = H) can be up to 20 times more reactive than the corresponding *trans*-alkene (R^1, R^4 = H). Several computational studies also indicate that a spiro transition state is favored for epoxidations with both DMDO[16,19] and fluorinated dioxiranes.[20]

Computational studies also indicate that it is possible for dioxiranes to act as either an electrophilic or a nucleophilic oxidant.[21] This can be rationalized by considering the frontier orbitals for both the dioxirane and the alkene (Figure 8.7). In the case of electron-rich alkenes, the π-orbital of the olefin can serve as a donor to the σ*-orbital of the dioxirane O–O bond. In the case of electron-deficient alkenes, an occupied π* lone pair from the dioxirane can engage in backbonding with the π*-orbital of the olefin.

8.1.3 Initial Forays into Asymmetric Catalysis

In 1984, Curci reported the first examples of asymmetric epoxidation in which a chiral ketone is used as the precursor to a chiral dioxirane (Figure 8.8).[22] Although the enantioselectivities afforded by ketones **13** and **14** were low, they were certainly not negligible. In addition, both ketones could be used in substoichiometric amounts, thereby setting the stage for future developments in the area of asymmetric catalysis with chiral ketones. Later, Curci reported that fluorinated ketones **15** and **16** delivered enantioenriched epoxides with improved reactivity but still less than desirable enantioselectivity (up to 20% ee).[23]

8.1.4 C_2-Symmetric Catalysts

Soon after Curci's report on the use of fluorinated ketones **15** and **16**, Song reported that C_2-symmetric ketones **17a** and **18** (Figure 8.9) afforded enantioenriched

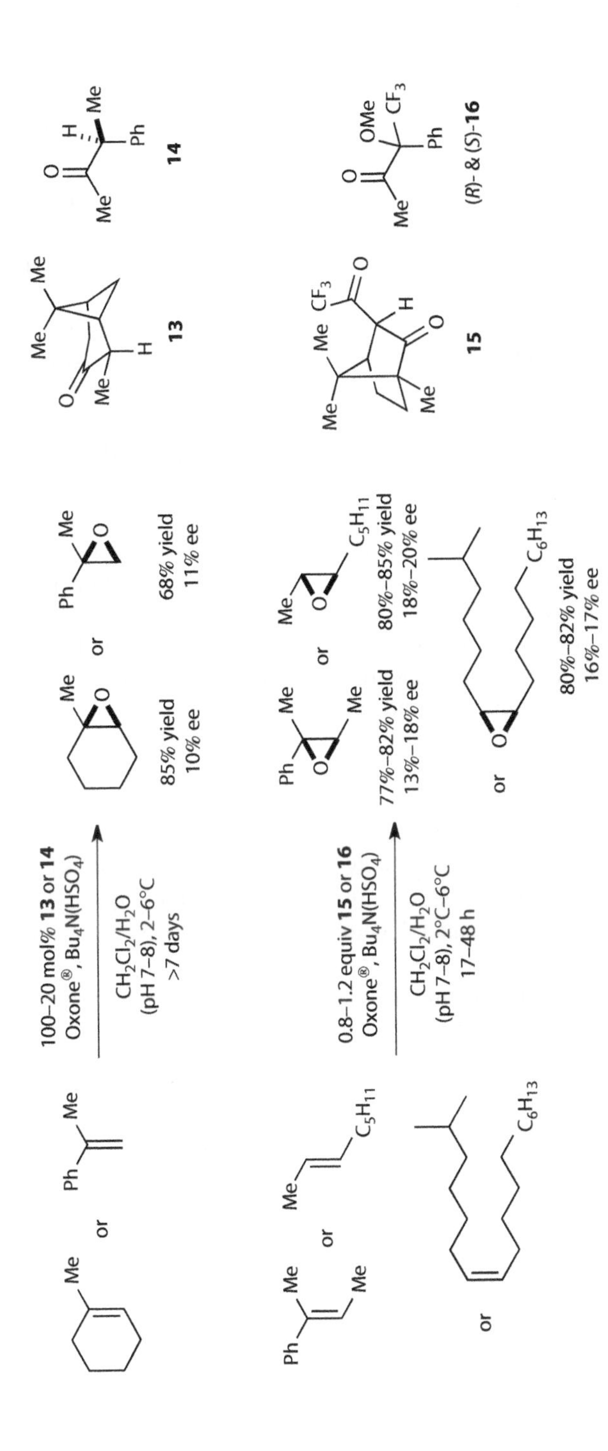

FIGURE 8.8 Curci's first examples of asymmetric epoxidation with a chiral ketone.

epoxides with improved selectivity.[24] Unfortunately, the level of selectivity was quite substrate dependent. For example, epoxidation of *trans*-stillbene with **18** (100 mol%) delivered the epoxide with 30% ee (59% ee at 0°C). But, when the same conditions were used with *trans*-β-methylstyrene, the epoxide was formed with 20% ee. Importantly, the 1,1′-bi-2-naphthol (BINOL)-derived catalysts **17a–17c** could be used in substoichiometric amounts (30 mol%) but could deliver epoxides with poor enantiomeric excess (2–20% ee).[25] At the same time, Adam and Zhao reported on the use of C_2-symmetric ketone **19** as a stoichiometric reagent for asymmetric epoxidations.[26] Although the observed enantioselectivities were higher than those reported by Curci and Song, there was considerable variability (14%–81% ee) between substrates. Interestingly, they also observed a significant pH dependence on the enantioselectivity, with the highest selectivities being generated at pH 10.5 (65%–81% ee) rather than pH 8.0 (14%–66% ee). The rate of reaction at pH 10.5 was also dramatically faster than that at pH 8.0.

At about the same time, Song and Adam reported their work on C_2-symmetric ketones, and Yang and coworkers reported their work with very similar catalysts (**20–21**, Figure 8.10).[27] These catalysts proved to be quite selective for the asymmetric epoxidation of *trans*-stillbenes. Unfortunately, when they were used as catalysts for oxidizing other alkenes, the enantioselectivity tended to be more moderate (50%–80% ee). Yang and coworkers also demonstrated that catalyst **20b** was effective for the kinetic resolution of certain silyl-protected allylic alcohols (Figure 8.11).[28]

Behar[29] and Denmark[30] independently reported work, evaluating the use of fluorinated C_2-symmetric catalysts **24** and **25**, respectively (Figure 8.12). With respect to alkene substrates, Behar's study was limited to just *trans*-β-methylstyrene, but catalyst **24** delivered the epoxide with a respectable 86% ee.[29] Denmark's study included several other alkenes. Catalyst **25** performed well with disubstituted styrene derivatives (88%–94% ee), but lower selectivities were observed with other substrates.[30]

To be sure, many other ketone catalysts have been prepared and evaluated in asymmetric epoxidation reactions. But, the story is essentially the same as with the catalysts that is already mentioned: low to moderate selectivity and/or limited demonstrated substrate scope.[31] However, the story does not end here. As will be seen in the next section, one family of chiral ketone catalysts has proven to be quite effective for the asymmetric epoxidation of a wide range of substrates.

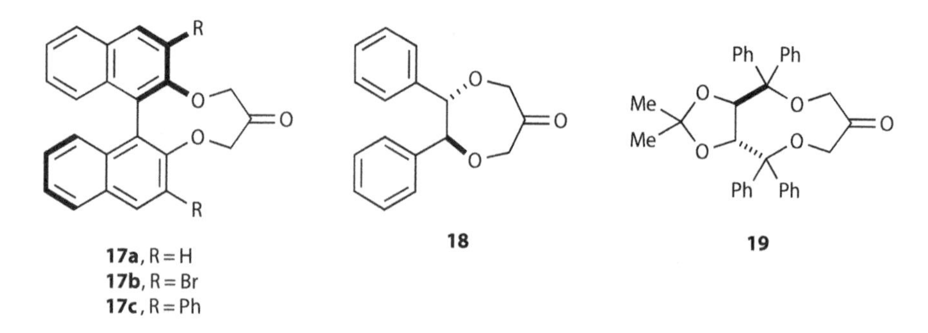

FIGURE 8.9 C_2-symmetric ketone catalysts reported by Song and Adam.

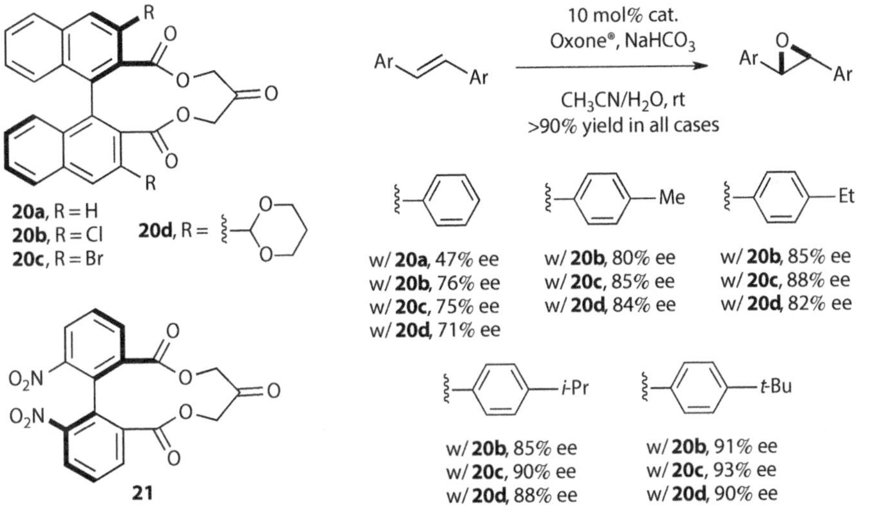

FIGURE 8.10 Asymmetric epoxidation of *trans*-stillbenes with Yang's catalysts.

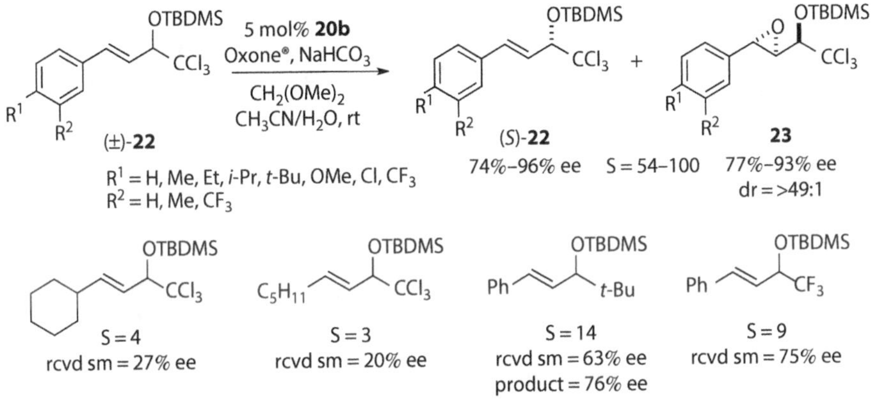

FIGURE 8.11 Kinetic resolution of silyl-protected allylic alcohols.

8.1.5 Fructose-Derived Ketones as General Asymmetric Epoxidation Catalysts

In 1996, Shi and coworkers introduced a new class of chiral ketone catalysts for use in asymmetric epoxidation reactions (Figure 8.13). The initially disclosed catalyst (**26**), which can be prepared in two steps from D-fructose,[32] was found to be quite effective for the asymmetric epoxidation of both *trans*-alkenes[33] (Figure 8.14) and trisubstituted alkenes[34] (Figure 8.15) with various substituents.[35] In order to widen the scope of this asymmetric epoxidation, Shi and coworkers have added several other catalysts to the family.[4] By using catalyst **27**, this methodology could be

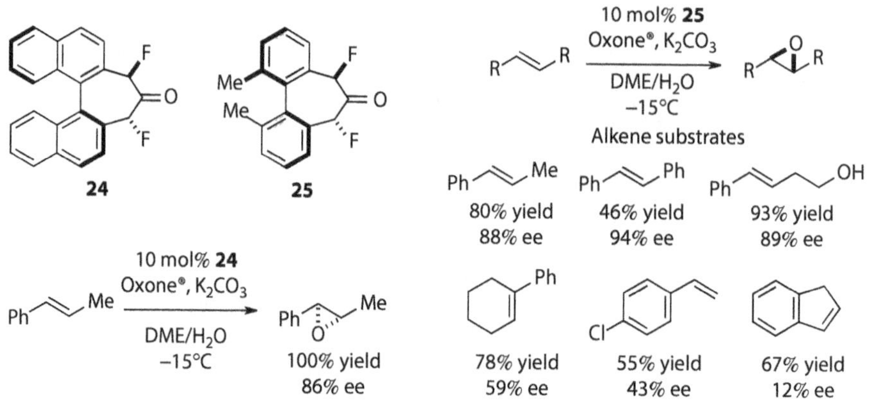

FIGURE 8.12 Asymmetric epoxidation with fluorinated C_2-symmetric ketones.

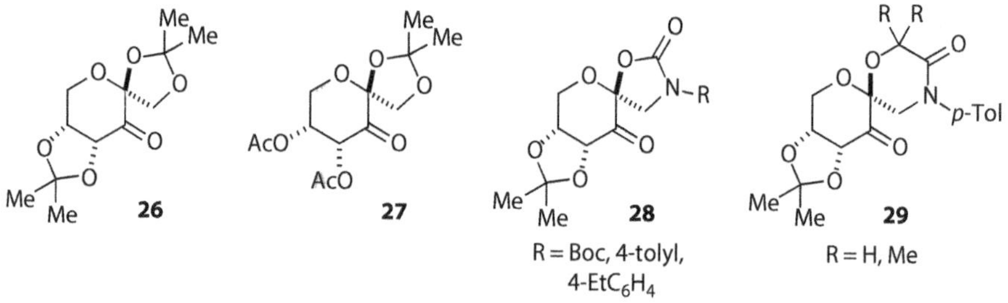

FIGURE 8.13 Fructose-derived catalysts developed by Shi and coworkers.

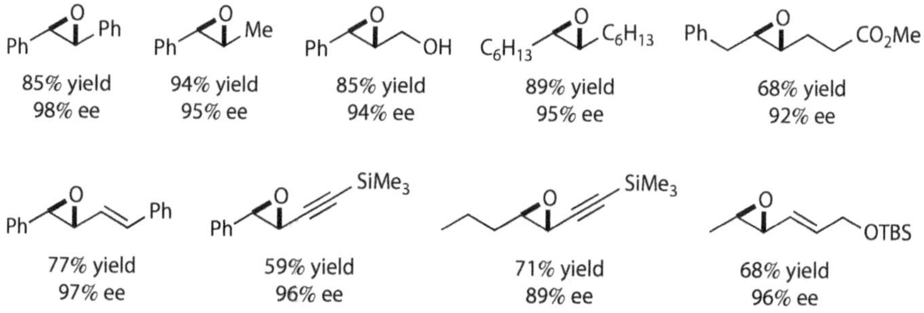

FIGURE 8.14 Examples of *trans*-disubstituted epoxides produced by catalyst **26**.

extended to α,β-unsaturated esters, enimides, and some *cis*-alkenes[36] (Figure 8.16). In order to perform asymmetric epoxidations on *cis*-alkenes (Figure 8.17), catalyst **28** was prepared.[37] Later, it was found that catalyst **29** could be used for the epoxidation of 1,1-disubstituted alkenes (Figure 8.18).[38] Finally, it should be mentioned that the enantiomer of these catalysts can be prepared from L-sorbose.[39]

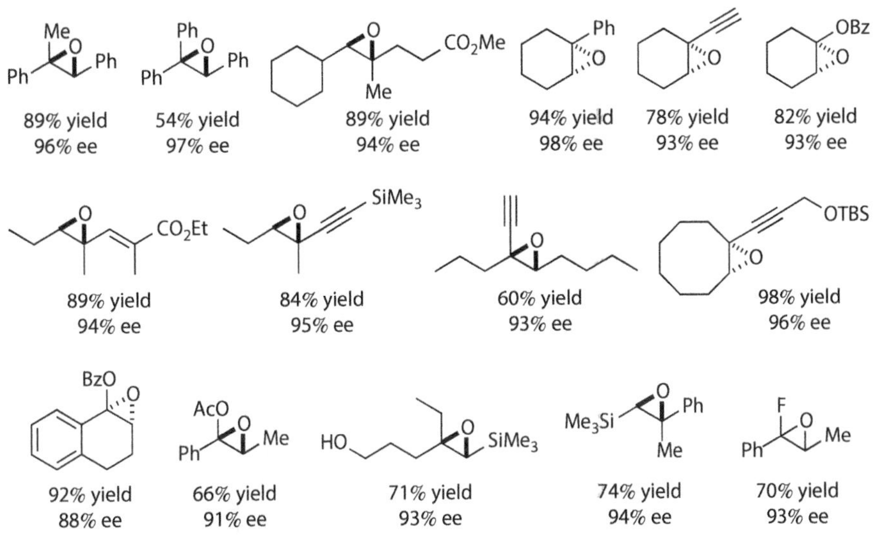

89% yield 96% ee; 54% yield 97% ee; 89% yield 94% ee; 94% yield 98% ee; 78% yield 93% ee; 82% yield 93% ee

89% yield 94% ee; 84% yield 95% ee; 60% yield 93% ee; 98% yield 96% ee

92% yield 88% ee; 66% yield 91% ee; 71% yield 93% ee; 74% yield 94% ee; 70% yield 93% ee

FIGURE 8.15 Examples of trisubstituted epoxides produced by catalyst **26**.

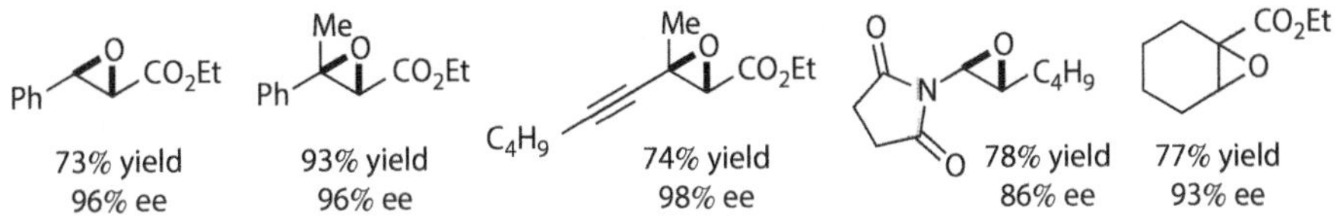

73% yield 96% ee; 93% yield 96% ee; 74% yield 98% ee; 78% yield 86% ee; 77% yield 93% ee

FIGURE 8.16 Examples of epoxides produced by catalyst **27**.

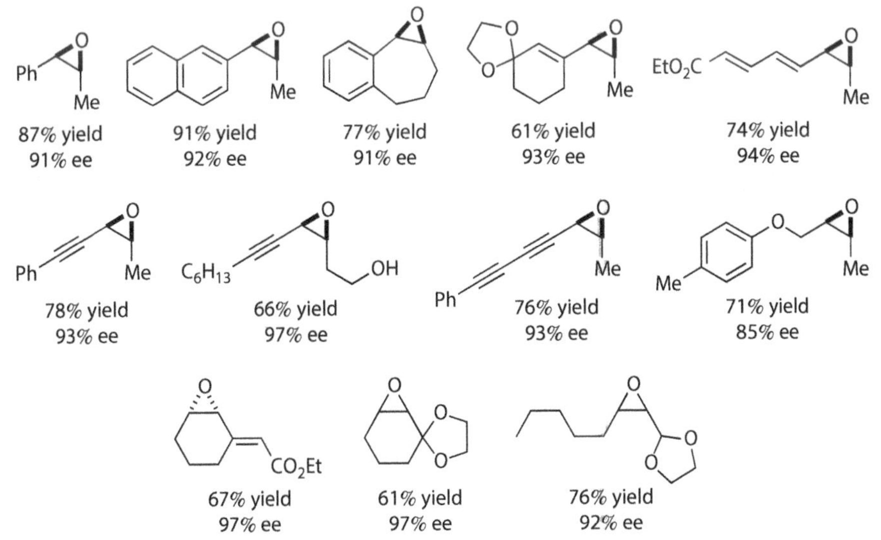

87% yield 91% ee; 91% yield 92% ee; 77% yield 91% ee; 61% yield 93% ee; 74% yield 94% ee

78% yield 93% ee; 66% yield 97% ee; 76% yield 93% ee; 71% yield 85% ee

67% yield 97% ee; 61% yield 97% ee; 76% yield 92% ee

FIGURE 8.17 Examples of *cis*-disubstituted epoxides produced by catalyst **28**.

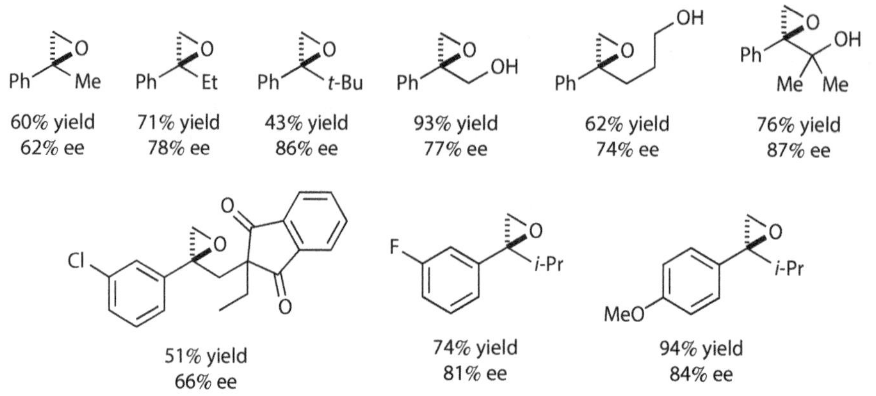

FIGURE 8.18 Examples of 1,1-disubstituted epoxides produced by catalyst **29**.

Mechanistically, epoxidations using the Shi catalysts are thought to proceed as shown in Figure 8.19. Initial reaction between catalyst **26** and monoperoxysulfate will give intermediate **30** after proton transfer. Deprotonation of the hemiketal hydroxyl group leads to dianion **31**, which cyclizes to form dioxirane **32**. Oxygen transfer from **32** to the alkene substrate affords the epoxide and regenerates the ketone catalyst. The available experimental and theoretical evidence points toward a spiro transition state similar to **34** that is being favored in epoxidations using the Shi catalysts.[16,40]

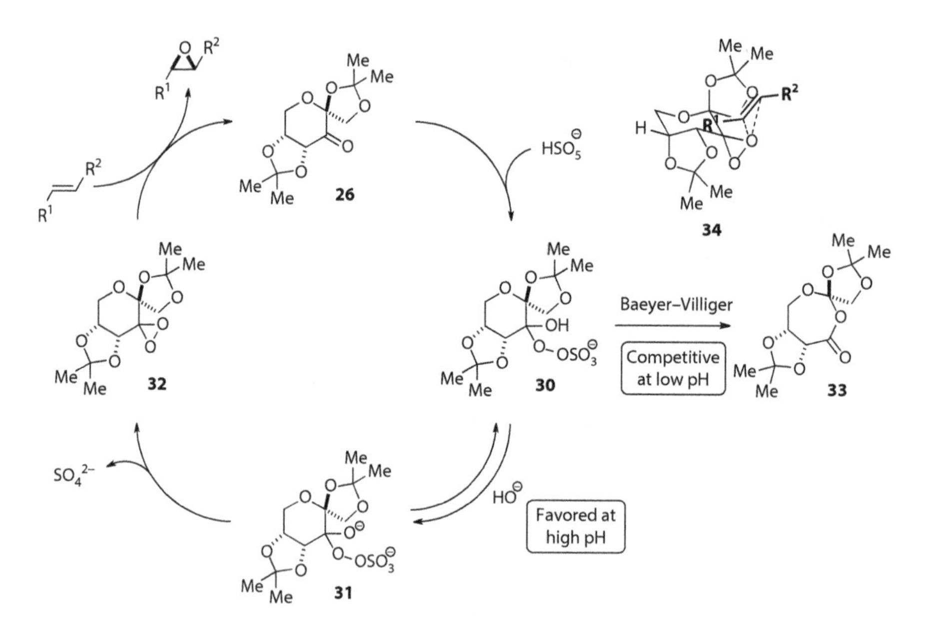

FIGURE 8.19 Proposed mechanism for the Shi epoxidation.

However, in some cases, the planar transition state can compete and appears to be responsible for generating the minor enantiomer.

Shi has found that these catalysts tend to operate more effectively at a pH of about 10.5.[41] This is in contrast to other ketone catalysts, where a somewhat lower pH (7–8) is preferred. The reason for this is that the higher pH will facilitate the formation of intermediate **31**, thereby thwarting the Baeyer–Villiger process that leads to **33**. In addition to pH control, the competing catalyst decomposition pathway can also be overcome to a large extent by modifying the catalyst structure.[42] It is interesting to note that the Shi group found that the terminal oxidant Oxone® can be replaced with H_2O_2 under certain conditions.[43]

Finally, a further testament to the generality of the Shi catalysts has been the use of these catalysts, especially **26**, by others for the synthesis of complex natural products.[31] It is outside the scope of this chapter to give a detailed analysis of these efforts, but several representative natural products whose syntheses made use of catalyst **26** or its enantiomer are highlighted in Figure 8.20.[44] Stereocenters whose construction is a direct result of either the asymmetric epoxidation or it products are indicated with a "*".

8.2 *N*-HETEROCYCLIC CARBENE CATALYSTS AS ACYL ANION EQUIVALENTS

8.2.1 Introduction

In Chapter 8, it was shown how imidazolium salts can serve as catalysts for the formation of homoenolates and related intermediates. This reactivity was predicated on the ability of imidazolium salts, and related compounds, to function as precursors for so-called *N*-heterocyclic carbenes (NHCs). This section will focus more on the use of NHC catalysts as a means of accessing the reactivity of acyl anion equivalents. The use of NHC catalysts has expanded rapidly over the last 10–15 years, making it difficult to include all this work in just two chapters. The reader is encouraged to seek out one or more reviews[45] on this topic for more in-depth coverage. Throughout this chapter, the terms imidazolium salt (or relevant analogs thereof) and NHC may be used interchangeably. It should be remembered that for the most part, the salt is the precatalyst that is actually added to the reaction, but the NHC is what is actually responsible for the reactivity.

8.2.2 Structural Features and General Reactivity of *N*-Heterocyclic Carbenes and Their Precursors

The literature is replete with different heterocyclic precursors to NHCs. Some of the more useful precatalysts for organocatalytic reactions are based on thiazole (**35**), imidazole (**36**), imidazoline (**38**), and triazole (**37**) scaffolds (Figure 8.21a). These cores are responsible for the reactivity and stability of the carbene itself and have been decorated by all manner of substituents in order to further tune the electronic and steric properties of the catalyst. In a recent review, Rovis has compiled a very comprehensive structural

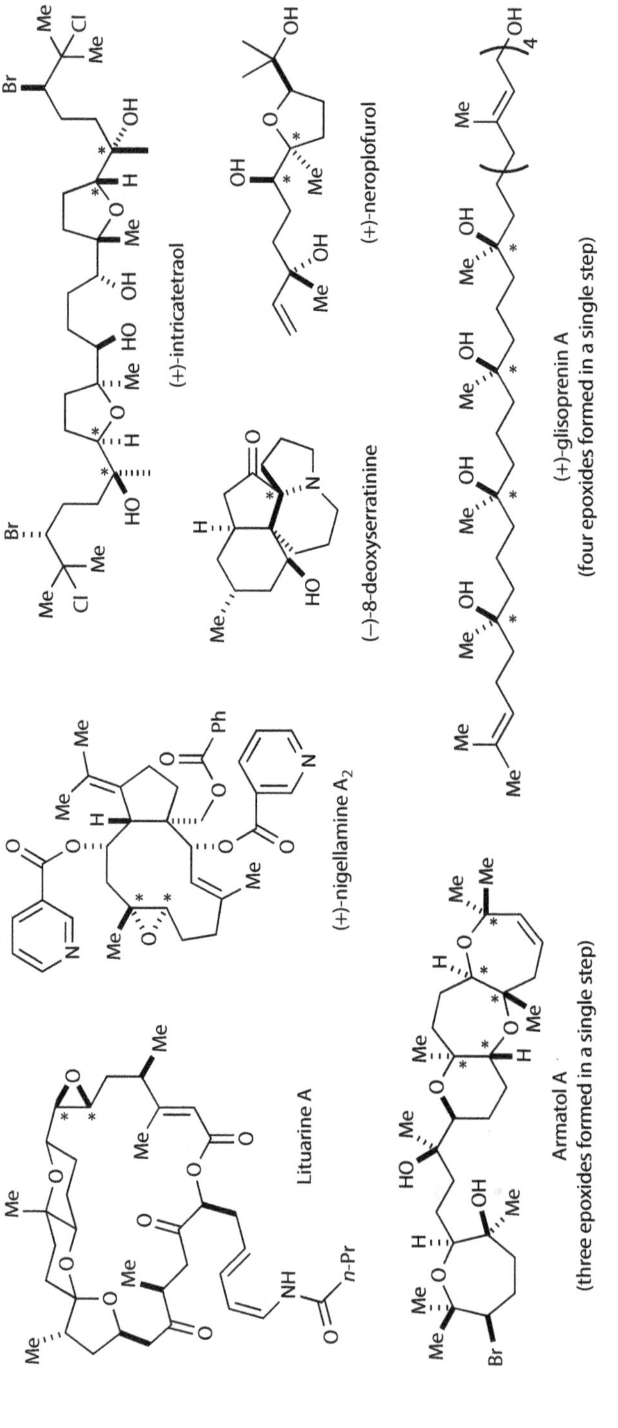

Lituarine A

(+)-nigellamine A_2

(+)-intricatetraol

(−)-8-deoxyserratinine

(+)-neroplofurol

Armatol A
(three epoxides formed in a single step)

(+)-glisoprenin A
(four epoxides formed in a single step)

FIGURE 8.20 Examples of natural products synthesized using catalyst **26** or its enantiomer.

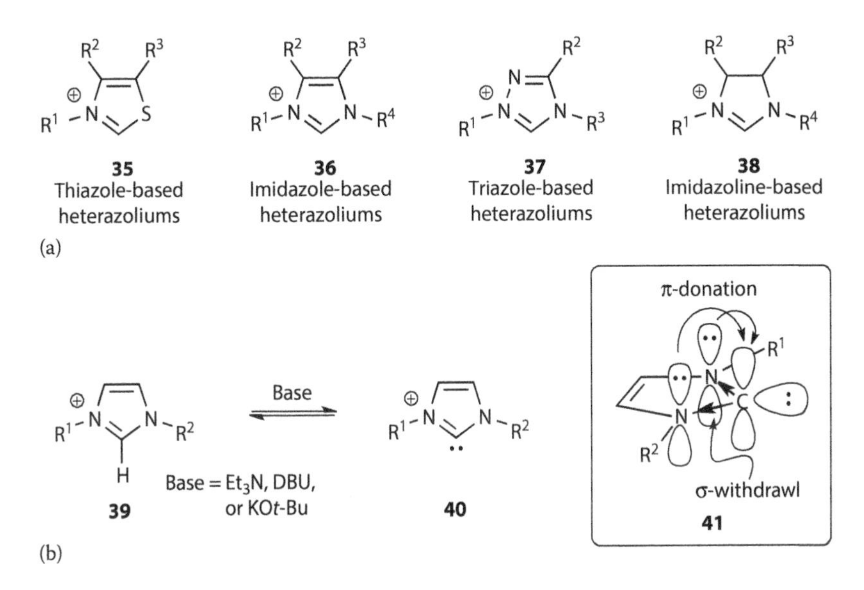

FIGURE 8.21 (a) Commonly used heterocyclic precatalysts. Counterions are not shown. (b) Generation of NHCs and their orbital description.

survey of these structures for both chiral and achiral catalysts.[46] Most of the organocatalytic reactions that involve NHCs generate the active carbene (e.g., **40**) *in situ* through deprotonating the precatalyst (e.g., **39**) with an appropriate base (Figure 8.21b). To be sure, there is some variance in the acidity of the different heterocyclic precursors, but for the most part, the pK_a values of the different precatalysts are between ~16 and ~24 in water. This is low enough that bases such as KO*t*-Bu, K_2CO_3, Et_3N, or 1,8-diazabicyclo(5.4.0)undec-7-ene (DBU) can be used to generate the carbene, but this does not need to be a complete deprotonation for catalysis to proceed.[47]

When compared to classical triplet carbenes, NHCs are remarkably stable. A large part of this stabilization can be ascertained by considering the molecular orbitals of the NHC, as shown in **41**.[48] The carbene itself is best thought of as a singlet carbene with a lone pair of electrons in an sp²-orbital and an empty p-orbital. The adjacent heteroatoms (N or S) stabilize the carbene in two ways. First, the electronegativity of the heteroatoms provides inductive stabilization of the lone pair. Second, the heteroatoms provide resonance stabilization of the carbene through backbonding into the empty p-orbital. Furthermore, the aromatic nature of precatalysts such as **35–37** provides further stabilization. These structural features combine to make NHCs potent nucleophiles with enhanced Lewis basicity.[49]

The key intermediate associated with much of the organocatalytic reactivity of NHCs is 2,2-diminoenol (**44a**) that is commonly referred to as the Breslow intermediate. This intermediate is quite analogous to intermediates proposed by Breslow[50] for the mechanism of action of thiamine (vitamin B1). It has been shown that the formation of the Breslow intermediate (Figure 8.22) can be reversible,[51] but certain substituents on the carbene may make this process irreversible.[47] Once formed, the Breslow intermediate can be the subject of several different transformations, depending on

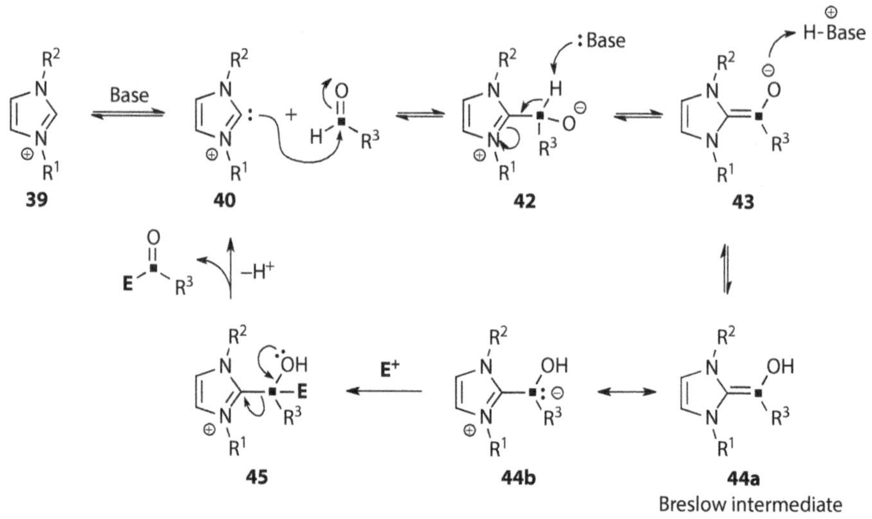

FIGURE 8.22 Formation of the Breslow intermediate and reaction with an electrophile.

the nature of both the reaction partner and the substituent of the original aldehyde (R^3). The primary function of the Breslow intermediate is as a nucleophile. This can be rationalized by considering resonance structure **44b** in which electron donation by the heterocycle results in a buildup of charge on what was the aldehyde carbon. In doing so, this has changed the reactivity of the original aldehyde from electrophilic to nucleophilic (umpolung reactivity). Thus, the Breslow intermediate will react with electrophiles to give tetrahedral intermediate **45**. Collapse of this intermediate will liberate the NHC catalyst and will generate the ketone product.

The umpolung reactivity shown in Figure 8.22 essentially means that the Breslow intermediate behaves as an acyl anion equivalent. Common reaction partners seen with this mode of reactivity are aldehydes and imines (in the benzoin and azabenzoin reaction) and electron-deficient alkenes (in the Stetter reaction) (Figure 8.23). Interestingly, using different aldehyde starting materials can modify the reactivity displayed by the Breslow intermediate. For example, the Breslow intermediate formed from an α,β-unsaturated aldehyde (**46**) has extended conjugation with the heterocyclic portion (Figure 8.23). Reaction between **46** and an electrophile occurs at the β-position of the original aldehyde and generates an acylazolium intermediate (**47**). Catalyst turnover occurs by nucleophilic attack on the acylazolium. Through this process, NHCs allow α,β-unsaturated aldehydes to function as homoenolate equivalents. More traditional enolate reactivity can be conferred to the starting aldehyde if the α-carbon of the aldehyde is substituted with a leaving group (Figure 8.23). Now, the electron-rich double bond of the Breslow intermediate is poised to expel the leaving group, as shown with **48**. This results in the formation of a new enol or enolate (**49**) that can react with electrophiles at the α-carbon of the original aldehyde.

In addition to the basic reactivity highlighted in Figure 8.23, researchers have uncovered other exciting ways to intercept the Breslow intermediate (e.g., oxidative processes), but the general theme remains: NHCs provide an umpolung reactivity

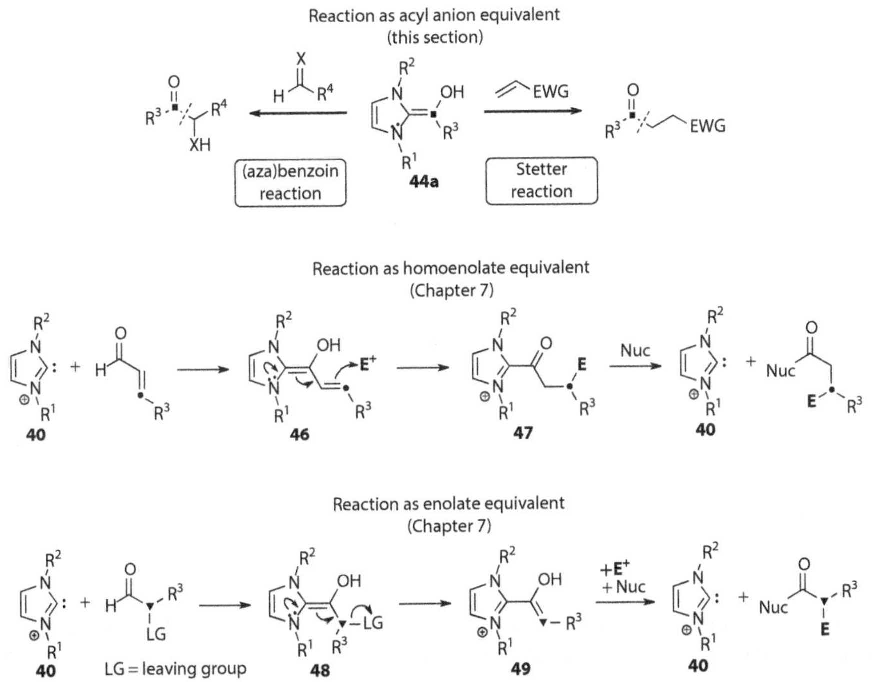

Reaction as acyl anion equivalent (this section)

(aza)benzoin reaction

Stetter reaction

Reaction as homoenolate equivalent (Chapter 7)

Reaction as enolate equivalent (Chapter 7)

LG = leaving group

FIGURE 8.23 Reaction pathways of differently substituted Breslow intermediates.

to differently substituted aldehydes, and this can occur at the α-carbon, β-carbon, or carbonyl carbon of the original aldehyde. This section will only cover the acyl anion reactivity conferred by NHC catalysts. The homoenolate and enolate reactivities were covered in Chapter 7.

8.2.3 Early Work with Achiral *N*-Heterocyclic Carbenes

In 1943, Ukai and coworkers found that the benzoin reaction of furfural and benzaldehyde could be promoted by various *N*-alkylthiazolium salts in the presence of a small amounts of added base (Figure 8.24).[52] At that time, they thought that the base promoted the ring opening of the thiazolium salt, and that a mercaptan similar to **50** was the actual promoter for this reaction. Since then, it has been recognized that the

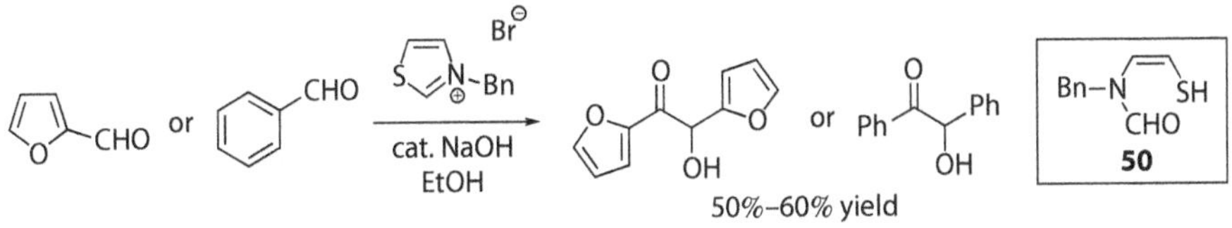

cat. NaOH EtOH

50%–60% yield

FIGURE 8.24 The first NHC-promoted benzoin reactions reported by Ukai.

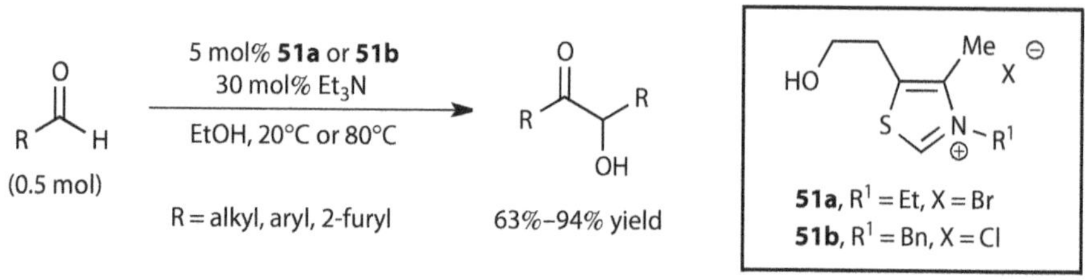

5 mol% **51a** or **51b**
30 mol% Et_3N
EtOH, 20°C or 80°C
(0.5 mol)
R = alkyl, aryl, 2-furyl
63%–94% yield
51a, R^1 = Et, X = Br
51b, R^1 = Bn, X = Cl

FIGURE 8.25 Stetter's NHC-catalyzed benzoin reaction.

R^2 - CHO
10 mol% **51b**
20 mol% Et_3N
EtOH
EWG = nitrile, ester, ketone
R^1, R^2 = alkyl, aryl, heteroaryl
30%–75% yield

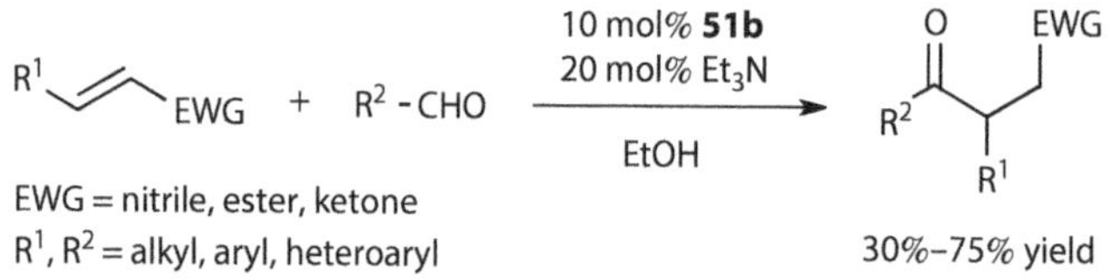

FIGURE 8.26 First examples of the Stetter reaction.

actual catalyst was the NHC formed from the thiazolium, and this has been credited as the first example of an NHC-catalyzed benzoin reaction. In 1973, Tagaki and Hara reported that alkyl aldehydes could be accommodated by incorporating a larger substituent on the nitrogen of the thiazolium catalyst.[53] Later, Stetter showed that thiazolium catalysts **51a** or **51b** allowed benzoin reactions of both alkyl and aryl aldehydes to be performed on a preparative scale (Figure 8.25).[54] Prior to their work on the benzoin reaction, Stetter and coworkers found that the same thiazolium salts could catalyze the conjugate addition of aldehydes to α,β-unsaturated ketones and esters (Figure 8.26).[55] Since these initial pioneering studies, the achiral NHC-catalyzed benzoin and Stetter reactions have been continuously developed and expanded to many other substrates.[56] In particular, the Stetter reaction has become a very reliable method for the generation of 1,4-dicarbonyl compounds, which themselves are useful intermediates for the construction of heterocycles.

8.2.4 Asymmetric Benzoin and Stetter Reactions

Since the mid-1990s, a plethora of chiral NHC precatalysts has been reported in the literature. It is not possible to describe all these advances in one section of one chapter. Some will be highlighted later as appropriate. The reader should seek out reviews by Enders[57] and Rovis[46] for a comprehensive description of these developments.

It is important to recognize that the first example of using a chiral NHC catalyst in an asymmetric benzoin reaction appeared less than 10 years after Breslow's determination that a carbene was responsible for the biological activity of vitamin B1.[58] Notably, this report was also among the first examples of nonenzymatic asymmetric catalysis. In 1966, Sheehan and Hunneman reported that chiral thiazolium salt catalyst **52** generated benzoin in low yield but with 22% optical purity (Figure 8.27).[59] Later, Sheehan reported that catalyst **53** offered improved enantioselectivity for benzoin, but the yield was still poor.[60] Almost 25 years later, Rawal reported that simply

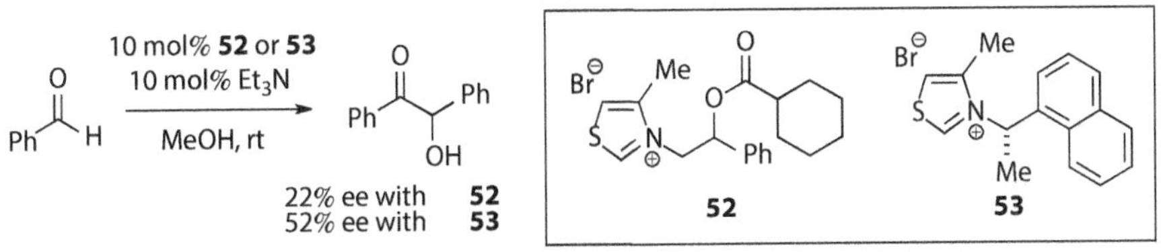

10 mol% **52** or **53**
10 mol% Et_3N
MeOH, rt

22% ee with **52**
52% ee with **53**

52 **53**

FIGURE 8.27 Sheehan's asymmetric benzoin reactions.

adding water to the reaction dramatically improved the performance of catalyst **53**, and benzoin could be obtained in 52% yield and with 48% ee.[61] This would remain as the highest enantioselectivity for the benzoin reaction for many years.[62]

In 1996, Enders and coworkers reported a breakthrough in the area of asymmetric benzoin reactions. This work was based on previous results that used achiral triazolium salts as catalysts for benzoin-type reactions.[63] Enders found that chiral triazolium catalyst **54** could perform asymmetric benzoin reactions with aromatic aldehydes, and the resulting products could be obtained with up to 86% ee (Figure 8.28).[64] Unfortunately, poor results were seen with aliphatic aldehydes. Later, it was found that the enantioselectivities could be further improved by using bicyclic triazolium catalyst **55** (Figure 8.29).[65]

Examples of asymmetric Stetter reactions were slower to appear. The first attempts at an intermolecular version were carried out in 1989 by Enders and coworkers (Figure 8.30).[57] They found that catalyst **56** could promote the addition of butanal into chalcone. The resulting 1,4-diketone was isolated with an encouraging 39% ee but with a very disappointing yield, which is likely due to low catalytic activity of the catalyst. It would be another 20 years before improved asymmetric control in the intermolecular Stetter reaction was realized. In 2008, Enders found that triazolium salt **57** could

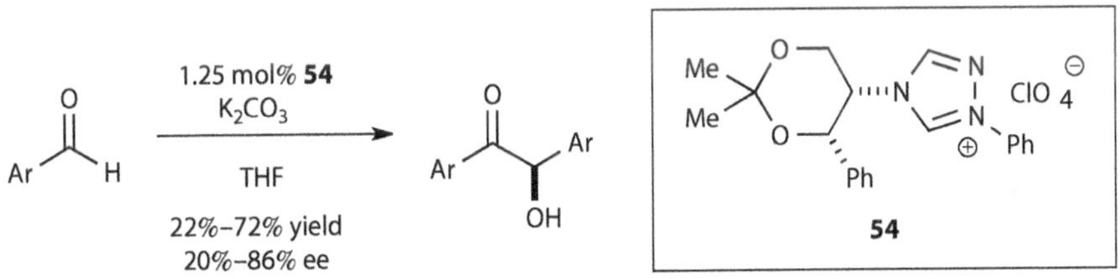

1.25 mol% **54**
K_2CO_3
THF

22%–72% yield
20%–86% ee

54

FIGURE 8.28 Enders' first report on the asymmetric benzoin reaction.

10 mol% **55**
10 mol% KO*t*-Bu
THF, 0 or 18°C

41%–100% yield
53%–93% ee

55

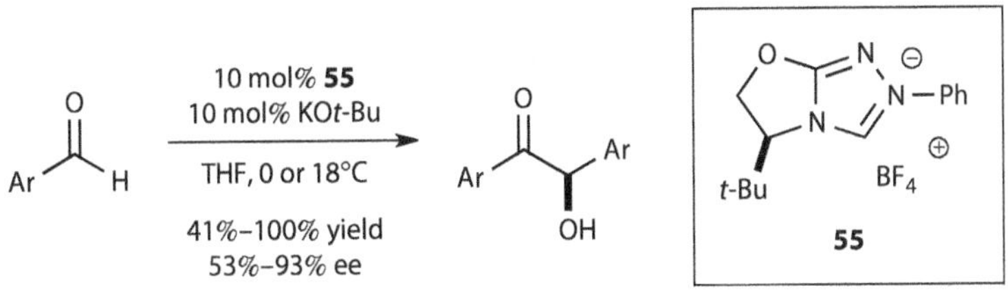

FIGURE 8.29 Enders' improved asymmetric benzoin reaction.

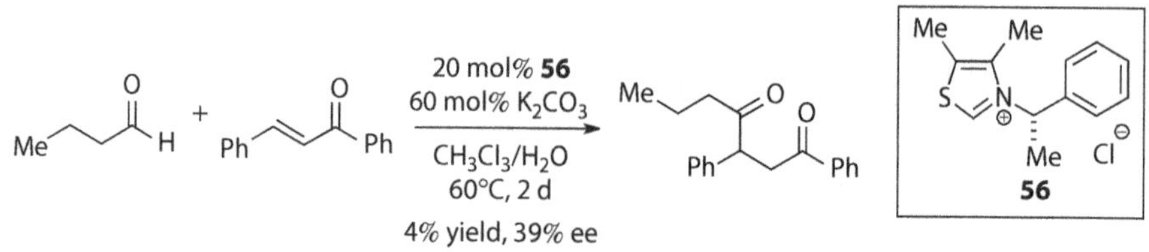

FIGURE 8.30 Initial asymmetric intermolecular Stetter reaction studies by Enders.

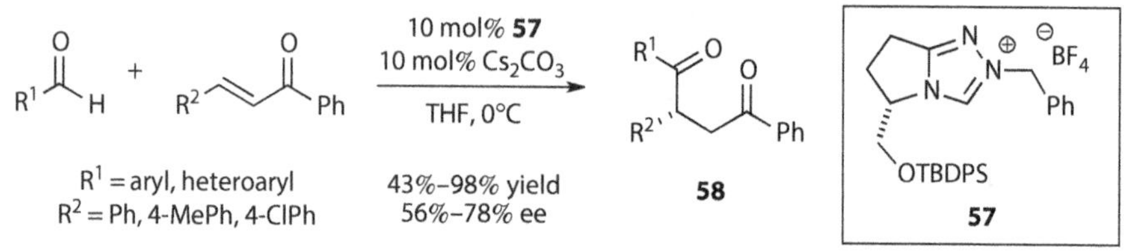

FIGURE 8.31 Asymmetric intermolecular Stetter reaction reported by Enders.

promote the addition of different aromatic and heteroaromatic aldehydes into chalcones (Figure 8.31).[66] The resulting 1,4-diketones (**58**) were obtained in moderate yields but with improved enantioselectivites, relative to their earlier work. Concurrently, Rovis and coworkers found that catalyst **59** promoted the addition of morpholine-derived glyoxamide **60** into different alkylidenemalonates (**61**, Figure 8.32).[67] The products were isolated in generally good yields and with high enantioselectivity. This result would serve as a springboard for continued exploration by the Rovis group.[68] Ryu and Yang,[69] Gravel,[70] Chi,[71] and Glorius[72] have also made contributions that aimed at further expanding the scope of the asymmetric intermolecular Stetter reaction.

The first detailed study of the intramolecular Stetter reaction was reported by Ciganek in 1995.[73] Before this work, the only example of an intramolecular Stetter reaction was in the context of a natural product synthesis.[74] Soon after Ciganek's report, Enders and coworkers reported the first attempts at an asymmetric variant (Figure 8.33).[75] It was found that triazolium catalyst **54** promoted the cyclization of aromatic aldehydes (**63**) to give 4-chromanones (**64**) with moderate levels of enantiocontrol. In 2002, Rovis and coworkers reported the first highly enantioselective intramolecular Stetter reaction (Figure 8.34).[76] They found that aminoindanol-derived catalyst **65** promoted the cyclization of appropriately substituted aldehydes **66**. The resulting products (**67**) could be isolated in high yields and with high enantioselectivity. Following this initial

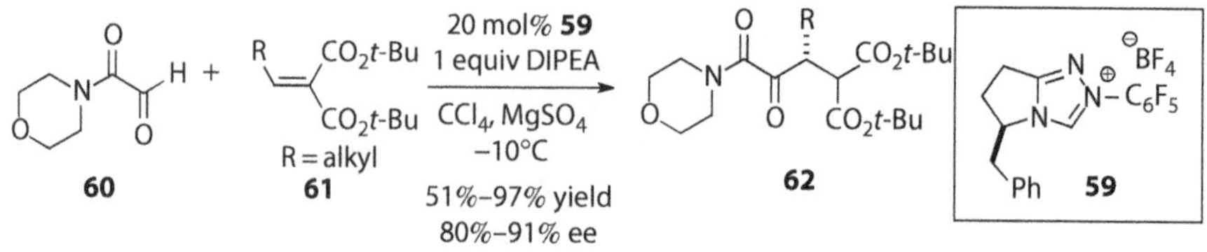

FIGURE 8.32 Asymmetric intermolecular Stetter reaction reported by Rovis.

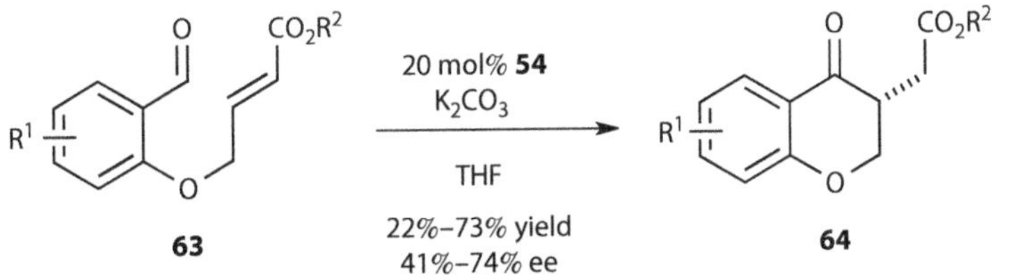

FIGURE 8.33 Asymmetric intramolecular Stetter reaction reported by Enders.

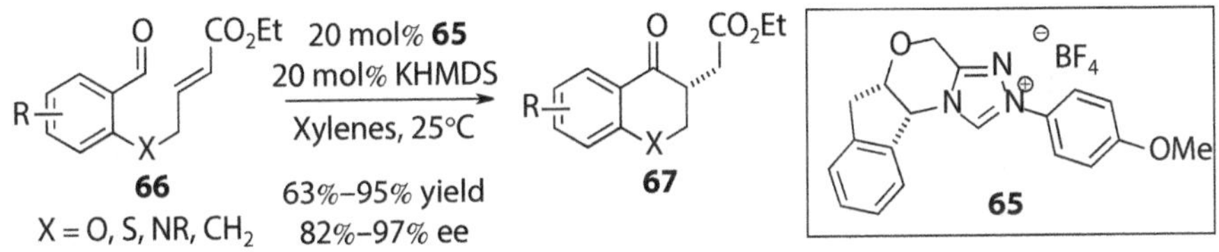

FIGURE 8.34 Asymmetric intramolecular Stetter reaction reported by Rovis.

disclosure by Rovis, the asymmetric intramolecular Stetter reaction would quickly be applied to the formation of quaternary stereocenters,[77] other trisubstituted acceptors,[78] and desymmetrization reactions[79] all with generally high enantioselectivity. Continued progress on asymmetric intramolecular Stetter reactions was rapid. Discussion of these developments can be found elsewhere.[46,57]

8.2.5 Crossed-Benzoin Reactions

One limitation of the classical benzoin reaction is that it is really just a dimerization of the starting aldehyde. A more useful scenario would be a crossed-benzoin reaction in which two different aldehydes are used (Figure 8.35). In practice, such a reaction often leads to a mixture of four different benzoin products: two homodimers (**68,69**) and two cross-products (**70,71**). Statistical mixtures are seen when two different aliphatic aldehydes are used, but good yields of one product can be obtained if an excess of one aldehyde is used.[80] Stetter reported that selectivity could be observed when a three-fold excess of an aliphatic aldehyde is reacted with an aromatic aldehyde partner (Figure 8.36).[81] Reactions using 2-chlorobenzaldehyde were found to be particularly selective. In these cases, it is thought that the aromatic aldehyde forms a more stable Breslow intermediate, and this preferentially reacts with the more reactive aliphatic

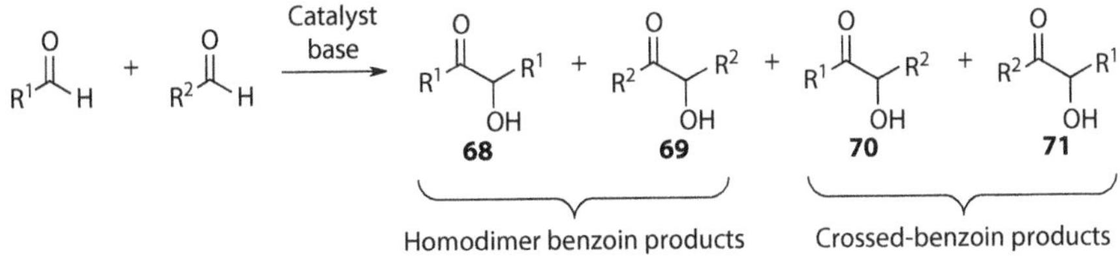

FIGURE 8.35 The crossed-benzoin reaction.

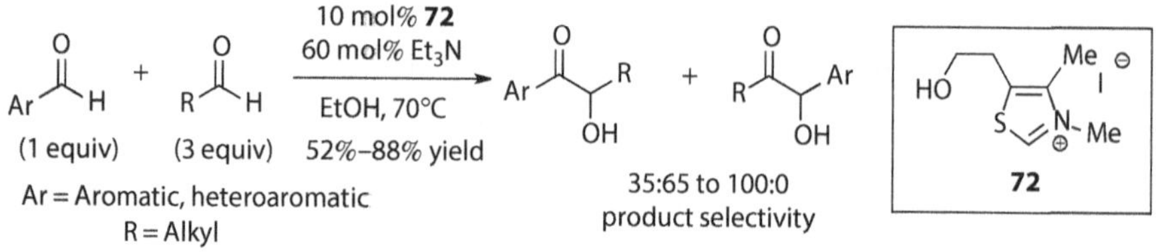

FIGURE 8.36 Crossed-benzoin reactions of aromatic and aliphatic aldehydes.

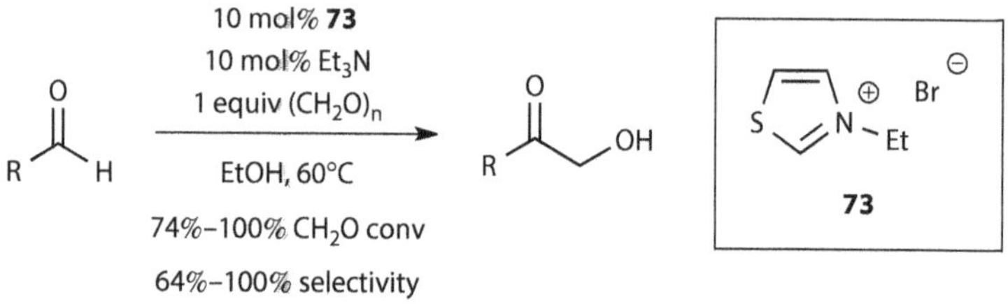

FIGURE 8.37 Crossed-benzoin reactions with formaldehyde.

aldehyde. Efficient crossed-benzoin reactions are also observed when formaldehyde is used as one of the reaction components (Figure 8.37).[82]

Intramolecular reactions offer another opportunity for cross-benzoin reactions. The first examples of intramolecular benzoin reactions appeared in 1976 (Figure 8.38).[83] A series of dialdehydes (**75**) were efficiently cyclized to isomeric 2-hydroxycyclopentanones **76** and **77**. Unfortunately, the triethylamine required for this reaction also resulted in product isomerization, so it was not possible to judge the selectivity inherent to the reaction. The scope of the intramolecular benzoin would remain unexplored for more than 20 years. In 2004, Enders[84] and Suzuki[85] independently reported the cyclization of various ketoaldehydes using thiazolium **72** as the source of the NHC. Some of the better results are shown in Figure 8.39. Similar conditions could be used to construct five-membered rings, but the yields were generally lower. By using ketoaldehyde starting materials, the authors were able to ensure that only one product was formed in contrast to the earlier work with dialdehyde substrates.

Following their initial work, Enders[86] and Suzuki[87] reported that chiral triazolium catalysts could be used for asymmetric cyclizations of ketoaldehydes. Suzuki found that catalyst **78** could be applied to several different substrate types, and the resulting products were generally isolated with high enantioselectivity (Figure 8.40).

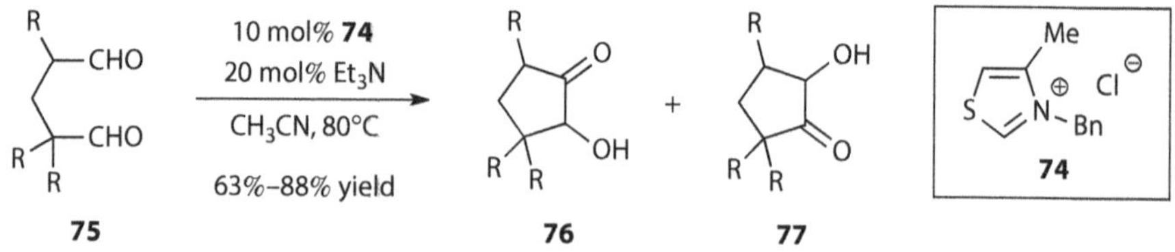

FIGURE 8.38 First examples of intramolecular benzoin reactions.

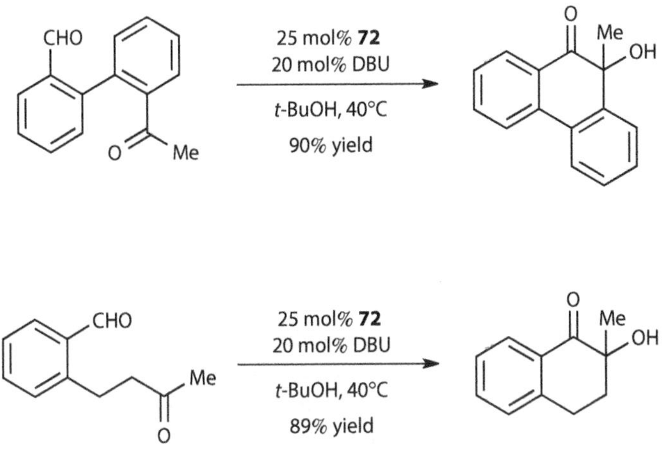

CHO
O
Me
25 mol% 72
20 mol% DBU
t-BuOH, 40°C
90% yield
O
Me
OH
CHO
Me
O
25 mol% 72
20 mol% DBU
t-BuOH, 40°C
89% yield
O
Me
OH

FIGURE 8.39 Thiazolium-catalyzed cyclization of ketoaldehydes.

BF4
78
44% yield
96% ee
70% yield
96% ee
69% yield
60% ee
Me
Et
i-Pr
Bn
OH
73% yield
39% ee
47% yield
90% ee
47% yield
90% ee
56% yield
88% ee
Ph
Me
MeO
N-O
91% yield
98% ee
73% yield
99% ee

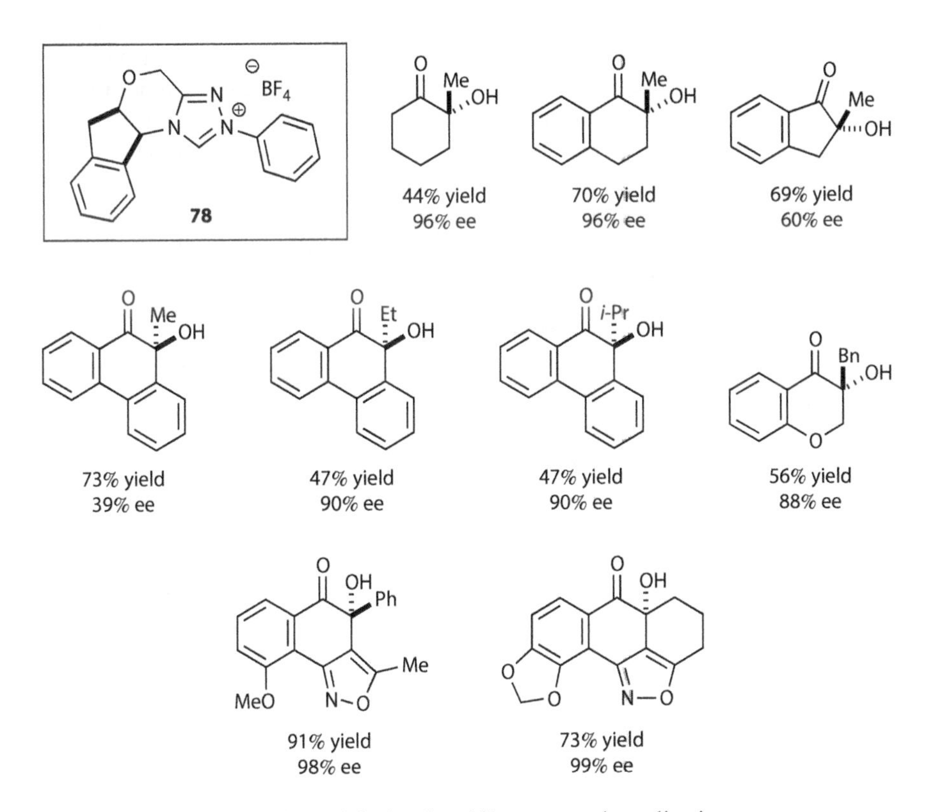

FIGURE 8.40 Products formed during Suzuki's asymmetric cyclization.

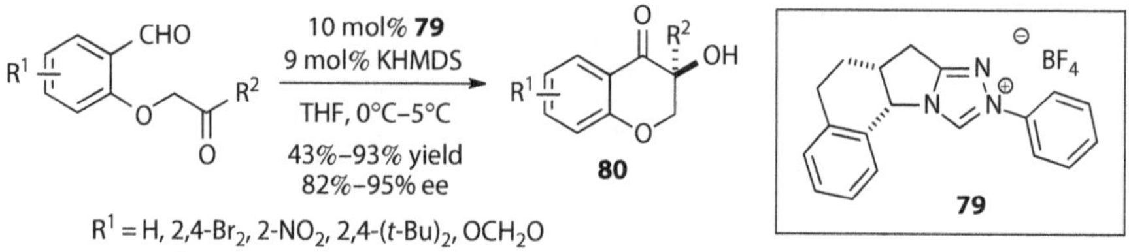

CHO
R1
R2
O
O
10 mol% 79
9 mol% KHMDS
THF, 0°C–5°C
43%–93% yield
82%–95% ee
O
R2
OH
80
BF4
79

R^1 = H, 2,4-Br_2, 2-NO_2, 2,4-(*t*-Bu)$_2$, OCH_2O
R^2 = Et, Me, *i*-Bu, *c*-Hex, *n*-Pr

FIGURE 8.41 Asymmetric synthesis of 3-hydroxy-4-chromanones by Enders.

Later, Enders was able to show that several different 4-chromanones (**80**) could be produced with high enantioselectivity using catalyst **79** (Figure 8.41).[88]

In recent years, several researchers have reinvestigated the problem of the cross-benzoin reaction.[89] Of particular importance were the findings by Connon and Zeitler[90] and Gravel[91] who found that bicyclic triazolium catalysts with a pentafluorophenyl substituent could perform highly selective cross-benzoin reactions, even without large excesses of one aldehyde (Figure 8.42). Connon and Zeitler reported that selectivity could be achieved when catalyst **81** is used in conjunction with an aromatic aldehyde with an electron-withdrawing substituent in the ortho position.[90] It was proposed that the ortho group plays an important role in modulating the electrophilicity of the aromatic aldehyde. A few years later, Gravel and coworkers found that highly selective cross-benzoin reactions could be achieved by using catalyst **83**.[91] In addition to the improved chemoselectivity, this catalyst did not require the use of ortho-substituted aromatic aldehydes. Subsequent mechanistic works by Gravel and Legault[92] suggest that the selectivity observed with catalyst **83** is kinetically controlled. In particular, the formation of the Breslow intermediate from the alkyl aldehyde is preferred and rapid. In addition, the attack of the Breslow intermediate onto the aromatic aldehyde is rate determining and appears to be stabilized

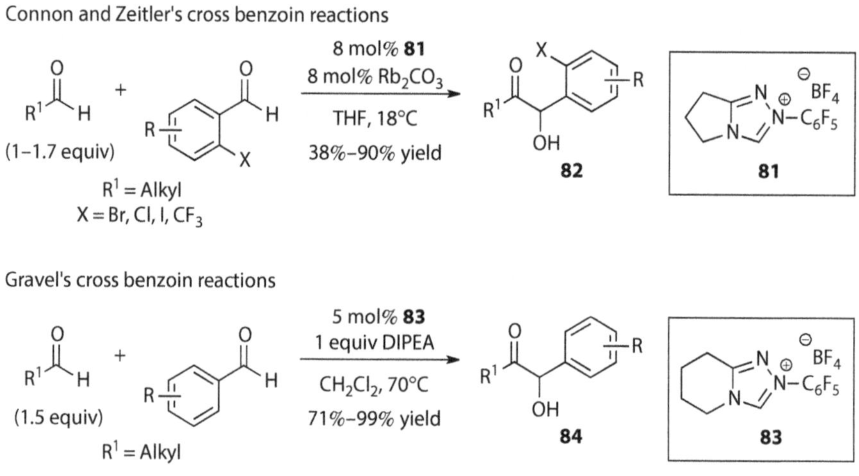

Connon and Zeitler's cross benzoin reactions
R1 H
(1–1.7 equiv)
R
X
8 mol% 81
8 mol% Rb2CO3
THF, 18°C
38%–90% yield
OH
82
BF4
N–C6F5
81
R1 = Alkyl
X = Br, Cl, I, CF3
Gravel's cross benzoin reactions
R1 H
(1.5 equiv)
5 mol% 83
1 equiv DIPEA
CH2Cl2, 70°C
71%–99% yield
OH
84
BF4
N–C6F5
83
R1 = Alkyl

FIGURE 8.42 Cross-benzoin reactions under catalyst control.

by π interactions between the triazole of the catalyst and the aromatic group of the electrophile. It should be mentioned that both Gravel and Connon reported some preliminary experiments in asymmetric cross-benzoin reactions, but these have yet to be explored to a greater extent.

Cross-benzoin reactions can be achieved if one of the aldehyde components is replaced with an activated ketone. Enders and Henseler reported the first example of using ketone electrophiles in benzoin-type reactions in 2009.[93] They found that triazolium catalyst **85** effectively promotes the addition of aromatic and heteroaromatic aldehydes into aromatic trifluoromethyl ketones (Figure 8.43). The CF_3 presumably increases the electrophilicity of the ketone acceptor. These conditions do form the classical benzoin adduct of the aldehyde, but its formation is reversible. In one case, they were even able to use benzoin as the source of the aldehyde. Soon thereafter, Enders reported an asymmetric variant of this reaction that employed chiral triazolium catalyst **86** (Figure 8.44).[94] Interestingly, the asymmetric version was most successful with heteroaromatic aldehydes. This was explained by invoking attractive stabilizing interactions in the Breslow intermediate between the heteroaromatic group of the aldehyde and the pentafluorophenyl group of the catalyst. More recently, Connon and Zeitler,[95] Gravel,[96] and Johnson[97] have shown that α-ketoesters can serve as effective acceptors in cross-benzoin reactions with aldehydes (Figure 8.45). Moreover, these acceptors are also compatible with nonaromatic (alkyl) aldehyde donors. Both Gravel and Johnson were able to show that these reactions can be rendered asymmetric when a chiral triazolium catalyst is used.

Cross-coupling can also be achieved by replacing one of the aldehyde components with an imine. This is sometimes referred to as an azabenzoin reaction. The first examples of the azabenzoin reaction were reported in 2001 by researchers at

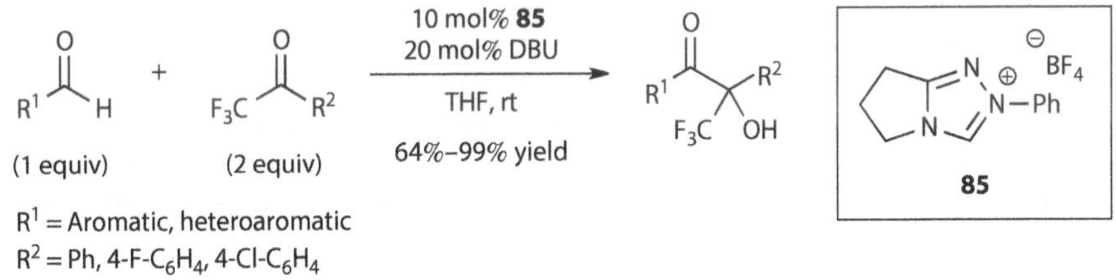

10 mol% **85**
20 mol% DBU
THF, rt
64%–99% yield

(1 equiv) (2 equiv)

R^1 = Aromatic, heteroaromatic
R^2 = Ph, 4-F-C_6H_4, 4-Cl-C_6H_4

85

FIGURE 8.43 Cross-benzoin reactions with trifluoromethyl ketone acceptors.

10 mol% **86**
1 equiv DIPEA
THF, 0°C
69%–96% yield
39%–84% ee

(1 equiv) (2 equiv)

R^1 = Heteroaromatic
R^2 = Aromatic, heteroaromatic

86

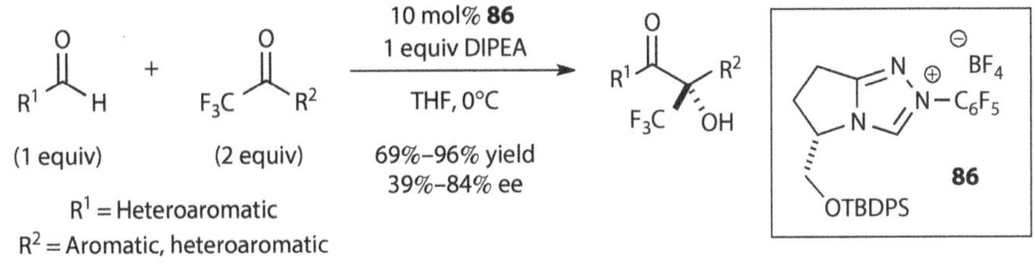

FIGURE 8.44 Asymmetric cross-benzoin reactions with trifluoromethyl ketone acceptors.

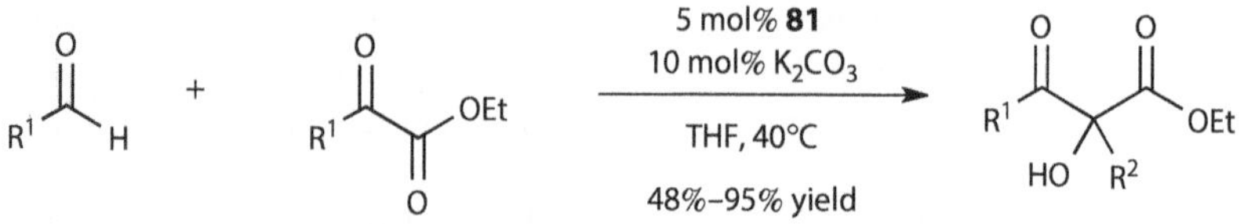

5 mol% **81**
10 mol% K_2CO_3
THF, 40°C
48%–95% yield

R^1 = Aromatic, heteroaromatic, alkyl
R^1 = Alkyl, phenyl

FIGURE 8.45 Connon's cross-benzoin reactions with α-ketoester acceptors.

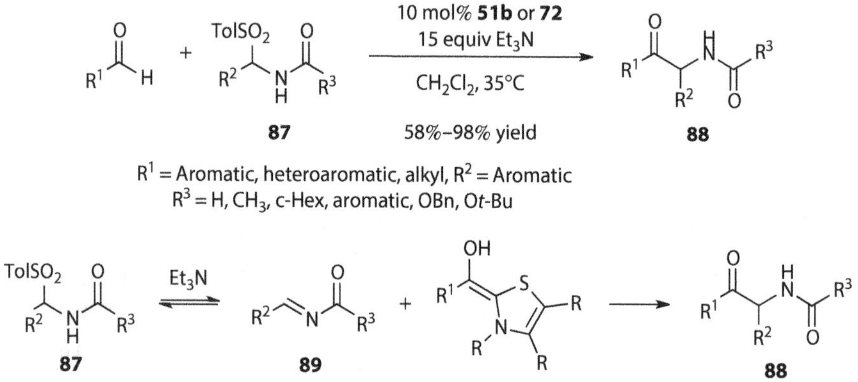

10 mol% **51b** or **72**
15 equiv Et_3N
CH_2Cl_2, 35°C
58%–98% yield

R^1 = Aromatic, heteroaromatic, alkyl, R^2 = Aromatic
R^3 = H, CH_3, c-Hex, aromatic, OBn, O*t*-Bu

FIGURE 8.46 Examples of azabenzoin reactions.

Merck (Figure 8.46).[98] They found that thiazolium catalysts such as **51a** and **72** promoted the reaction of different aldehydes with arylsulfonylamides (**87**). The resulting α-amino ketones (**88**) were isolated in good-to-high yields. It was shown that under the reaction conditions, the starting arylsulfonylamides were in equilibrium with the corresponding *N*-acyl imines (**89**). The product would therefore be generated by reaction of the *N*-acyl imine with the Breslow intermediate. Importantly, the corresponding benzoin product was not observed under these conditions. In addition, crossover products were not observed when the reaction was carried out in the presence of an added α-amino ketone. Similarly, the corresponding benzoin products do not serve as starting materials for these reactions. Later, Miller and coworkers reported an asymmetric version of this reaction that employed peptide-based thiazolium catalyst **90** (Figure 8.47).[99]

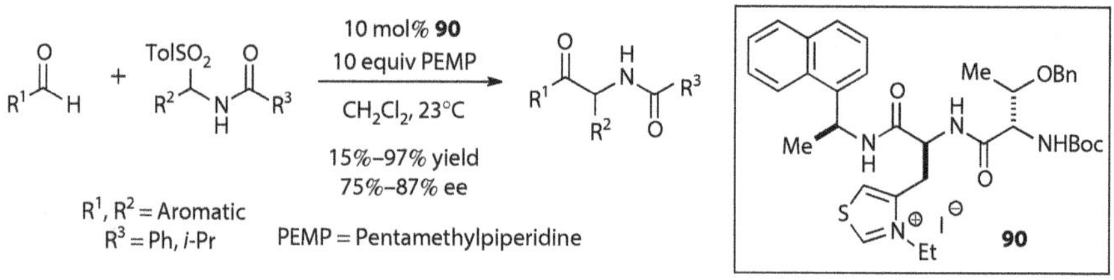

10 mol% **90**
10 equiv PEMP
CH_2Cl_2, 23°C
15%–97% yield
75%–87% ee

R^1, R^2 = Aromatic
R^3 = Ph, *i*-Pr
PEMP = Pentamethylpiperidine

FIGURE 8.47 Miller's asymmetric azabenzoin reaction.

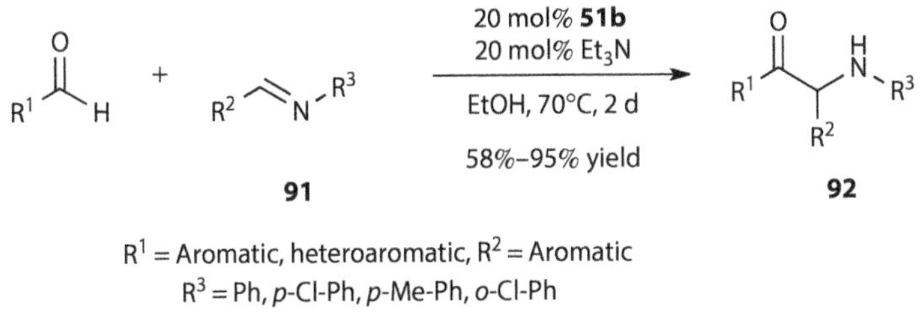

20 mol% **51b**
20 mol% Et_3N
EtOH, 70°C, 2 d
58%–95% yield

91 **92**

R^1 = Aromatic, heteroaromatic, R^2 = Aromatic
R^3 = Ph, *p*-Cl-Ph, *p*-Me-Ph, *o*-Cl-Ph

FIGURE 8.48 Azabenzoin reaction with *N*-arylimines.

Later, You and coworkers found that α-amino ketones **92** could be produced in good-to-high yields when aldehydes were reacted with *N*-arylimines (**91**) in the presence of thiazolium catalyst **51b** (Figure 8.48).[100] Mechanistically, these conditions appear to be somewhat different from those reported by the Merck group. This was ascertained by performing several experiments. First, it was found that the aldehyde could be replaced by the corresponding benzoin with little diminution in yield. Second, a small amount of crossover product was detected in a control experiment, suggesting that the azabenzoin reaction may be reversible under these conditions. Since these important first examples, the azabenzoin reaction has been extended to use ketamine substrates.[101] Enantioselective variants have been reported by Rovis[102] and Ye.[103] In general thought, this area has not been explored to the extent that is seen by the classic benzoin reaction.

8.2.6 Use of Acylsilanes as Acyl Anion Equivalents

A recurring problem in both Stetter reactions and azabenzoin reactions is the preferential dimerization of the aldehyde component through the benzoin pathway. In some cases, this dimerization is not a significant problem as the benzoin reaction is reversible, but this does lead to a diminution of the overall reaction rate. In other cases, the competitive benzoin reaction is not reversible. This scenario would obviously have a significant negative impact on the efficiency of the process. In 2004, Scheidt and Johnson reported an interesting solution to this problem. They found that acylsilanes could, under appropriate reaction conditions, serve as a surrogate for the aldehyde during the formation of the Breslow intermediate. As will be seen, Scheidt and coworkers found that NHC catalysts could promote these reactions. Johnson and coworkers have found that cyanide[104] or phosphite[105] catalysts can also promote reactions that involve acylsilanes, but these efforts will not be discussed here.

Scheidt's initial report concerned using acylsilanes as donors in Stetter-type reactions (Figure 8.49).[106] Under the optimized conditions, it was found that both aryl and alkyl acylsilanes (**93**) could be added to different chalcones (**94**) in order to provide 1,4-diketones **95**. The reaction generally proceeded with good-to-high yields. Most examples used acylsilanes with a $SiMe_3$ group, but other silanes were also tolerated. The use of thiazolium catalyst **51a** was critical for the success of these reactions as imidazolium and triazolium catalysts gave poor results.[107] In addition to chalcones, the sila-Stetter reaction could also be extended to α,β-unsaturated esters and ketones. Later, Scheidt

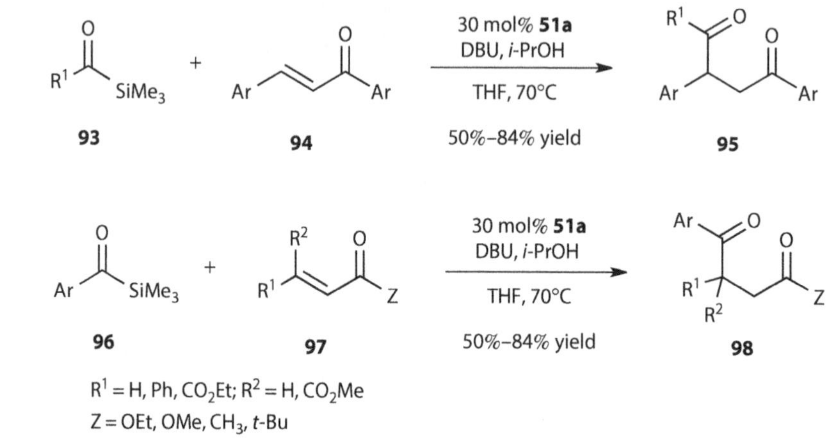

R[1]=H, Ph, CO_2Et; R^2 = H, CO_2Me
Z = OEt, OMe, CH_3, *t*-Bu

FIGURE 8.49 Sila-Stetter reactions of chalcones and other electron-deficient alkenes.

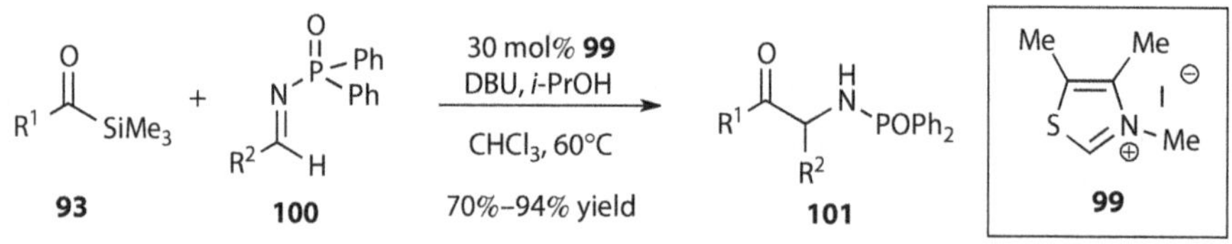

FIGURE 8.50 Scheidt's use of acylsilanes in azabenzoin reactions.

would use this sila-Stetter reaction in a one-pot synthesis of pyrroles.[108] The use of acylsilanes was also extended to the formation of α-aminoketones **101** though an azabenzoin-type reaction that involve *N*-phosphorylimines (**100**, Figure 8.50).[109]

Scheidt and coworkers proposed the following mechanism for the sila-Stetter reaction (Figure 8.51) and, by extension, the azabenzoin reaction. Initial reaction between the *in situ* generated carbene (**102**) and the acylsilane produces a tetrahedral intermediate (**103**) that can undergo a Brook rearrangement to give **104**. Transfer of the silyl group from this intermediate to the alcohol additive generates the Breslow intermediate (**105**) that is required for conjugate addition into the chalcone. The fact that an alcohol additive is required for efficient reaction supports the notion that the Breslow intermediate is generated by the proposed silyl transfer step.

8.2.7 Use of Acyl Anion Equivalents in Natural Product Synthesis

The synthetic potential of the acyl anion equivalents generated by NHC catalysts is demonstrated in the numerous natural product syntheses that make use of this technology.[110] Again, it is outside the scope of this chapter to give a detailed analysis of these efforts, but several representative natural products, and one blockbuster drug, whose syntheses made use of an NHC are highlighted in Figure 8.52.[111] Bonds whose construction was made possible by an NHC, either chiral or achiral, are indicated with a "*".

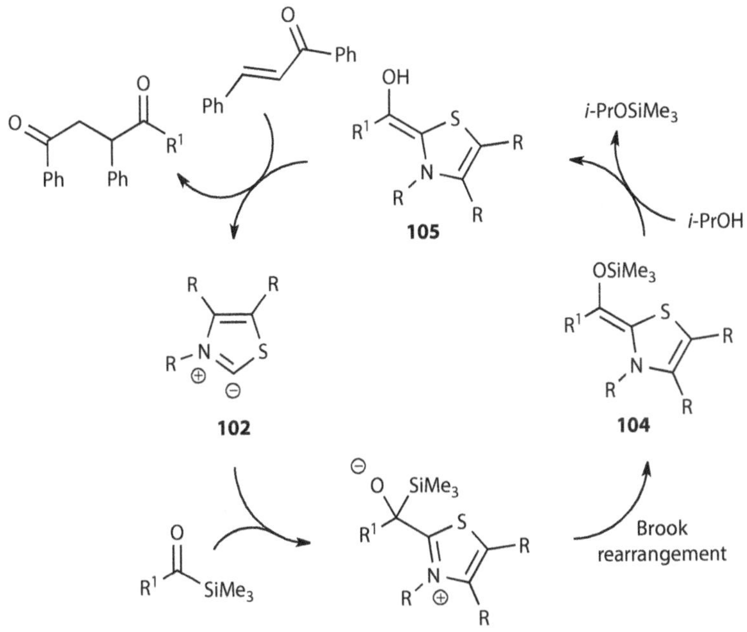

FIGURE 8.51 Proposed mechanism of the sila-Stetter reaction.

8.3 TERTIARY PHOSPHINES AS VERSATILE NUCLEOPHILIC CATALYSTS

8.3.1 Introduction to Phosphine Catalysis

Much like *N*-heterocyclic carbenes, the use of tertiary phosphines as catalysts for organic transformations can be traced back several decades. The genesis appears to be the independent work by Rauhut and Currier,[112] Baizer and Anderson,[113] and McClure[114]; all of whom found that electron-deficient alkenes, such as acrylates or acrylonitrile, could be dimerized in the presence of a catalytic amount of a tertiary phosphine (Figure 8.53). At about the same time, Morita described a similar reaction between an electron-deficient alkene and an aldehyde (Figure 8.54).[115] Curiously, despite these promising first steps, little subsequent work would be reported until the 1990s. In the case of the Morita–Baylis–Hillman (MBH) reaction, this lack of activity can probably be attributed to the notoriously poor reaction rates. In the case of the Rauhut–Currier (RC) reaction, the lack of control in cross-coupling reactions is the likely culprit. The apparent lack of interest in these reactions by the synthetic community may also be due to the air sensitivity and relatively high cost of the tertiary phosphine catalysts. Not surprisingly, therefore, what little activity that was reported concerned the use of tertiary amines as catalysts.[116,117] Although tertiary amines can be active catalysts for these reactions, they are often much less active than phosphines. This is likely due to the softer nature of the phosphine nucleophile

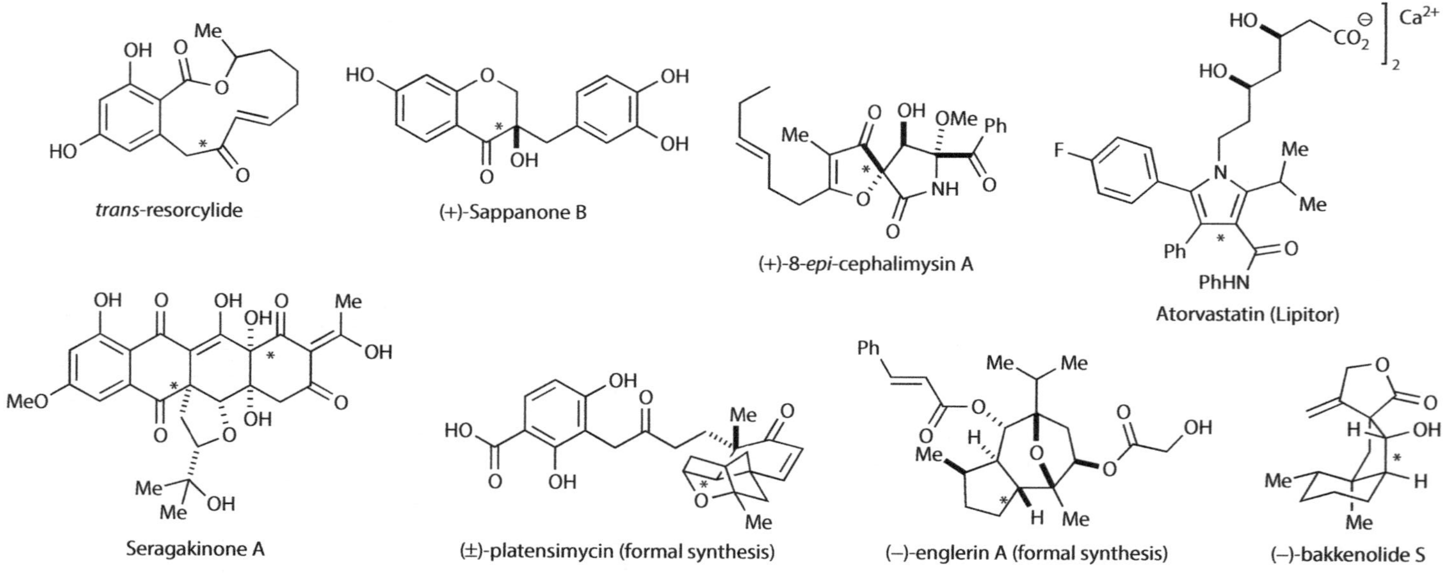

trans-resorcylide

(+)-Sappanone B

(+)-8-*epi*-cephalimysin A

Atorvastatin (Lipitor)

Seragakinone A

(±)-platensimycin (formal synthesis)

(−)-englerin A (formal synthesis)

(−)-bakkenolide S

FIGURE 8.52 Examples of natural products synthesized using acyl anion equivalents generated with an NHC catalyst.

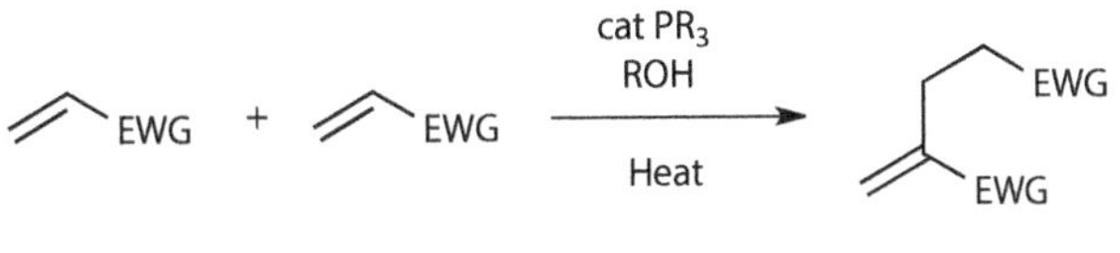

FIGURE 8.53 The Rauhut–Currier reaction.

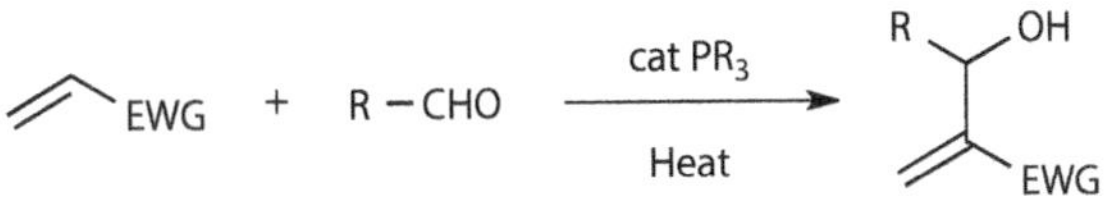

FIGURE 8.54 The Morita–Baylis–Hillman reaction.

being a better match for the soft alkene electrophile. In addition, the amines tend to be stronger bases than the corresponding phosphine, and this may give rise to complications that are not seen with phosphine catalysts.

Interest in phosphine-catalyzed reactions was reignited in the 1990s following reports on several new transformations. The first were the reports by Trost[118] and Lu[119] that concern the phosphine-catalyzed isomerization of electron-deficient alkynes into the corresponding 1,3-dienes (Figure 8.55). Following these results, Trost and coworkers reported that these internal redox reactions could also involve the addition of internal[120] or external[121] nucleophiles (Figure 8.56). Meanwhile, Lu reported that phosphines could catalyze an annulation reaction between allenoates and electron-deficient alkenes (Figure 8.57).[122] All of these reactions share some similarities with the RC and MBH reactions, and together would serve as the basis for subsequent phosphine-catalyzed reactions. At the same time, Trost and Lu were investigating the reactions mentioned earlier; Vedejs and coworkers were

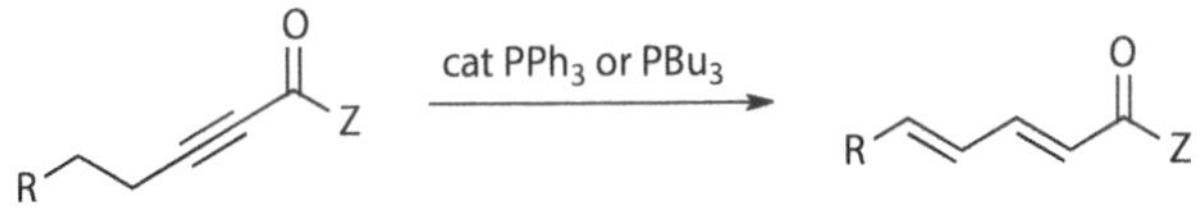

FIGURE 8.55 Phosphine-catalyzed isomerization of alkynes.

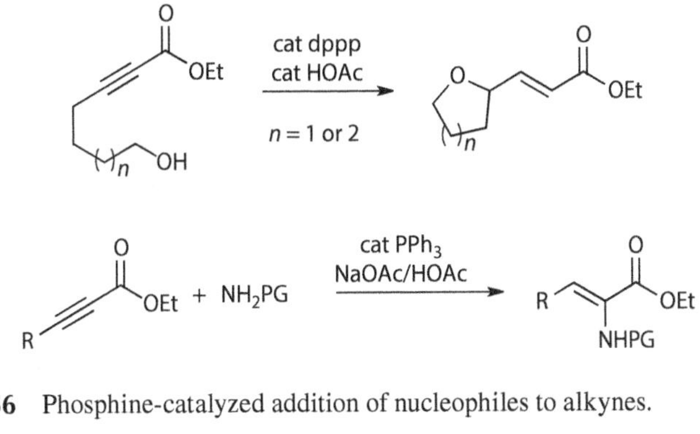

FIGURE 8.56 Phosphine-catalyzed addition of nucleophiles to alkynes.

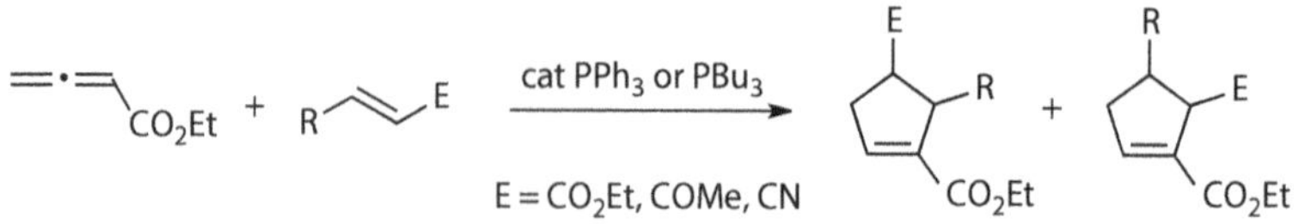

FIGURE 8.57 Phosphine-catalyzed annulation of allenoates.

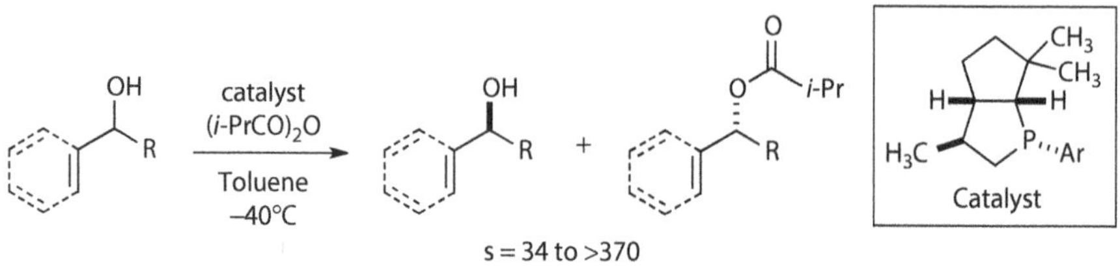

FIGURE 8.58 Phosphine-catalyzed resolution of secondary alcohols.

investigating the use of chiral phosphine catalysts for the kinetic resolution of secondary alcohols (Figure 8.58).[123] Although these resolution reactions would not catch on in a general way, they serve as an important demonstration that chiral phosphines can catalyze asymmetric transformations in the absence of a supporting metal.

Much similar to the other catalyst classes discussed in this chapter, the use of phosphines as organocatalysts has been extensively reviewed.[124] The use of chiral phosphines in asymmetric organocatalysis has also grown sufficiently to warrant several recent reviews.[125] The MBH[126] and RC[127] reactions have also been the subject of thorough reviews and will not be covered here. Instead, the discussion below will focus more on the reactivity that is seen in Figures 8.55 through 8.57. Some coverage of asymmetric catalysis will appear as appropriate. As before, the reader is encouraged to seek out previous reviews for more detailed discussions.

8.3.2 Isomerization of Electron-Deficient Alkynes

Investigations into the phosphine-catalyzed isomerization of electron-deficient alkynes started in 1988 with independent reports by Trost, Lu, and Inoue.[128] These researchers found that organometallic complexes based on Pd and Ru catalyzed the isomerization of alkynones into conjugated dienones (e.g., **106–107**). Subsequent work by Lu[129] revealed that excess phosphine was beneficial. Eventually, it was determined that if the metal was left out of the reaction altogether, the phosphine alone would promote the reaction. Two similar reactions were reported in 1992. Trost found that PPh_3 could function as the catalyst if the reaction is carried out at elevated temperatures.[118] Alternatively, Lu found that PBu_3 was an effective catalyst at ambient temperatures.[119]

This reaction has proven useful for the isomerization of all manner of electron-deficient alkynes, with the reactivity trend as follows: alkynones > alkynoates > alkynamides. Allenoates afford the same product as the corresponding alkyne.[118] Nonconjugated alkynes are inert under these conditions. Both alkynoates and

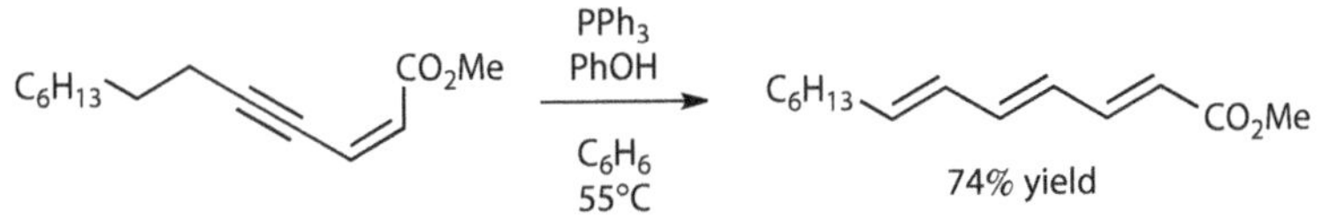

FIGURE 8.59 Synthesis of conjugated trienes by alkyne isomerization.

alkynamides generally require an acidic cocatalyst in order to achieve efficient reactivity. Trost and Kazmaier employed acetic acid for their studies. Later, Rychnovsky found that phenol could also be used.[130] This reaction can also be extended to the synthesis of conjugated trienes as shown in Figure 8.59.

The mechanism that has been proposed for this isomerization reaction is shown in Figure 8.60. Conjugate addition of the phosphine into the alkyne produces a resonance-stabilized anion that can be protonated to give vinyl phosphonium salt **108**. This intermediate can then be isomerized into **109** and **110** by a series of proton transfer steps. It is thought that the conjugate base of any acid cocatalyst facilitates this isomerization.[130] Once formed, intermediate **111** provides diene **107** by elimination of phosphine. As will be seen, this mechanism has proved quite inspirational for developing other phosphine-catalyzed reactions, and several of the intermediates can be intercepted under appropriate conditions. For example, intermediate **110** can be intercepted by aldehydes during stoichiometric olefination reactions.[131]

Recently, Harmata and Hampton have shown that a similar phosphine-catalyzed isomerization of allenic sulfones **112** to 2-arylsulfonyl-1,3-dienes **113** can be performed (Figure 8.61). Interestingly, when the same starting materials were reacted with $Pd(PPh_3)_4$, the isomerization proceeded to give 1-arylsulfonyl-1,3-dienes **114** instead.[132] Subsequent

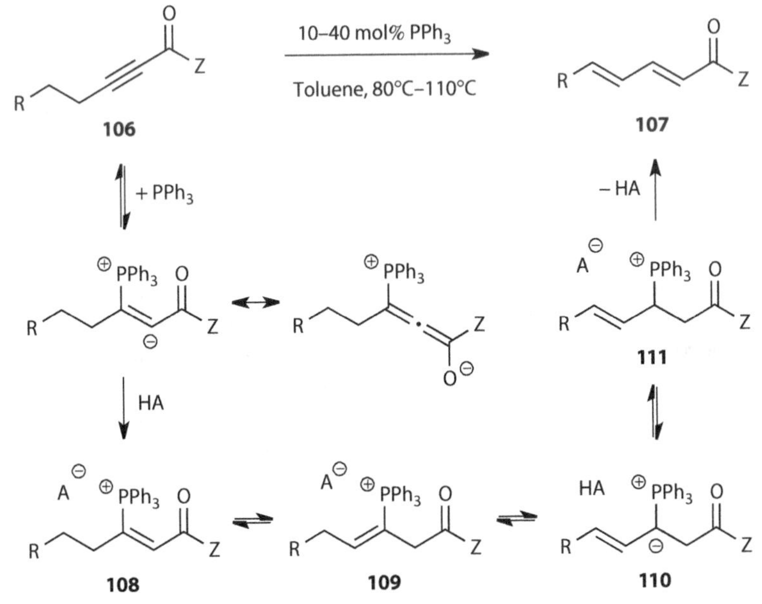

FIGURE 8.60 Isomerization of electron-deficient alkynes as reported by Trost.

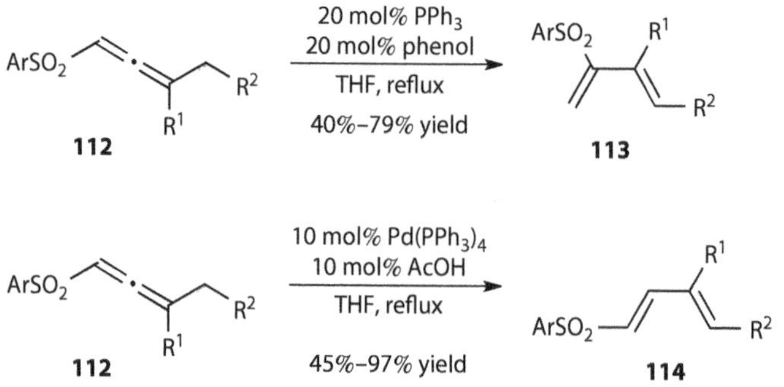

FIGURE 8.61 Harmata's phosphine-catalyzed and palladium-catalyzed isomerization of allenic sulfones.

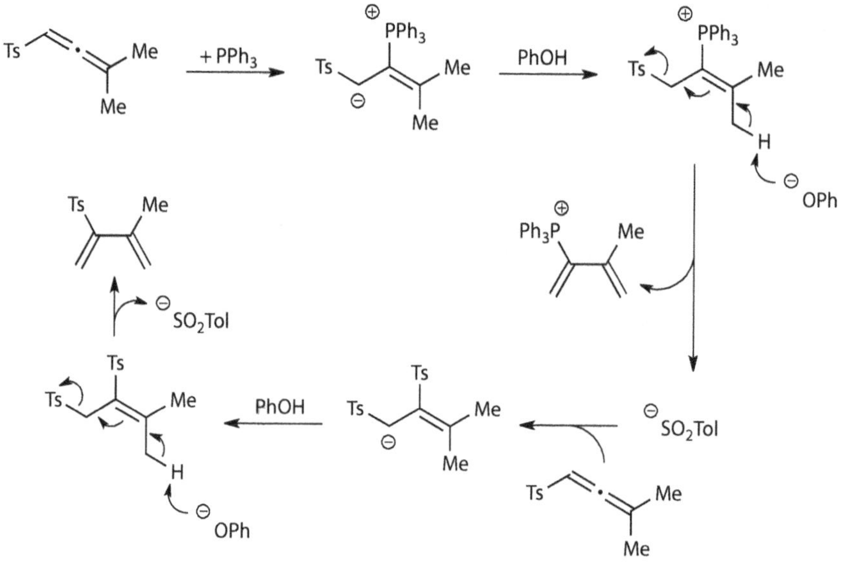

FIGURE 8.62 Proposed mechanism for Harmata's isomerization of allenic sulfones.

mechanistic work revealed that the phosphine only serves as a means to generate sulfinate ions, which are the actual isomerization catalyst (Figure 8.62).[133] This mechanism was supported by a number of crossover experiments and control experiments in which sodium toluenesulfinate was used as the catalyst without added phosphine.

The phosphine-catalyzed isomerization of alkynes is quite synthetically useful owing to its mildness and the useful nature of the resulting electronically differentiated 1,3-dienes. Campagne and coworkers have used this reaction as the final step in their construction of the C4–C_{24} fragment of macrolactin A (Figure 8.63).[134] O'Doherty and coworkers have also used this isomerization to great success in the synthesis of several natural products, including cladospolides B–D,[135] cryptocaryols A and B,[136] apicularen,[137] milbemycin β_3,[138] and northern portion of RK-397.[139] Much of this work has taken advantage of a strategy developed in their laboratory (Figure 8.64)[140] that involves

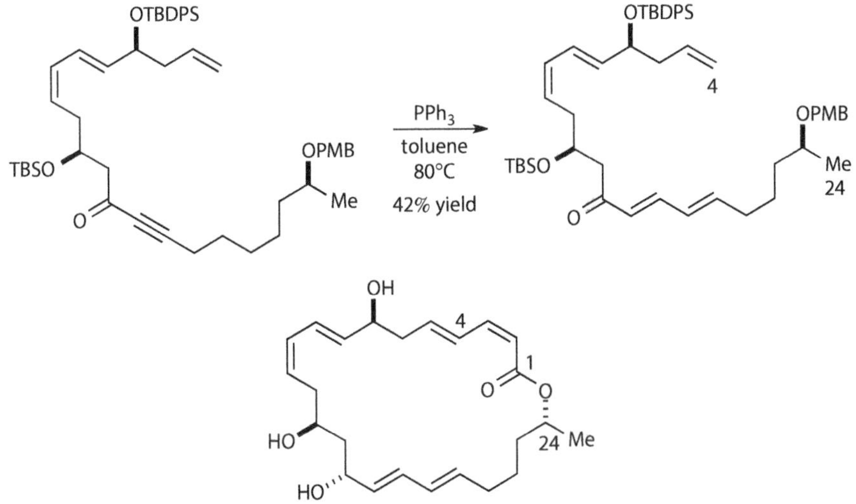

FIGURE 8.63 Synthesis of C_4–C_{24} fragment of macrolactin A.

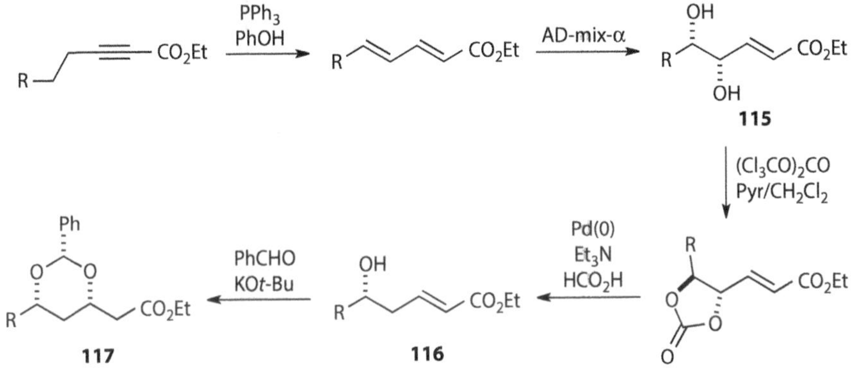

FIGURE 8.64 O'Doherty's strategy for elaboration of the alkynoate isomerization products.

asymmetric dihydroxylation of the dienoate products (to give **115**), Pd-catalyzed reductive deoxygenation (to give **116**), and base-catalyzed benzylidene acetal formation (to give **117**). More recently, Trost and Biannic have shown that with appropriate substitution, this isomerization can be used to initiate a 6π-electrocyclization.[141]

8.3.3 Phosphine-Catalyzed Nucleophilic Additions to Alkynes

Following their work on phosphine-catalyzed alkyne isomerizations, the Trost group investigated reactions that could intercept one or more intermediates that were generated during the isomerization mechanism. In particular, it was envisioned that vinyl phosphonium intermediates **108** and **109** might be electrophilic enough to accept a nucleophile and generate **118** or **120** (Figure 8.65). Elimination of the phosphine from these

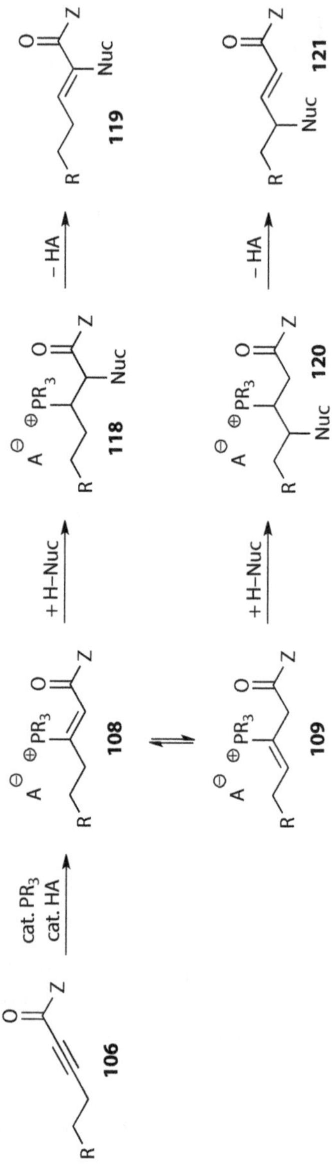

FIGURE 8.65 Proposed pathways for phosphine-catalyzed addition of nucleophiles to alkynes.

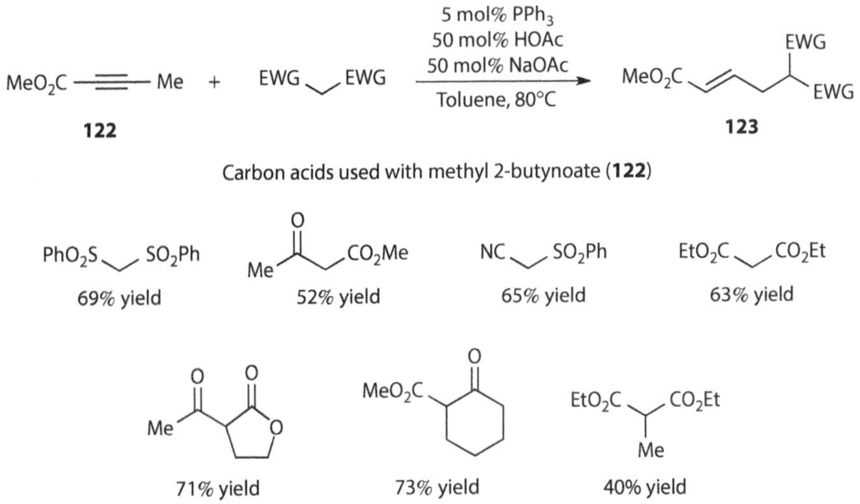

FIGURE 8.66 Phosphine-catalyzed addition of carbon acids to alkynes.

intermediates would then give rise to products in which the nucleophile has been added to the α- or γ-position of the starting alkyne (**119** and **121**, respectively).

The first success came when soft carbon nucleophiles (active methylene compounds) were used as the nucleophilic component (Figure 8.66).[142] Several different carbon acids could be used with methyl 2-butynoate (**122**) as the acceptor. The reaction was also successful with amide and ketone acceptors. In all the cases, the isolated product (**123**) was the one in which the nucleophile was added to the γ-position of the starting alkyne. This suggests that the phosphine plays a more important role than just a simple base, as deprotonation of the nucleophile would likely lead to a simple Michael addition. Later, Lu reported that similar nucleophilic additions could be carried out using allenoates as the acceptor.[143]

Following their success with carbon acids, Trost and Li turned their attention to the use of hydroxyl groups as nucleophiles (Figure 8.67).[120] In exploratory work, it was

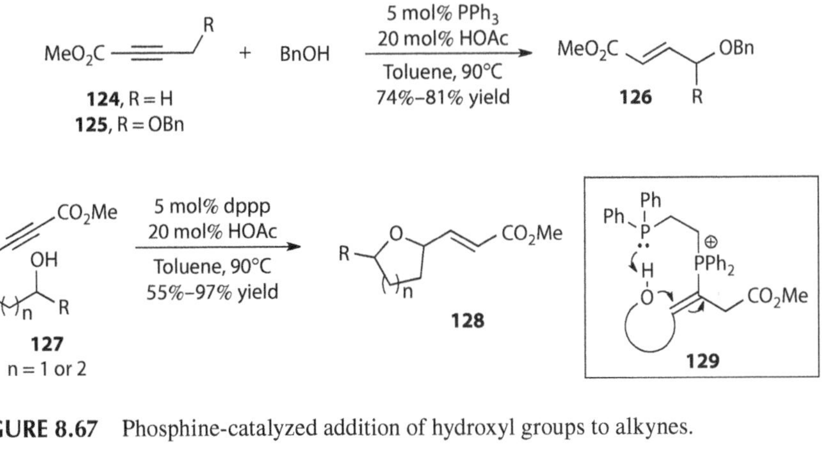

FIGURE 8.67 Phosphine-catalyzed addition of hydroxyl groups to alkynes.

found that PPh_3 catalyzed the addition of benzyl alcohol to the γ-position of alkynes **124** and **125**. In contrast, no product was obtained when a carbon acid was reacted with alkyne **125**. A secondary alcohol was a competent nucleophile for alkyne **124**, but the yield was lower (48% yield) and the rate was diminished. Following these proof-of-principle reactions, the cyclization of a series of alkynyl alcohols (**127**) was investigated. Interestingly, although PPh_3 was a competent catalyst for the addition of benzyl alcohol, significant amounts of the conjugated diene were observed when this catalyst was used for the cyclization. The full alkyne isomerization process could be suppressed almost entirely by changing it to a bidentate phosphine catalyst [1,3-bis(diphenylphosphino)propane (dppp)]. In order to explain the superiority of the bidentate phosphine, it was proposed that the cyclization event is facilitated by the second phosphine that serves as a general base as shown by **129**. The reaction was successful in forming both five- and six-membered rings. The reaction could also be extended to the construction of spirocyclic and fused bicyclic products (not shown).

Nonbasic nitrogen nucleophiles can also be used in these reactions. As seen earlier, when methyl 2-butynoate, or another similar electrophile, is used, the reaction occurs exclusively at the γ-position of the starting alkyne (Figure 8.68).[144] In all these cases, it is thought that the nucleophile prefers to react with an intermediate phosphonium salt similar to **109**, rather than **108**. Presumably, this is due to the rapid isomerization of these intermediates by proton transfer processes, but steric factors cannot be ruled out. However, if the isomerization of **108** is prevented, then the nucleophile will add to the α-position of the starting alkyne. This is seen during the addition of nitrogen nucleophiles to alkyne **134** (Figure 8.69).[121]

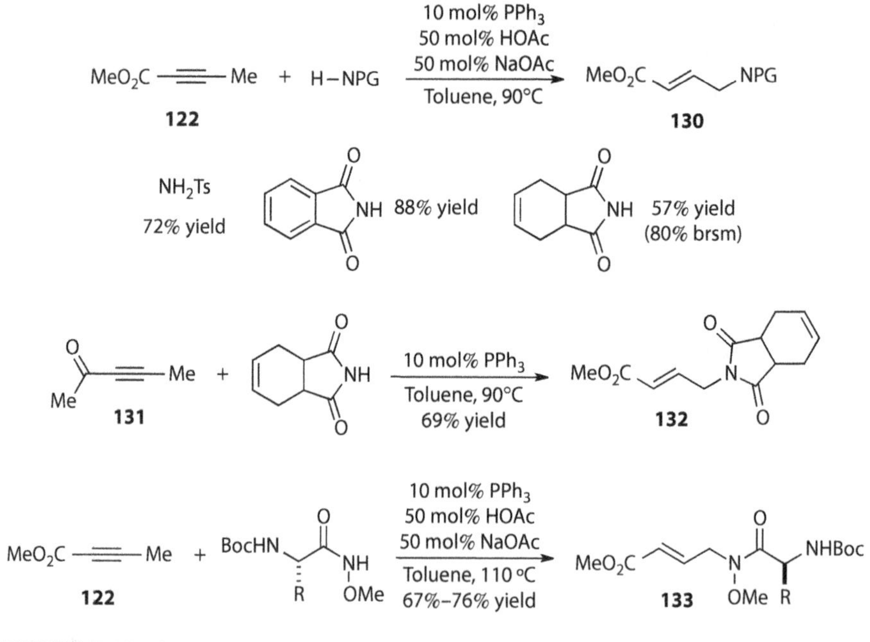

FIGURE 8.68 Phosphine-catalyzed addition of nitrogen nucleophiles to the γ-position of alkynes.

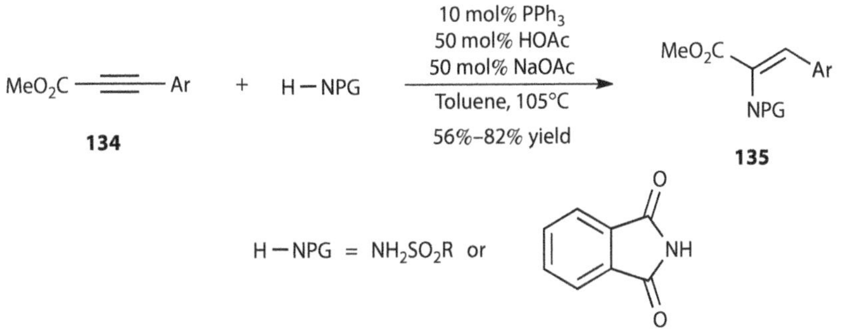

FIGURE 8.69 Phosphine-catalyzed addition of nitrogen nucleophiles to the α-position of alkynes.

In addition to the work described earlier, Alvarez-Ibarra has shown that carboxylates can function as the nucleophilic component in this phosphine-catalyzed coupling reaction.[145] Lu has also shown that various heterocycles can be formed by sequential addition at the γ- and β-positions or at the α- and β-positions when bifunctional nucleophiles are employed (Figure 8.70).[146] Related reactions that form heterocycles have also been reported by Kwon.[147]

New stereocenters are formed in many of the addition reactions that are described earlier, albeit in racemic fashion. Moreover, the proposed mechanism involves the temporary formation of a covalent C–P bond. This suggests that asymmetric induction may be observed if a chiral phosphine is used as the catalyst. Zhang and coworkers reported the first attempt at enantioselective catalyst in this manner.[148]

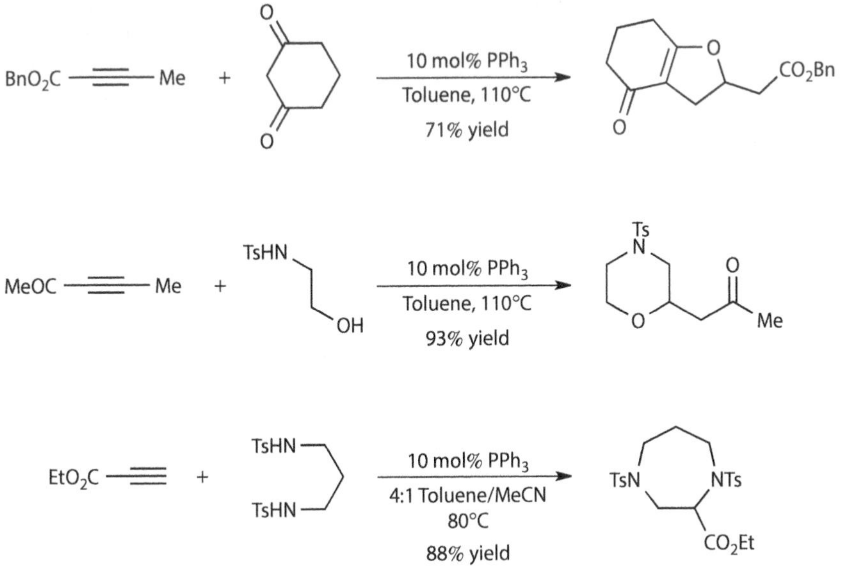

FIGURE 8.70 Examples of heterocycles formed by phosphine-catalyzed addition of nucleophiles.

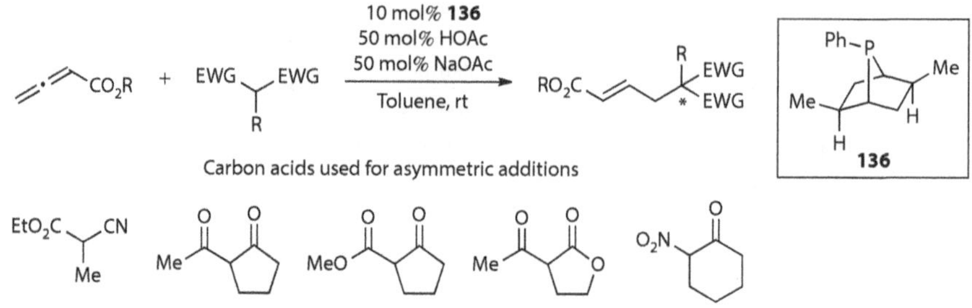

FIGURE 8.71 First attempts at enantioselective additions of nucleophiles to alkynes.

After some experimentation, it was found that bicyclic phosphine **136** afforded the highest selectivity for the addition of carbon acid nucleophiles to allenoate **137** (Figure 8.71). Although the enantioselectivity was only moderate, this was an important first step and demonstrated that asymmetric induction could be achieved for these reactions. It should also be noted that the stereogenic carbon atom formed in **138** originates on the nucleophile and not on the allene. The use of an allenoate electrophile was also quite important. Lower asymmetric induction is observed when the allene is substituted for ethyl 2-butynoate. This was attributed to the higher temperatures (80°C–110°C) that were required with the alkyne electrophile. In contrast, the higher electrophilicity of the allenoate allows the reaction to be carried out at ambient temperatures. This would be a key piece of information for the continued development of asymmetric reactions that involve phosphine catalysts.

In 2014, Lu[149] and Zhao[150] independently reported similar solutions to the problem of forming δ-stereocenters. They found that high enantioselectivity could be achieved during the alkylation of substituted oxindoles (**139**) with allenoates (Figure 8.72). Both groups reported that amino acid-derived phosphines were optimal for this transformation. More specifically, Lu and coworkers found that alanine-derived catalyst **141** was optimal for oxindoles with alkyl substituents at the 3-position, but

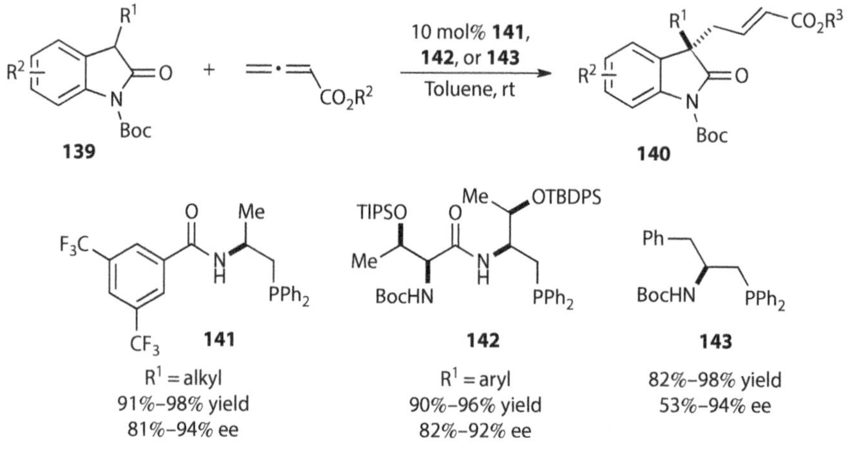

FIGURE 8.72 Phosphine-catalyzed alkylation of oxindoles.

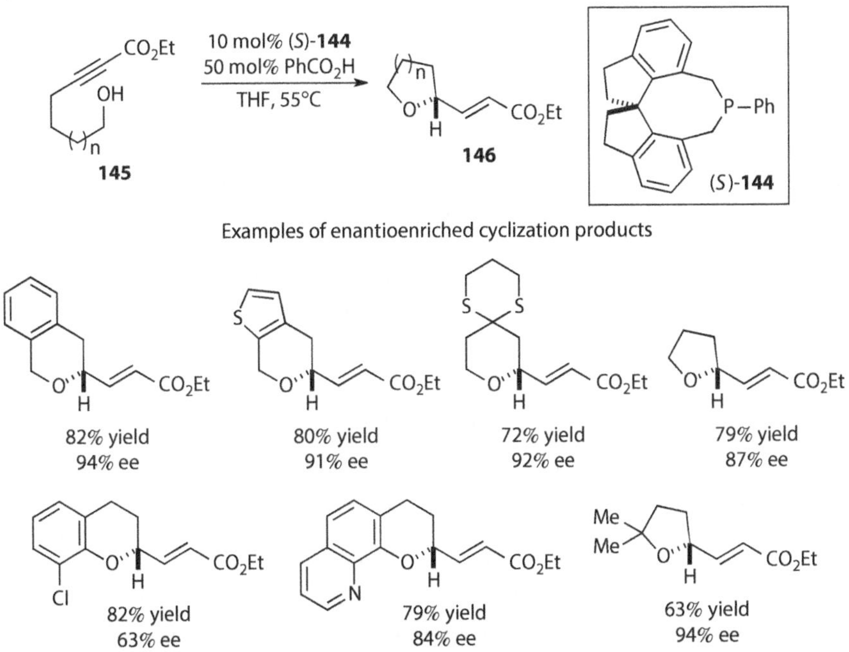

FIGURE 8.73 Fu's asymmetric phosphine-catalyzed cyclization.

threonine-derived catalyst **142** was superior when aromatic substituents were present.[149] Phenylalanine-derived catalyst **143** was the best catalyst to emerge from the screening efforts that were made by Zhao and coworkers,[150] but the enantioselectivities tended to be somewhat lower compared to those reported by Lu.

In 2009, Fu and Chung reported that Trost's cyclization of alkynyl alcohols[120] could be rendered asymmetric with the use of catalyst (*S*)-**144** (Figure 8.73).[151] The observed asymmetric induction was high in most cases. The reaction could be applied to the cyclization of primary alcohols, tertiary alcohols, and phenols, though the enantioselectivity with phenolic substrates was somewhat lower. Both five- and six-membered rings could be formed. Several years later, Fu and coworkers would be able to extend this methodology to include the formation of nitrogen heterocycles.[152]

Later in 2009, Fu and Smith reported that a modified catalyst, (*S*)-**147**, was capable of promoting the highly enantioselective addition of nitromethane to the γ-position of racemic allenamides **148** (Figure 8.74).[153] Both yield and enantioselectivity were

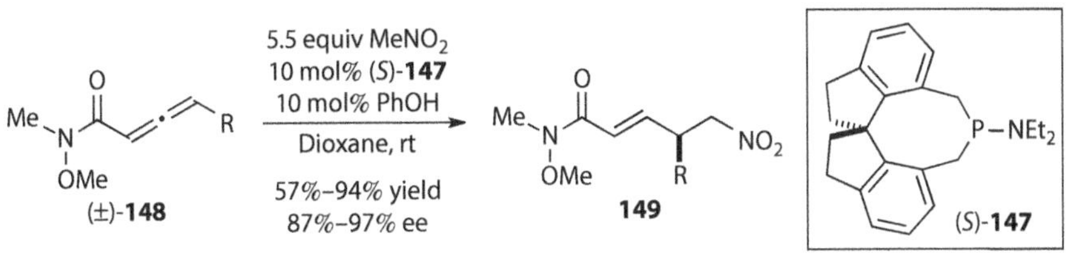

FIGURE 8.74 Fu's asymmetric addition of nitromethane to racemic allenes.

high in all the cases. In addition to the allenamides, the reaction could be extended to allenoates with little detriment to the yield or selectivity. The use of allenyl phosphonate resulted in somewhat lower enantioselectivity, but the reaction remained high yielding. It is interesting to note that there was no apparent kinetic resolution of the starting allene, suggesting that the chiral catalyst reacts with both enantiomers of the starting material with equal affinity. In addition, the authors reported little-to-no formation of the conjugated diene, suggesting that catalyst (*S*)-**147** is able to make this alternative pathway noncompetitive. Since these reports, the scope of this methodology has been extended to include other external nucleophiles such as malonic esters,[154] thiols,[155] trifluoroacetamide,[152] and suitably acidic heterocycles.[156,157]

8.3.4 Phosphine-Catalyzed Ring-Forming Reactions

One area of research that has seen extensive exploration over the last 20 years has been the use of phosphine catalysts for the formation of heterocycles and carbocycles. Interest in these reactions was kindled by a 1995 report by Lu and Zhang, which reported that PPh_3 catalyzed the [3 + 2] cycloaddition between allenoate **150** and electron-deficient alkenes **151** (Figure 8.75).[122] There was some preference for the formation of cycloadduct **152a** over **152b**, but the selectivity was generally low. Spirocycles can also be formed when *exo*-methyleneketones and amides are used as the dipolarophile (Figure 8.76).[158] The reaction could also be extended to include alkynoates (not shown), but those reactions required the use of PBu_3 as catalyst.

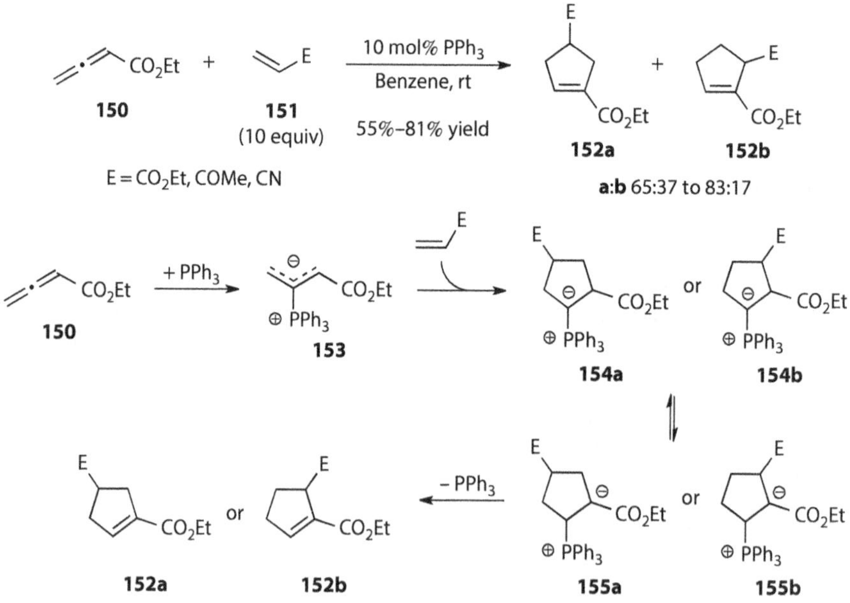

FIGURE 8.75 Phosphine-catalyzed annulation of allenoates.

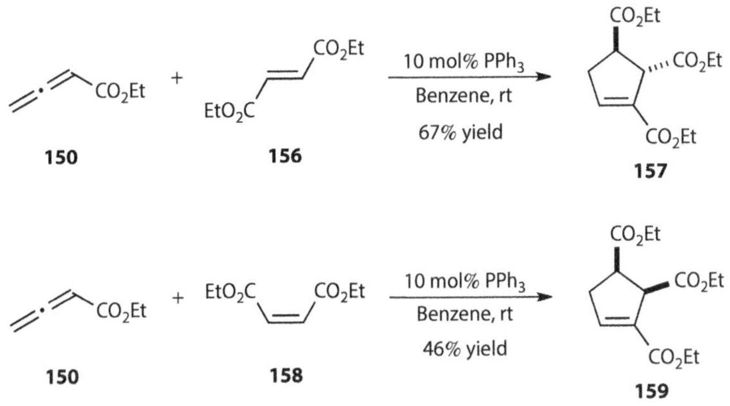

FIGURE 8.76 Phosphine-catalyzed spirocycle formation.

The authors proposed that cycloaddition between the alkene component and zwitterion **153** would form ylide **154a** and **154b** (Figure 8.75). Proton transfer would then give **155a** and **155b**, which give rise to the products by elimination of phosphine. The nature of the cycloaddition step was probed by using isomeric alkenes **156** and **158** (Figure 8.77). Interestingly, these reactions proceeded in a stereospecific manner to give **157** and **159**, respectively. This suggests that the cycloaddition step proceeds by a concerted mechanism.

After their initial report,[122] Lu and coworkers began investigating the use of other dipolarophiles in phosphine-catalyzed [3 + 2] cycloadditions. In 1997, they reported that *N*-sulfonylimines **161** could be used to form dihydropyrroles **162** in high yield (Figure 8.78).[159] As earlier, alkynoates could be used in place of the allenoate, provided PBu_3 was used as the catalyst. The synthetic utility of both the allene–alkene

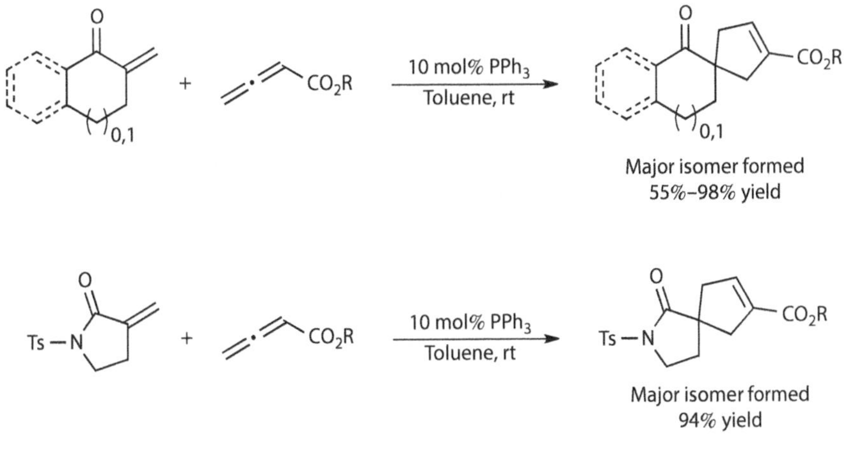

FIGURE 8.77 Stereospecificity of Lu's phosphine-catalyzed cycloaddition.

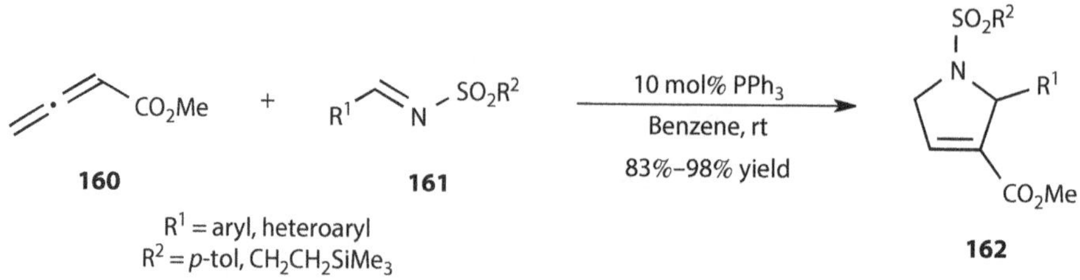

FIGURE 8.78 Phosphine-catalyzed [3 + 2] cycloaddition with *N*-sulfonylimines.

and allene–imine [3 + 2] cycloaddition reactions has been demonstrated through their use as the key step in several total syntheses.[160]

Intramolecular phosphine-catalyzed [3 + 2] cycloadditions have also been reported. The first examples were reported by the Krische group in 2003.[161] They found that 1,7-enynes **163** could be converted into diquinanes **164** (Figure 8.79). The products were all obtained as essentially a single diastereomer (>95:5 dr). Unfortunately, attempts to cyclize 1,8-enynes gave poor results. Shortly thereafter, the Krische group would use this cycloaddition as the key step in an efficient synthesis of (±)-hirsutene (Figure 8.80).[162] Later, Kwon and Henry would show that substituted dihydrocoumarins (**166**) could be constructed using an intramolecular phosphine-catalyzed [3 + 2] cycloaddition of aryl allenoates **165** (Figure 8.81).[163] Interestingly, when the cyclization was attempted with nitroalkene **167**, nitronate **168** was obtained as the major product rather than the expected cyclopentene (e.g., **166**). Recently, Fu and coworkers reported that the cyclizations shown in

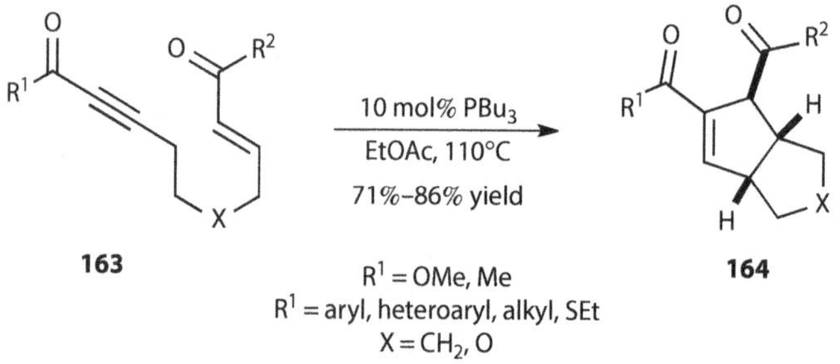

FIGURE 8.79 Krische's intramolecular [3 + 2] cycloaddition.

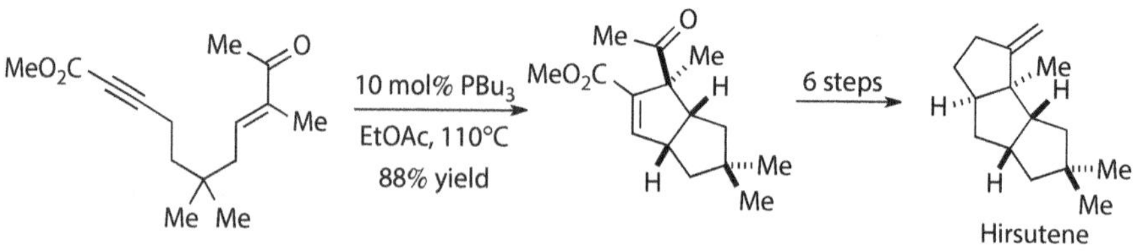

FIGURE 8.80 Krische's synthesis of (±)-hirsutene.

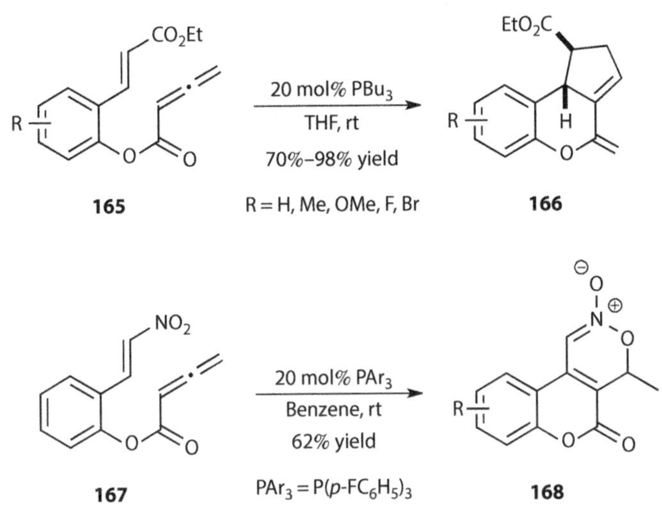

20 mol% PBu_3
THF, rt
70%–98% yield
R = H, Me, OMe, F, Br

165 166

20 mol% PAr_3
Benzene, rt
62% yield
$PAr_3 = P(p\text{-}FC_6H_5)_3$

167 168

FIGURE 8.81 Kwon's synthesis of substituted dihydrocoumarins.

Figures 8.79 and 8.81 both could be carried out in an asymmetric manner with high selectivity (up to 98% ee).[164]

In 2003, Kwon and coworkers reported that phosphine-catalyzed cycloadditions could be extended to [4 + 2] reactions. Their first efforts in this area focused on the reaction between allenoates **169** and *N*-tosylimines **170** (Figure 8.82).[165] In contrast

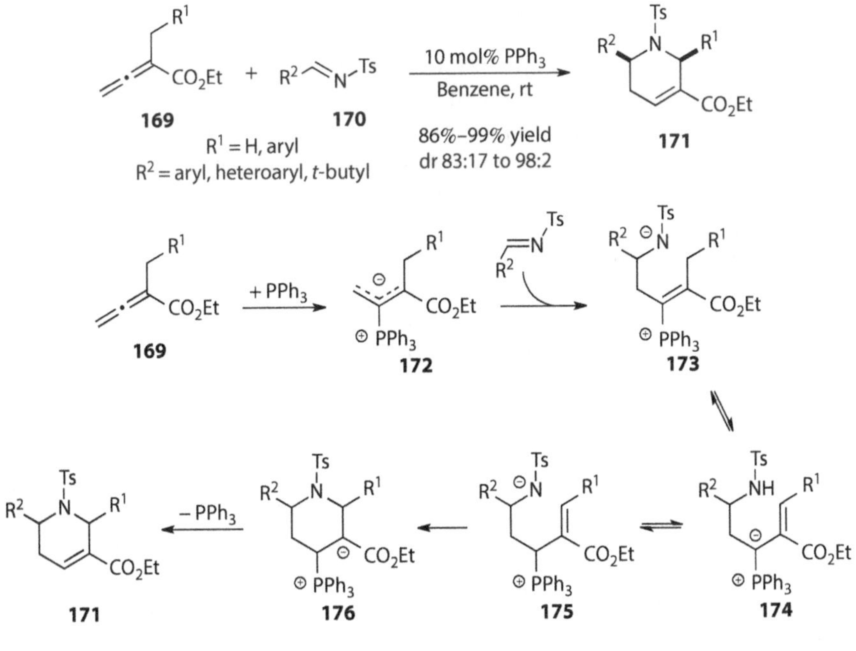

10 mol% PPh_3
Benzene, rt
86%–99% yield
dr 83:17 to 98:2

169 170 171

R^1 = H, aryl
R^2 = aryl, heteroaryl, *t*-butyl

+ PPh_3 169 172 173 174 175 176 – PPh_3 171

FIGURE 8.82 Phosphine-catalyzed [4 + 2] cycloaddition with *N*-sulfonylimines.

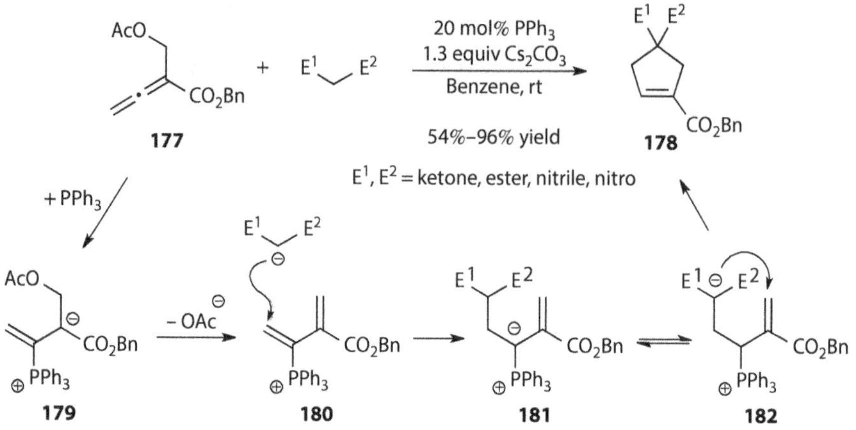

FIGURE 8.83 Synthesis of cyclopentenes by phosphine catalyzed [4 + 1] reactions.

to the [3 + 2] reactions of Lu, it was proposed that the [4 + 2] cycloadditions proceed by a stepwise mechanism. Thus, addition of **172** to the imine gives rise to **173**, which is then converted into **174** through several proton-transfer steps. Conjugate addition of the sulfonamide anion in **175** generates a cyclic intermediate (**176**) that eliminates phosphine and gives rise to tetrahydropyridine **171**. Later, this phosphine-catalyzed [4 + 2] reaction would be extended to include electron-deficient alkenes[166] and ketones.[167]

In 2010, Tong and coworkers reported a new pathway for performing [4 + *n*] reactions under phosphine catalysis. They found that cyclopentenes **178** could be formed in high yield when acetoxy allenoate **177** was reacted with various active methylene compounds in the presence of Cs_2CO_3 and a catalytic amount of PPh_3 (Figure 8.83).[168] The authors proposed that the initial reaction between the phosphine and the allenoate produced an initial intermediate (**179**), which expels acetate and forms phosphonium diene **180**. Nucleophilic addition of the carbanion into the vinyl phosphonium produces ylide **181**, which can undergo tautomerization to **182**. Subsequent cyclization and elimination of the phosphine would then afford the cyclopentene product. In addition to carbon acids, Tong and coworkers demonstrated that sulfonyl hydrazines and sulfonamides were also competent nucleophiles for this reaction (Figure 8.84). In 2014, Lu and coworkers reported that threonine-derived catalyst **183** could promote asymmetric [4 + 1] reactions between allenoate **177** and pyrazolones **184** (Figure 8.85).[169] The resulting spiropyrazolones (**185**) were formed with high enantioselectivity.

Of course, there has been a great deal of interest in performing these phosphine-catalyzed ring-forming reactions in an asymmetric manner. The first report of this came just 2 years after Lu's initial report on the phosphine-catalyzed [3 + 2] cycloaddition.[122] Zhang and coworkers reported that this cycloaddition could be rendered asymmetric by using bicyclic phosphines **136** and **186** (Figure 8.86).[170] In 2010, Loh reported that the chiral bisphosphine (*R*,*R*)-1,2-bis[(2-methoxyphenyl)(phenylphosphino)]ethane (DIPAMP) (**190**) was a highly effective catalyst for the asymmetric cycloaddition between 3-butynoates (**191**) and α,β-unsaturated ketones

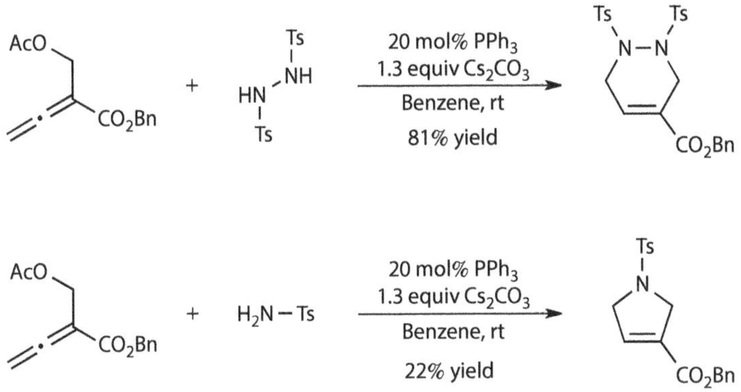
AcO
CO2Bn
+
Ts
NH
HN
Ts
20 mol% PPh3
1.3 equiv Cs2CO3
Benzene, rt
81% yield
Ts
Ts
N–N
CO2Bn
AcO
CO2Bn
+
H2N–Ts
20 mol% PPh3
1.3 equiv Cs2CO3
Benzene, rt
22% yield
Ts
N
CO2Bn

FIGURE 8.84 Other [4 + *n*] reactions reported by Tong.

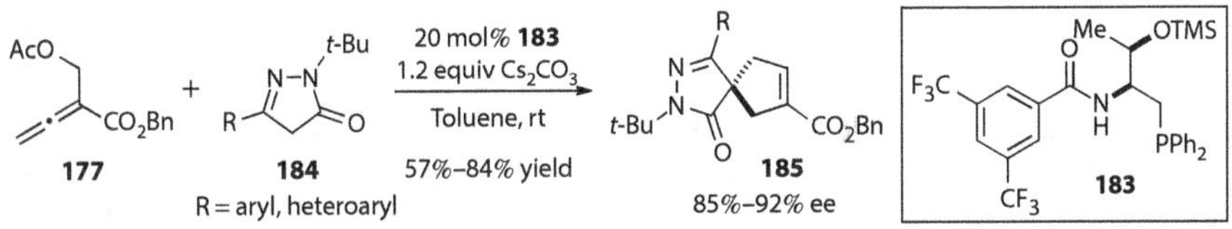
AcO
CO2Bn
177
+
t-Bu
N–N
R
O
184
R = aryl, heteroaryl
20 mol% 183
1.2 equiv Cs2CO3
Toluene, rt
57%–84% yield
R
N
N
t-Bu
O
CO2Bn
185
85%–92% ee
Me
OTMS
F3C
O
N
H
PPh2
CF3
183

FIGURE 8.85 Asymmetric [4 + 1] reactions reported by Lu.

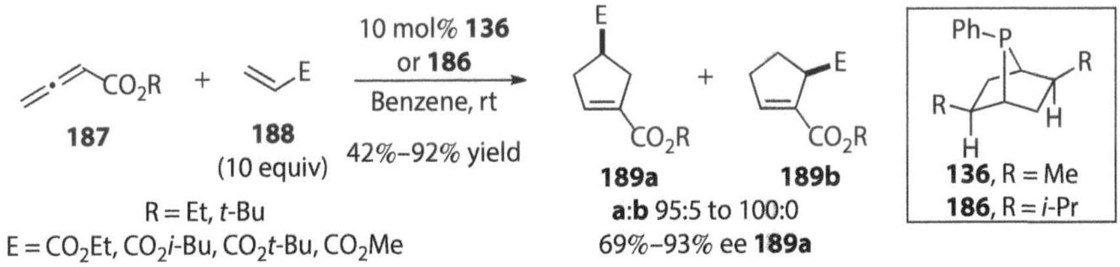
CO2R
187
+
E
188
(10 equiv)
10 mol% 136
or 186
Benzene, rt
42%–92% yield
R = Et, t-Bu
E = CO2Et, CO2i-Bu, CO2t-Bu, CO2Me
E
CO2R
189a
+
E
CO2R
189b
a:b 95:5 to 100:0
69%–93% ee 189a
Ph–P
R
R
H
H
136, R = Me
186, R = i-Pr

FIGURE 8.86 Zhang's asymmetric [3 + 2] cycloaddition reaction.

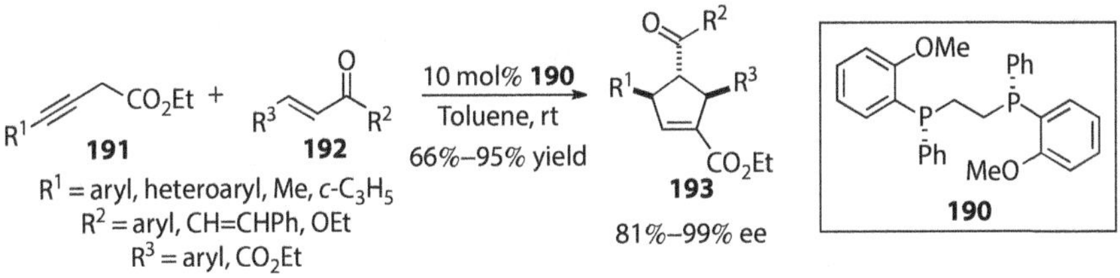
R1
CO2Et +
191
O
R3
R2
192
R1 = aryl, heteroaryl, Me, c-C3H5
R2 = aryl, CH=CHPh, OEt
R3 = aryl, CO2Et
10 mol% 190
Toluene, rt
66%–95% yield
O
R2
R1
R3
CO2Et
193
81%–99% ee
OMe
Ph
P
P
Ph
MeO
190

FIGURE 8.87 Loh's asymmetric [3 + 2] cycloaddition reaction.

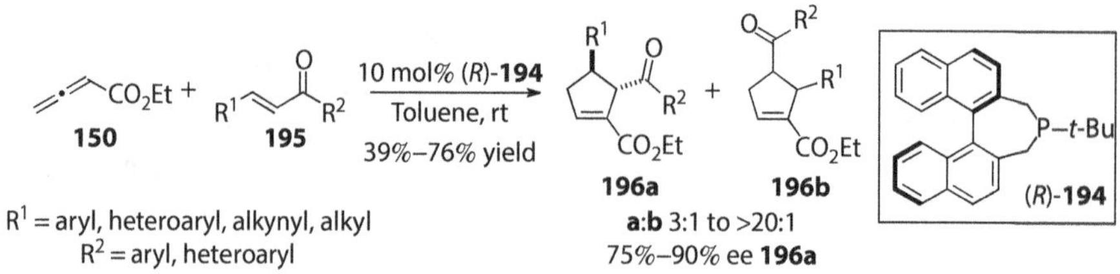

FIGURE 8.88 Fu's asymmetric [3 + 2] cycloaddition reaction.

and esters (**192**, Figure 8.87).[171] Interestingly, Fu reported that when phosphine (*R*)-**194** was used to catalyze the reaction between allenoate **150** and α,β-unsaturated ketones **195** (Figure 8.88), the obtained product (**196a**) was of the opposite regioselectivity from that observed by Zhang and Loh (**193**).[172] It is unclear why this is the case.

Since Zhang's report on asymmetric [3 + 2] cycloadditions, just about all the ring-forming reactions described earlier have been rendered asymmetric with various degrees of success. Much of this has been reviewed.[125] There are, however, a few interesting results that have recently emerged. Some of these have been discussed earlier. In 2014, Shi reported that catalyst **197** promoted asymmetric [3 + 2] cycloadditions between azomethine imines **199** and substituted allenoates **198** (Figure 8.89). Interestingly, the product (**200**) was formed by a cycloaddition that involve the γ- and δ-positions of the allenoate. To account for this, the authors proposed a mechanism that involve a proton transfer (or H-shift) and the initially formed vinyl phosphonium intermediate **201**. Addition of **202** to the azomethine imine gives rise to **203**, which can cyclize to form ylide **204**. Another proton transfer (or H-shift) generates **205**, which expels the catalyst and forms the product.

Finally, Fu and Kramer have reported that phosphine (*R*)-**206** catalyzed the enantioselective synthesis of dihydropyrroles (**208**) from a suitably substituted allene (**207**) and an aromatic sulfonamide (Figure 8.90).[173] This transformation is quite similar to the [4 + 1] reaction reported by Tong,[168] described earlier, and the PPh_3-catalyzed reaction reported by Kwon.[174] Again the products were all formed with uniformly high enantioselectivity and generally high yields.[173] Fu proposes that the key intermediate for this reaction is dienyl phosphonium salt **210**. This could be formed by initial addition of the phosphine to the allene, and subsequent elimination of acetate from **209**. Once formed, phosphonium **210** can react with the sulfonamide to give either (or both) **211** or **212**. Both of these intermediates would be poised for cyclization to the observed product. Interestingly, a crossover experiment suggests that the addition of the sulfonamide to intermediate **210** may be reversible.

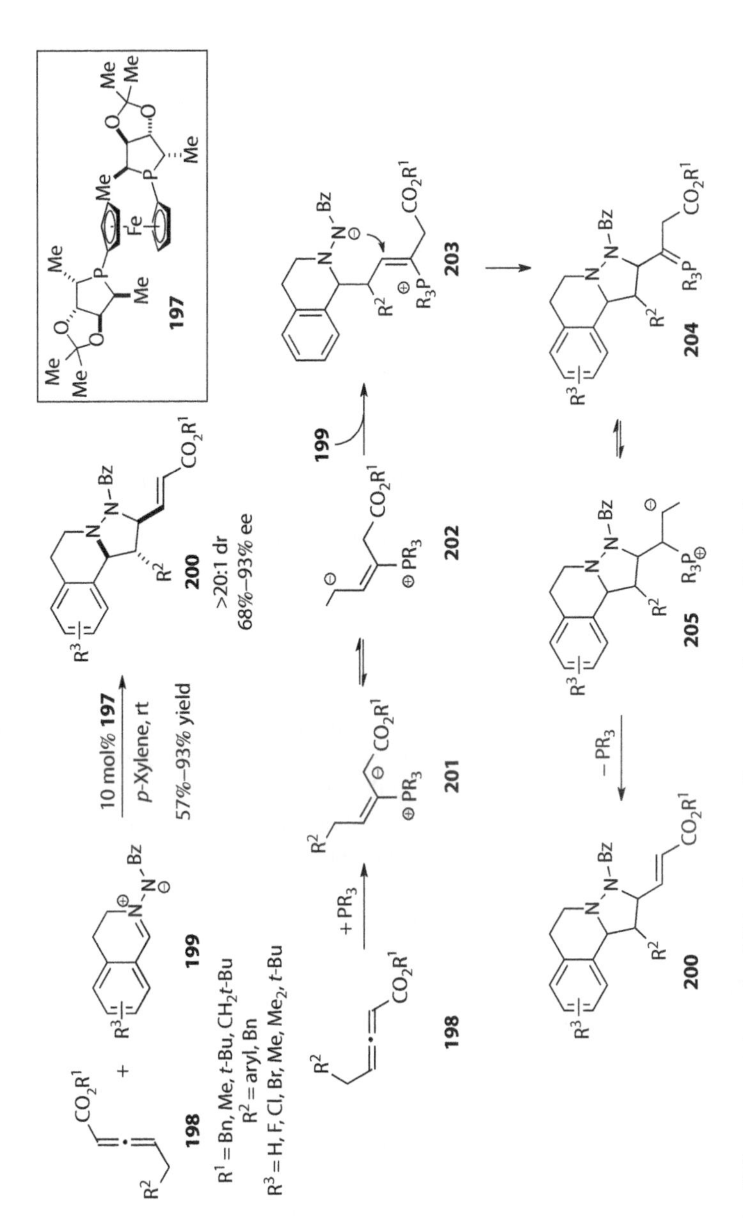

FIGURE 8.89 Shi's asymmetric cycloaddition between allenoates and azomethine imines.

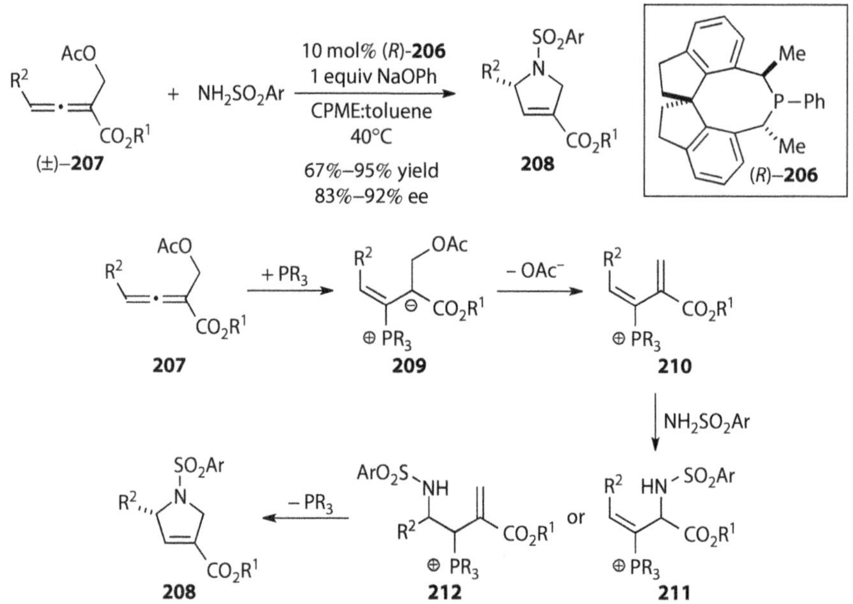

FIGURE 8.90 Asymmetric phosphine-catalyzed formation of dihydropyrroles.

8.4 CONCLUSION

This chapter has discussed three catalyst classes with very different reactivity profiles. All three have a rich history that can be traced back several decades or longer. Undoubtedly, the coming years will see even more exciting transformations that involve these catalysts. It is hoped that the present coverage of the known reactivity of these catalysts will inspire the development of new modes of reactivity. Also important will be the continued application of these catalysts, both in existing reactions and new reactions, in the synthesis of complex molecules. Only in these scenarios will the true utility and power of organocatalysis be realized.

REFERENCES

1. Burns, N. Z., Baran, P. S., Hoffmann, R. W. *Angew. Chem. Int. Ed.* **2009**, *48*, 2854–2867.
2. Denmark, S. E., Wu, Z. *Synlett* **1999**, 847–859.
3. (a) Yang, D. *Acc. Chem. Res.* **2004**, *37*, 497–505. (b) Wong, M. K., Yip, Y. C., Yang, D. *Top. Organomet. Chem.* **2011**, *36*, 123–152.
4. (a) Shi, Y. *Acc. Chem. Res.* **2004**, *37*, 488–496. (b) Wong, O. A., Shi, Y. *Top. Curr. Chem.* **2010**, *291*, 201–232.
5. (a) Murray, R. W. *Chem. Rev.* **1989**, *89*, 1187–1201. (b) Adam, W., Curci, R., Edwards, J. O. *Acc. Chem. Res.* **1989**, *22*, 205–211. (c) Curci, R., D'Accolti, L., Fusco, C. *Acc. Chem. Res.* **2006**, *39*, 1–9.
6. Baeyer, A. V., Villiger, V. *Chem. Ber.* **1899**, *32*, 3625–3633.
7. Talbot, R. I., Thompson, P. G. Fluorinated organic cyclic peroxides U.S. Patent 3,632,606, January 4, 1972.

8. Montgomery, R. E. *J. Am. Chem. Soc.* **1974**, *96*, 7820.
9. (a) Edwards, J. O., Pater, B. H., Curci, R., DiFuria, F. *Photochem. Photobiol.* **1979**, *30*, 63–70. (b) Gallopo, A. R., Edwards, J. O. *J. Org. Chem.* **1981**, *46*, 1684–1688.
10. (a) Curci, R., Fiorentino, M., Troisi, L., Edwards, J. O., Pater, R. H. *J. Org. Chem.* **1980**, *45*, 4758–4760. (b) Cicala, G., Curci, R., Fiorentino, M., Laricchiuta, O. *J. Org. Chem.* **1982**, *47*, 2670–2673.
11. Mello, R., Fiorentino, M., Sciacovelli, O., Curci, R. *J. Org. Chem.* **1988**, *53*, 3890–3891.
12. Adam, W., Paredes, R., Smerz, A. K., Veloza, L. A. *Eur. J. Org. Chem.* **1998**, *2*, 349–354.
13. Yang, D., Wong, M.-K., Yip, Y.-C. *J. Org. Chem.* **1995**, *60*, 3887–3889.
14. (a) Denmark, S. E., Forbes, D. C., Hays, D. S., DePue, J. S., Wilde, R. G. *J. Org. Chem.* **1995**, *60*, 1391–1407. (b) Denmark, S. E., Wu, Z. *J. Org. Chem.* **1998**, *63*, 2810–2811.
15. Denmark, S. E., Wu, Z., Crudden, C. M., Matsuhashi, H. *J. Org. Chem.* **1997**, *62*, 8288–8289.
16. Houk, K. N., Liu, J., DeMello, N. C., Condroski, K. R. *J. Am. Chem. Soc.* **1997**, *119*, 10147–10152.
17. Singleton, D. A., Wnag, Z. *J. Am. Chem. Soc.* **2005**, *127*, 6679–6685.
18. (a) Baumstark, A. L., McCloskey, C. J. *Tetrahedron Lett.* **1987**, *28*, 3311–3314. (b) Baumstark, A. L., Vasquez, P. C. *J. Org. Chem.* **1988**, *53*, 3437–3439.
19. (a) Bach, R. D., Andrés, J. L., Owensby, A. L., Schlegel, H. B., McDouall, J. J. W. *J. Am. Chem. Soc.* **1992**, *114*, 7207–7217. (b) Jensen, C., Liu, J., Houk, K. N., Jorgensen, W. L. *J. Am. Chem. Soc.* **1997**, *119*, 12982–12983.
20. Armstrong, A., Washington, I., Houk, K. N. *J. Am. Chem. Soc.* **2000**, *122*, 6297–6298.
21. Deubel, D. V. *J. Org. Chem.* **2001**, *66*, 3790–3796.
22. Curci, R., Fiorentino, M., Serio, M. R. *J. Chem. Soc. Chem. Commun.* **1984**, 155–156.
23. Curci, R., D'Accolti, L., Fiorentino, M., Rosa, A. *Tetrahedron Lett.* **1995**, *36*, 5831–5834.
24. Song, C. E., Kim, Y. H., Lee, K. C., Lee, S.-G., Jin, B. W. *Tetrahedron: Asymmetry* **1997**, *8*, 2921–2926.
25. Kim, Y. H., Lee, K. C., Chi, D. Y., Lee, S.-G., Song, C. E. *Bull. Korean Chem. Soc.* **1999**, *20*, 831–834.
26. Adam, W., Zhao, C.-G. *Tetrahedron: Asymmetry* **1997**, *8*, 3995–3998.
27. (a) Yang, D., Yip, Y.-C., Tang, M.-W., Wong, M.-K., Zheng, J.-H., Cheung, K.-K. *J. Am. Chem. Soc.* **1996**, *118*, 491–492. (b) Yang, D., Wang, X.-C., Wong, M.-K., Yip, Y.-C., Tang, M.-W. *J. Am. Chem. Soc.* **1996**, *118*, 11311–11312.
28. Yang, D., Jiao, G.-S., Yip, Y.-C., Lai, T.-H., Wong, M.-K. *J. Org. Chem.* **2001**, *66*, 4619–4624.
29. Stearman, C. J., Behar, V. *Tetrahedron Lett.* **2002**, *43*, 1943–1946.
30. Denmark, S. E., Matsuhashi, H. *J. Org. Chem.* **2002**, *67*, 3479–3486.
31. Zhu, Y., Wang, Q., Cornwall, R. G., Shi, Y. *Chem. Rev.* **2014**, *114*, 8199–8256.
32. Tu, Y., Frohn, M., Wang, Z.-X., Shi, Y. *Org. Synth.* **2003**, *80*, 1–8.
33. Tu, Y., Wang, Z.-X., Shi, Y. *J. Am. Chem. Soc.* **1996**, *118*, 9806–9807.
34. Wang, Z.-X., Tu, Y., Frohn, M., Zhang, J.-R., Shi, Y. *J. Am. Chem. Soc.* **1997**, *119*, 11224–11235.
35. (a) Frohn, M., Dalikiewicz, M. Tu, Y., Wang, Z.-X., Shi, Y. *J. Org. Chem.* **1998**, *63*, 2948–2953. (b) Wang, Z.-X., Shi, Y. *J. Org. Chem.* **1998**, *63*, 3099–3104. (c) Wang, Z.-X., Cao, G.-A., Shi, Y. *J. Org. Chem.* **1999**, *64*, 7646–7650. (d) Zhu, Y., Manske, K. J., Shi, Y. *J. Am. Chem. Soc.* **1999**, *121*, 4080–4081. (e) Zhu, Y., Shu, L., Tu, Y., Shi, Y. *J. Org. Chem.* **2001**, *66*, 1818–1826. (f) Warren, J. D., Shi, Y. *J. Org. Chem.* **1999**, *64*, 7675–7677. (g) Wong, O. A., Shi, Y. *J. Org. Chem.* **2009**, *74*, 8377–8380.

36. (a) Wu, X.-Y., She, X., Shi, Y. *J. Am. Chem. Soc.* **2002**, *124*, 8792–8793. (b) Wang, B., Wu, X.-Y., Wong, O. A., Nettles, B., Zhao, M.-X., Chen, D., Shi, Y. *J. Org. Chem.* **2009**, *74*, 3986–3989.
37. (a) Tian, H., She, X., Yu, H., Shu, L., Shi, Y. *J. Org. Chem.* **2002**, *67*, 2435–2446. (b) Wong, O. A., Shi, Y. *J. Org. Chem.* **2006**, *71*, 3973–3976. (c) Burke, C. P., Shi, Y. *Angew. Chem. Int. Ed.* **2006**, *45*, 4475–4478. (d) Burke, C. P., Shi, Y. *J. Org. Chem.* **2007**, *72*, 4093–4097. (e) Burke, C. P., Shi, Y. *Org. Lett.* **2009**, *11*, 5150–5153.
38. Wang, B., Wong, O. A., Zhao, M.-X., Shi, Y. *J. Org. Chem.* **2008**, *73*, 9539–9543.
39. Zhao, M.-X., Shi, Y. *J. Org. Chem.* **2006**, *71*, 5377–5379.
40. Schneebeli, S. T., Hall, M. L., Breslow, R., Friesner, R. *J. Am. Chem. Soc.* **2009**, *131*, 3965–3973.
41. Wang, Z.-X., Tu, Y., Frohn, M., Shi, Y. *J. Org. Chem.* **1997**, *62*, 2328–2329.
42. Tian, H., She, X., Shi, Y. *Org. Lett.* **2001**, *3*, 715–718.
43. Shu, L., Shi, Y. *Tetrahedron* **2001**, *57*, 5213–5218.
44. (a) Smith, A. B., III, Walsh, S. P., Frohn, M., Duffey, M. O. *Org. Lett.* **2005**, *7*, 139–142. (b) Bian, J., Van Wingerden, M., Ready, J. M. *J. Am. Chem. Soc.* **2006**, *128*, 7428–7429. (c) Adams, C. M., Ghosh, I., Kishi, Y. *Org. Lett.* **2004**, *6*, 4723–4726. (d) Morimoto, Y., Okita, T., Takaishi, M., Tanaka, T. *Angew. Chem. Int. Ed.* **2007**, *46*, 1132–1135. (e) Yang, Y.-R., Lai, Z.-W., Shen, L., Huang, J.-Z., Wu, X.-D., Yin, J.-L., Wei, K. *Org. Lett.* **2010**, *12*, 3430–3433. (f) Huo, X., Pan, X., Huang, G., She, X. *Synlett* **2011**, 1149–1150. (g) Tanuwidjaja, J., Ng, S.-S., Jamison, T. F. *J. Am. Chem. Soc.* **2009**, *131*, 12084–12085.
45. Selected reviews: (a) Bugaut, X., Glorius, F. *Chem. Soc. Rev.* **2012**, *41*, 3511–3522. (b) Cohen, D. T., Scheidt, K. A. *Chem. Sci.* **2012**, *3*, 53–57. (c) Biju, A. T., Kuhl, N., Glorius, F. *Acc. Chem. Res.* **2011**, *44*, 1182–1195. (d) Philips, E. M, Chan, A. Scheidt, K. A. *Aldrichimica Acta* **2009**, *42*, 55–66. (e) Johnson, J. S. *Angew. Chem. Int. Ed.* **2004**, *43*, 1326–1328.
46. Flanigan, D. M., Romanov-Michailidis, F., White, N. A., Rovis, T. *Chem. Rev.* **2015**, *115*, 9307–9387.
47. Mahatthananchai, J., Bode, J. W. *Chem. Sci.* **2012**, *3*, 192–197.
48. Hopkinson, M. N., Richter, C., Schedler, M., Glorius, F. *Nature* **2014**, *510*, 485–496.
49. Maji, B., Breugst, M., Mayr, H. *Angew. Chem. Int. Ed.* **2011**, *50*, 6915–6919.
50. Breslow, R. *J. Am. Chem. Soc.* **1958**, *80*, 3719–3726.
51. Berkessel, A., Elfert, S., Yatham, V. R., Neudörfl, J.-M., Schlörer, N. E., Teles, J. H. *Angew. Chem. Int. Ed.* **2012**, *51*, 12370–12374.
52. Ukai, T., Tanaka, R., Dokawa, T. *J. Pharm. Soc. Jpn.* (*Yakugaku Zasshi*) **1943**, *63*, 296–300, *Chem. Abstr.* **1951**, *45*, 5148.
53. Tagaki, W., Hara, H. *J. Chem. Soc. Chem. Commun.* **1973**, 891.
54. Stetter, H., Rämsch, R. Y., Kuhlmann, H. *Synthesis* **1976**, 733–735.
55. Stetter, H., Kuhlmann, H. *Angew. Chem. Int. Ed. Engl.* **1974**, *13*, 539.
56. Stetter, H., Kuhlmann, H. *Org. React.* **1976**, *40*, 407–496.
57. Enders, D., Niemeier, O., Hanseler, A. *Chem. Rev.* **2007**, *107*, 5606–5655.
58. Breslow, R. *J. Am. Chem. Soc.* **1958**, *80*, 3719–3726.
59. Sheehan, J. C., Hunneman, D. H. *J. Am. Chem. Soc.* **1966**, *88*, 3666–3667.
60. Sheehan, J. C., Hara, T. *J. Org. Chem.* **1974**, *39*, 1196–1199.
61. Dvorak, C. A., Rawal, V. H. *Tetrahedron Lett.* **1998**, *39*, 2925–2928.
62. (a) Tagaki, W., Tamura, Y., Yano, Y. *Bull. Chem. Soc. Jpn.* **1980**, *53*, 478–480. (b) Marti, J., Castellas, J., López-Calahorra, F. *Tetrahedron Lett.* **1993**, *34*, 521–524.
63. (a) Teles, J. H., Melder, J.-P., Ebel, K., Schneider, R., Gehrer, E., Harder, W., Brode, S., Enders, D., Breuer, K., Raabe, G. *Helv. Chim. Acta* **1996**, *79*, 61–83. (b) Dietrich, E., Lubell, W. D. *J. Org. Chem.* **2003**, *68*, 6988–6996.
64. Enders, D., Breuer, K., Teles, J. H. *Helv. Chim. Acta* **1996**, *79*, 1217–1221.

65. Enders, D., Kallfass, U. *Angew. Chem. Int. Ed.* **2002**, *41*, 1743–1745.
66. (a) Enders, D., Han, J., Henseler, A. *Chem. Commun.* **2008**, 3989–3991. (b) Enders, D., Han, J. *Synthesis* **2008**, 3864–3868.
67. Liu, Q., Perreault, S., Rovis, T. *J. Am. Chem. Soc.* **2008**, *130*, 14066–14067.
68. (a) Liu, Q., Rovis, T. *Org. Lett.* **2009**, *11*, 2856–2859. (b) DiRocco, D. A., Oberg, K. M., Dalton, D. M., Rovis, T. *J. Am. Chem. Soc.* **2009**, *131*, 10872–10874. (c) DiRocco, D. A., Rovis, T. *J. Am. Chem. Soc.* **2011**, *133*, 10402–10405. (d) Moore, J. L., Silvestri, A. P., Read de Alaniz, J., DiRocco, D. A., Rovis, T. *Org. Lett.* **2011**, *13*, 1742–1745. (e) DiRocco, D. A., Noey, E. L., Houk, K. N., Rovis, T. *Angew. Chem. Int. Ed.* **2012**, *51*, 2391–2394.
69. Kim, S. M., Jin, M. Y., Cui, Y., Kim, Y. S., Zhang, L., Song, C. E., Ryu, D. H., Yang, J. W. *Org. Biomol. Chem.* **2011**, *9*, 2069–2071.
70. Sánchez-Larios, E., Thai, K., Bilodeau, F., Gravel, M. *Org. Lett.* **2011**, *13*, 4942–4945.
71. Fang, X., Chen, X., Lv, H., Chi, Y. R. *Angew. Chem. Int. Ed.* **2011**, *50*, 11782–11785.
72. (a) Jousseaume, T., Wurz, N. E., Glorius, F. *Angew. Chem. Int. Ed.* **2011**, *50*, 1410–1414. (b) Wurz, N. E., Daniliuc, C. G., Glorius, F. *Chem.—Eur. J.* **2012**, *18*, 16297–16301.
73. Ciganek, E. *Synthesis* **1995**, 1311–1314.
74. Trost, B. M., Shuey, C. D., DiNinno, F., Jr., Mcelvain, S. S. *J. Am. Chem. Soc.* **1979**, *101*, 1284–1285.
75. Enders, D., Breuer, K., Runsink, J., Teles, J. H. *Helv. Chin. Acta* **1996**, *79*, 1899–1902.
76. Kerr, M. S., Read de Alaniz, J., Rovis, T. *J. Am. Chem. Soc.* **2002**, *124*, 10298–10299.
77. (a) Kerr, M. S., Rovis, T. *J. Am. Chem. Soc.* **2004**, *126*, 8876–8877. (b) Nakamura, T., Hara, O., Tamura, T., Makino, K., Hamada, Y. *Synlett* **2005**, 155–157.
78. Read de Alaniz, J., Rovis, T. *J. Am. Chem. Soc.* **2005**, *127*, 6284–6289.
79. Liu, Q., Rovis, T. *J. Am. Chem. Soc.* **2006**, *128*, 2552–2553.
80. Heck, R., Henderson, A. P., Köhler, B., Rétey, J., Golding, B. T. *Eur. J. Org. Chem.* **2001**, 2623–2627.
81. Stetter, H., Dämbkes, G. *Synthesis* **1977**, 403–404.
82. Matsumoto, T., Ohishi, M., Inoue, S. *J. Org. Chem.* **1985**, *50*, 603–606.
83. Cookson, R., Lane, R. M. *J. Chem. Soc. Chem. Commun.* **1976**, 804–805.
84. Enders, D., Niemeier, O. *Synlett* **2004**, 2111–2114.
85. (a) Hachisu, Y., Bode, J. W., Suzuki, K. *J. Am. Chem. Soc.* **2003**, *125*, 8432–8433. (b) Hachisu, Y., Bode, J. W., Suzuki, K. *Adv. Synth. Catal.* **2004**, *346*, 1097–1100.
86. Enders, D., Niemeier, O., Balensiefer, T. *Angew. Chem. Int. Ed.* **2006**, *45*, 1463–1467.
87. Takikawa, H., Hachisu, H., Bode, J. W., Suzuki, K. *Angew. Chem. Int. Ed.* **2006**, *45*, 3492–3494.
88. Enders, D., Niemeier, O., Raabe, G. *Synlett* **2006**, 2431–2434.
89. (a) Jin, M. Y., Kim, S. M., Han, H., Ryu, D. H., Yang, J. W. *Org. Lett.* **2011**, *13*, 880–883. (b) Jin, M. Y., Kim, S. M., Mao, H., Ryu, D. H., Song, C. E., Yang, J. W. *Org. Biomol. Chem.* **2014**, *12*, 1547–1550. (c) Piel, I., Pawelczyk, M. D., Hirano, K., Fröhlich, R., Glorius, F. *Eur. J. Org. Chem.* **2011**, 5475–5484.
90. (a) O'Toole, S. E., Rose, C. A., Gundala, S., Zeitler, K., Connon, S. J. *J. Org. Chem.* **2011**, *76*, 347–357. (b) Rose, C. A., Gundala, S., Connon, S. J., Zeitler, K. *Synthesis* **2011**, 190–198.
91. Langdon, S. M., Wilde, M. M. D., Thai, K. Gravel, M. *J. Am. Chem. Soc.* **2014**, *136*, 7539–7542.
92. Langdon, S. M., Legault, C. Y., Gravel, M. *J. Org. Chem.* **2015**, *80*, 3597–3610.
93. Enders, D., Henseler, A. *Adv. Synth. Catal.* **2009**, *351*, 1749–1752.
94. Enders, D., Grossmann, A., Fronert, J., Raabe, G. *Chem. Commun.* **2010**, *46*, 6282–6284.
95. Rose, C. A., Gundala, S., Fagan, C.-L., Franz, J. F., Connon, S. J., Zeitler, K. *Chem. Sci.* **2012**, *3*, 735–740.

96. Tahi, K., Langdon, S. M., Bilodeau, F., Gravel, M. *Org. Lett.* **2013**, *15*, 2214–2217.
97. (a) Goodman, C. G., Johnson, J. S. *J. Am. Chem. Soc.* **2014**, *136*, 14698–14701. (b) Goodman, C. G., Walker, M. M., Johnson, J. S. *J. Am. Chem. Soc.* **2015**, *137*, 122–125.
98. Murry, J. A., Frantz, D. E., Soheili, A., Tillyer, R., Grabowski, E. J. J., Reider, P. J. *J. Am. Chem. Soc.* **2001**, *123*, 9696–9697.
99. Mennen, S. M., Gipson, J. D., Kim, Y. R., Miller, S. J. *J. Am. Chem. Soc.* **2005**, *127*, 1654–1655.
100. Li, G.-Q., Dai, L.-X., You, S.-L. *Chem. Commun.* **2007**, 852–854.
101. Enders, D., Henseler, A., Lowins, S. *Synthesis* **2009**, 4125–4128.
102. DiRocco, D. A., Rovis, T. *Angew. Chem. Int. Ed.* **2012**, *51*, 5904–5906.
103. Sun, L.-H., Liang, Z.-Q., Jia, W.-Q., Ye, S. *Angew. Chem. Int. Ed.* **2013**, *52*, 5803–5806.
104. (a) Linghu, X., Johnson, J. S. *Angew. Chem. Int. Ed.* **2003**, *42*, 2534–2536. (b) Bausch, C. C., Johnson, J. S. *J. Org. Chem.* **2004**, *69*, 4283–4285. (c) Linghu, X., Bausch, C. C., Johnson, J. S. *J. Am. Chem. Soc.* **2005**, *127*, 1833–1840.
105. Linghu, X., Potnick, J. R., Johnson, J. S. *J. Am. Chem. Soc.* **2004**, *126*, 3070–3071.
106. Mattson, A. E., Bharadwaj, A. R., Scheidt, K. A. *J. Am. Chem. Soc.* **2004**, *126*, 2314–2315.
107. Mattson, A. E., Bharadwaj, A. R., Zuhl, A. M., Scheidt, K. A. *J. Org. Chem.* **2006**, *71*, 5715–5724.
108. Bharadwaj, A. R., Scheidt, K. A. *Org. Lett.* **2004**, *6*, 2465–2468.
109. Mattson, A. E., Scheidt, K. A. *Org. Lett.* **2004**, *6*, 4363–4366.
110. Izquierdo, J., Hutson, G. E., Cohen, D. T., Scheidt, K. A. *Angew. Chem. Int. Ed.* **2012**, *51*, 11686–11698.
111. (a) Couladouros, E. A., Mihou, A. P., Bouzas, E. A. *Org. Lett.* **2004**, *6*, 977–980. (b) Baumann, K. L., Butler, D. E., Deering, C. F., Mennen, K. E., Millar, A., Nanninga, T. N., Palmer, C. W., Roth, B. D. *Tetrahedron Lett.* **1992**, *33*, 2283–2284. (c) Takada, A., Hashimoto, Y., Takikawa, H., Hikita, K., Suzuki, K. *Angew. Chem. Int. Ed.* **2011**, *50*, 2297–2301. (d) Takikawa, H., Suzuki, K. *Org. Lett.* **2007**, *9*, 2713–2716. (e) Nicolaou, K. C., Tang, Y., Wang, J. *Chem. Commun.* **2007**, 1922–1923. (f) Xu, J., Caro-Diaz, E. J. E., Theodorakis, E. A. *Org. Lett.* **2010**, *12*, 3708–3711. (g) Phillips, E. M., Roberts, J. M., Scheidt, K. A. *Org. Lett.* **2010**, *12*, 2830–2833. (h) Lathrop, S. P., Rovis, T. *Chem. Sci.* **2013**, *4*, 1668–1673.
112. Rauhut, M. M., Currier, H. Preparation of dialkyl-2-methylene glutamates. U.S. Patent 3,074,999, January 22, 1963.
113. Baizer, M. M., Anderson, J. D. *J. Org. Chem.* **1965**, *30*, 1357–1360.
114. (a) McClure, J. D. Dimerization process of preparing 1,4-dicyano-1-butene from acrylonitrile. U.S. Patent 3,225,083, December 21, 1965. (b) McClure, J. D. *J. Org. Chem.* **1970**, *35*, 3045–3048.
115. Morita, K., Suzuki, Z., Hirose, H. *Bull. Chem. Soc. Jpn.* **1968**, *41*, 2815.
116. Baylis, A. B., Hillman, M. E. D. Acrylic compounds. German Patent DE 2,155,113, May 10, 1972.
117. (a) Basavaiah, D., Gowriswari, V. V. L. *Tetrahedron Lett.* **1986**, *27*, 2031–2032. (b) Amri, H., Villieras, J. *Tetrahedron Lett.* **1986**, *27*, 4307–4308. (b) Basavaiah, D., Gowriswari, V. V. L., Bharathi, T. K. *Tetrahedron Lett.* **1987**, *28*, 4591–4592. (c) Drewes, S. E., Emslie, N. D., Karodia, N. *Synth. Commun.* **1990**, *20*, 1915–1921.
118. Trost, B. M., Kazmaier, U. *J. Am. Chem. Soc.* **1992**, *114*, 7933–7935.
119. Guo, C., Lu, X. *J. Chem. Soc. Perkin Trans. 1* **1993**, 1921–1923.
120. Trost, B. M., Li, C.-J. *J. Am. Chem. Soc.* **1994**, *116*, 10819–10820.
121. Trost, B. M., Dake, G. R. *J. Am. Chem. Soc.* **1997**, *119*, 7595–7596.
122. Zhang, C., Lu, X. *J. Org. Chem.* **1995**, *60*, 2906–2908.
123. (a) Vedejs, E., Daugulis, O., Diver, S. T. *J. Org. Chem.* **1996**, *61*, 430–431. (b) Vedeja, E., Daugulis, O., MacKay, J. A., Rozners, E. *Synlett* **2001**, 1499–1505.

124. (a) Voituriez, A., Marinetti, A., Gicquel, M. *Synlett* **2015**, *26*, 142–166. (b) Wang, Z., Xu, X., Kwon, O. *Chem. Soc. Rev.* **2014**, *43*, 2927–2940. (c) Fan, Y. C., Kwon, O. *Chem. Commun.* **2013**, *49*, 11588–11619. (d) Lu, X., Du, Y., Lu, C. *Pure Appl. Chem.* **2005**, *77*, 1985–1990. (e) Methot, J. L., Roush, W. R. *Adv. Synth. Catal.* **2004**, *346*, 1035–1050.
125. (a) Xiao, Y., Sun, Z., Guo, H., Kwon, O. *Beilstein J. Org. Chem.* **2014**, *10*, 2089–2121. (b) Wei, Y., Shi, M. *Acc. Chem. Res.* **2010**, *43*, 1005–1018. (c) Marinetti, A., Voituriez, A. *Synlett* **2010**, 174–194. (d) Wei, Y., Shi, M. *Chem. Asian. J.* **2014**, *9*, 2720–2734.
126. (a) Basavaiah, D., Rao, A. J., Satyanarayana, T. *Chem. Rev.* **2003**, *103*, 811–892. (b) Masson, G., Housseman, C., Zhu, J. *Angew. Chem. Int. Ed.* **2007**, *46*, 4614–4628. (c) Declerck, V., Martinez, J., Lamaty, F. *Chem. Rev.* **2009**, *109*, 1–48. (d) Basavaiah, D., Reddy, B. S., Badsara, S. S. *Chem. Rev.* **2010**, *110*, 5447–5674. (e) Wei, Y., Shi, M. *Chem. Rev.* **2013**, *113*, 6659–6690.
127. Aroyan, C. E., Dermenci, A., Miller, S. J. *Tetrahedron* **2009**, *65*, 4069–4084.
128. (a) Ma, D., Lin, Y., Lu, X., Yu, L. *Tetrahedron Lett.* **1988**, *29*, 1045–1048. (b) Trost, B. M., Schmidt, T. A. *J. Am. Chem. Soc.* **1988**, *110*, 2301–2303. (c) Inoue, Y., Imaizumi, S. *J. Mol. Catal.* **1988**, *49*, L19–L21.
129. Lu, X., Zhang, C., Xu, Z. *Acc. Chem. Res.* **2001**, *34*, 535–544.
130. Rychnovsky, S. D., Kim, J. *J. Org. Chem.* **1994**, *59*, 2659–2660.
131. (a) Jacobsen, M. J., Funder, E. D., Cramer, J. R., Gothelf, K. V. *Org. Lett.* **2011**, *13*, 3418–3421. (b) Xu, S., Zhou, L., Zeng, S., Ma, R., Wang, Z., He, Z. *Org. Lett.* **2009**, *11*, 3498–3501.
132. Hampton, C. S., Harmata, M. *Org. Lett.* **2014**, *16*, 1256–1259.
133. Hampton, C. S., Harmata, M. *J. Org. Chem.* **2015**, *80*, 12151–12158.
134. Georgy, M., Lesot, P., Campagne, J.-M. *J. Org. Chem.* **2007**, *72*, 3543–3549.
135. Xing, Y., O'Doherty, G. A. *Org. Lett.* **2009**, *11*, 1107–1110.
136. Wang, Y., O'Doherty, G. A. *J. Am. Chem. Soc.* **2013**, *135*, 9334–9337.
137. Li, M., O'Doherty, G. A. *Org. Lett.* **2006**, *8*, 6087–6090.
138. Li, M., O'Doherty, G. A. *Org. Lett.* **2006**, *8*, 3987–3990.
139. Guo, H., Mortensen, M. S., O'Doherty, G. A. *Org. Lett.* **2008**, *10*, 3149–3152.
140. Hunter, T. J., O'Doherty, G. A. *Org. Lett.* **2001**, *3*, 1049–1052.
141. Trost, B. M., Biannic, B. *Org. Lett.* **2015**, *17*, 1433–1436.
142. Trost, B. M., Li, C.-J. *J. Am. Chem. Soc.* **1994**, *116*, 3167–3168.
143. Zhang, C., Lu, X. *Synlett* **1995**, 645–646.
144. Trost, B. M., Dake, G. R. *J. Org. Chem.* **1997**, *62*, 5670–5671.
145. Alvarez-Ibarra, C., Csákÿ, A. G., Gómez de la Oliva C. *Tetrahedron Lett.* **1999**, *40*, 8465–8467.
146. Lu, C., Lu, X. *Org. Lett.* **2002**, *4*, 4677–4679.
147. (a) Sriramurthy, V., Barcan, G. A., Kwon, O. *J. Am. Chem. Soc.* **2007**, *129*, 12928–12929. (b) Sriramurthy, V., Kwon, O. *Org. Lett.* **2010**, *12*, 1084–1087.
148. Chen, Z., Zhu, G., Jiang, Q., Xiao, D., Cao, P., Zhang, X. *J. Org. Chem.* **1998**, *63*, 5631–5635.
149. Wang, T., Yao, W., Zhong, F., Pang, G. H., Lu, Y. *Angew. Chem. Int. Ed.* **2014**, *53*, 2964–2968.
150. Chen, J., Cai, Y., Zhao, G. *Adv. Synth. Catal.* **2014**, *356*, 359–363.
151. Chung, Y. K., Fu, G. C. *Angew. Chem. Int. Ed.* **2009**, *48*, 2225–2227.
152. Lundgren, R. J., Wilsily, A., Marion, N., Ma, C., Chung, Y. K., Fu, G. C. *Angew. Chem. Int. Ed.* **2013**, *52*, 2525–2528.
153. Smith, S. W., Fu, G. C. *J. Am. Chem. Soc.* **2009**, *131*, 14231–14233.
154. Sinisi, R., Sun, J., Fu, G. C. *Proc. Natl. Acad. Sci. U.S.A.* **2010**, *107*, 20652–20654.
155. (a) Sun, J., Fu, G. C. *J. Am. Chem. Soc.* **2010**, *132*, 4568–4569. (b) Fujiwara, Y., Sun, J., Fu, G. C. *Chem. Sci.* **2011**, *2*, 2196–2198.
156. Kalek, M., Fu, G. C. *J. Am. Chem. Soc.* **2015**, *137*, 9438–9442.

157. (a) Wang, T., Yu, Z., Hoon, D. L., Huang, K.-W., Lan, Y., Lu, Y. *Chem. Sci.* **2015**, *6*, 4912–4922. (b) Wang, T., Hoon, D. L., Lu, Y. *Chem. Commun.* **2015**, *51*, 10186–10189.
158. Du, Y., Lu, X., Yu, Y. *J. Org. Chem.* **2002**, *67*, 8901–8905.
159. (a) Xu, Z., Lu, X. *Tetrahedron Lett.* **1997**, *38*, 3461–3464. (b) Xu, Z., Lu, X. *J. Org. Chem.* **1998**, *63*, 5031–5041.
160. (a) Du, Y., Lu, X. *J. Org. Chem.* **2003**, *68*, 6463–6465. (b) Jones, R. A., Krische, M. J. *Org. Lett.* **2009**, *11*, 1849–1851. (c) Sampath, M., Lee, P.-Y. B., Loh, T.-P. *Chem. Sci.* **2011**, *2*, 1988–1991. (d) Han, X., Zhong, F., Wang, Y., Lu, Y. *Angew. Chem. Int. Ed.* **2012**, *51*, 767–770. (e) Andrews, I. P., Kwon, O. *Chem. Sci.* **2012**, *3*, 2510–2514. (f) Tran, Y. S., Kwon, O. *Org. Lett.* **2005**, *7*, 4289–4291. (g) Barcan, G. A., Patel, A., Houk, K. N., Kwon, O. *Org. Lett.* **2012**, *14*, 5388–5391. (h) Villa, R. A., Xu, Q., Kwon, O. *Org. Lett.* **2012**, *14*, 4634–4637.
161. Wang, J.-C., Ng, S.-S., Krische, M. J. *J. Am. Chem. Soc.* **2003**, *125*, 3682–3683.
162. Wang, J.-C., Krische, M. J. *Angew. Chem. Int. Ed.* **2003**, *42*, 5855–5857.
163. Henry, C. E., Kwon, O. *Org. Lett.* **2007**, *9*, 3069–3072.
164. Lee, S. Y., Fujiwara, Y., Nishiguchi, A., Kalek, M., Fu, G. C. *J. Am. Chem. Soc.* **2015**, *137*, 4587–4591.
165. Zhu, X.-F., Lan, J., Kwon, O. *J. Am. Chem. Soc.* **2003**, *125*, 4716–4717.
166. Tran, Y. S., Kwon, O. *J. Am. Chem. Soc.* **2007**, *129*, 12632–12633.
167. Wang, T, Ye, S. *Org. Lett.* **2010**, *12*, 4168–4171.
168. Zhang, Q., Yang, L., Tong, X. *J. Am. Chem. Soc.* **2010**, *132*, 2550–2551.
169. Han, X., Yao, W., Wang, T., Tan, Y. R., Yan, Z., Kwiatkowski, J., Lu, Y. *Angew. Chem. Int. Ed.* **2014**, *53*, 5643–5647.
170. Zhu, G., Chen, Z., Jiang, Q., Xiao, D., Cao, P., Zhang, X. *J. Am. Chem. Soc.* **1997**, *119*, 3836–3837.
171. Sampath, M., Loh, T.-P. *Chem. Sci.* **2010**, *1*, 739–742.
172. Wilson, J. E., Fu, G. C. *Angew. Chem. Int. Ed.* **2006**, *45*, 1426–1429.
173. Kramer, S., Fu, G. C. *J. Am. Chem. Soc.* **2015**, *137*, 3803–3806.
174. Andrews, P., Blank, B. R., Kwon, O. *Chem. Commun.* **2012**, *48*, 5373–5375.

Index

Note: Page numbers followed by f and t refer to figures and tables respectively.

C

D

E

J

K

L

M

N

S

T

V

X

Z